# Intelligent Biomechatronics in Neurorehabilitation

# Intelligent Biomechatronics in Neurorehabilitation

*Edited by*
Xiaoling Hu

ACADEMIC PRESS
An imprint of Elsevier
ELSEVIER

Academic Press is an imprint of Elsevier
125 London Wall, London EC2Y 5AS, United Kingdom
525 B Street, Suite 1650, San Diego, CA 92101, United States
50 Hampshire Street, 5th Floor, Cambridge, MA 02139, United States
The Boulevard, Langford Lane, Kidlington, Oxford OX5 1GB, United Kingdom



**Library of Congress Cataloging-in-Publication Data**
A catalog record for this book is available from the Library of Congress

**British Library Cataloguing-in-Publication Data**
A catalogue record for this book is available from the British Library

ISBN: 978-0-12-814942-3

For information on all Academic Press publications visit our website at https://www.elsevier.com/books-and-journals

*Publisher:* Mara Conner
*Acquisition Editor:* Chris Katsaropaulos
*Editorial Project Manager:* Andrae Akeh
*Production Project Manager:* Selvaraj Raviraj
*Cover Designer:* Christian Bilbow

Typeset by TNQ Technologies

# Contents

# Contributors

**Sliman J. Bensmaia** Department of Organismal Biology and Anatomy, University of Chicago, Chicago, IL, United States

**Stuart Biggar** Beijing Advanced Innovation Center for Biomedical Engineering, Beihang University, Beijing, China

**Jack Brooks** Department of Organismal Biology and Anatomy, University of Chicago, Chicago, IL, United States

**Ka Leung Marko Chan** Health Care (Health Technology) (HKPolyU); Stroke and Clinical Neuroscience (CUHK); Biomedical Engineering (CUHK)

**Weidong Chen** Qiushi Academy for Advanced Studies, Zhejiang University, Hangzhou, China; Zhejiang Provincial Key Laboratory of Cardio-Cerebral Vascular Detection Technology and Medicinal Effectiveness Appraisal, Hangzhou, China; Innovation Joint Research Center for iCPS, Zhejiang University, Hangzhou, China; College of Biomedical Engineering and Instrument Science, Zhejiang University, Hangzhou, China

**Xin Chu** Qiushi Academy for Advanced Studies, Zhejiang University, Hangzhou, China; College of Computer Science and Technology, Zhejiang University, China

**John E. Downey** Department of Organismal Biology and Anatomy, University of Chicago, Chicago, IL, United States

**Yubo Fan** Key Laboratory for Biomechanics and Mechanobiology of Ministry of Education, School of Biological Science and Medical Engineering, Beihang University, Beijing, China; Beijing Advanced Innovation Center for Biomedical Engineering, Beihang University, Beijing, China; National Research Center for Rehabilitation Technical Aids, Beijing, China

**Zhefeng Gong** Department of Neurobiology, Zhejiang University School of Medicine, Hangzhou, Zhejiang, China

**Ziqi Guo** Department of Biomedical Engineering, The Hong Kong Polytechnic University, Kowloon, Hong Kong

**Yaoyao Hao** Qiushi Academy for Advanced Studies, Zhejiang University, Hangzhou, China; Key Laboratory of Biomedical Engineering of Ministry of Education, Hangzhou, China; Zhejiang Provincial Key Laboratory of Cardio-Cerebral Vascular Detection Technology and Medicinal Effectiveness Appraisal, Hangzhou, China; College of Biomedical Engineering and Instrument Science, Zhejiang University, Hangzhou, China

**Xiaoling Hu** Department of Biomedical Engineering, The Hong Kong Polytechnic University, Kowloon, Hong Kong

**Chaoyi Hu** Key Laboratory for Biomechanics and Mechanobiology of Ministry of Education, School of Biological Science and Medical Engineering, Beihang University, Beijing, China

**Yanhuan Huang** Department of Biomedical Engineering, The Hong Kong Polytechnic University, Kowloon, Hong Kong

**Yao Huang** Biomedical Engineering School, Faculty of Engineering and Information Technology, University of Technology Sydney, Sydney, Australia

**Michael TSUI. Kam-fai** Chief Executive Officer, Deltason Medical Ltd., Ma Liu Shui, Hong Kong

**Cliff Klein** Guangdong Work Injury Rehabilitation Center, Guangzhou, China

**Will Poyan Lai** Jockey Club Rehabilitation Engineering Clinic, Department of Biomedical Engineering, The Hong Kong Polytechnic University, Kowloon, Hong Kong

**Yue Li** Zhejiang Lab, Hangzhou, China; Qiushi Academy for Advanced Studies, Zhejiang University, Hangzhou, China; Key Laboratory of Biomedical Engineering of Ministry of Education, Hangzhou, China

**Le Li** Department of Rehabilitation Medicine, Guangdong Engineering and Technology Research Center for Rehabilitation Medicine and Translation, The First Affiliated Hospital, Sun Yat-sen University, Guangzhou, Guangdong Province, China

**Waiming Li** Department of Biomedical Engineering, The Hong Kong Polytechnic University, Kowloon, Hong Kong

**Xiaoyan Li** Department of Physical Medicine and Rehabilitation, University of Texas Health Science Center at Houston, TIRR Memorial Hermann Research Center, Houston, TX, United States

**Sheng Li** Department of Physical Medicine and Rehabilitation, University of Texas Health Science Center at Houston, TIRR Memorial Hermann Research Center, Houston, TX, United States

**Wai Leung Ambrose Lo** Department of Rehabilitation Medicine, Guangdong Engineering and Technology Research Center for Rehabilitation Medicine and Translation, The First Affiliated Hospital, Sun Yat-sen University, Guangzhou, Guangdong Province, China

**Chingyi Nam** Department of Biomedical Engineering, The Hong Kong Polytechnic University, Kowloon, Hong Kong

**Sin-Wa Ng** Community Rehabilitation Service Support Center, Hospital Authority, Hong Kong

**Zhenhuan Ouyang** Qiushi Academy for Advanced Studies, Zhejiang University, Hangzhou, China; College of Computer Science and Technology, Zhejiang University, China

**Gang Pan** Qiushi Academy for Advanced Studies, Zhejiang University, Hangzhou, China; Key Laboratory of Biomedical Engineering of Ministry of Education, Hangzhou, China; College of Computer Science and Technology, Zhejiang University, Hangzhou, China

**Guangshuai Peng** Key Laboratory for Biomechanics and Mechanobiology of Ministry of Education, School of Biological Science and Medical Engineering, Beihang University, Beijing, China; Beijing Advanced Innovation Center for Biomedical Engineering, Beihang University, Beijing, China

**Waisang Poon** Department of Surgery, Prince of Wales Hospital, The Chinese University of Hong Kong, Shatin, Hong Kong

**Yu Qi** Qiushi Academy for Advanced Studies, Zhejiang University, Hangzhou, China; Key Laboratory of Biomedical Engineering of Ministry of Education, Hangzhou, China; College of Computer Science and Technology, Zhejiang University, Hangzhou, China

**Qiuyang Qian** Department of Biomedical Engineering, The Hong Kong Polytechnic University, Kowloon, Hong Kong

**Wei Rong** Department of Biomedical Engineering, The Hong Kong Polytechnic University, Kowloon, Hong Kong

**Rong Song** School of Biomedical Engineering, Sun Yat-Sen University, Guangzhou, PR China

**Steven W. Su** Biomedical Engineering School, Faculty of Engineering and Information Technology, University of Technology Sydney, Sydney, Australia

**Eric W.C. Tam** Department of Biomedical Engineering, The Hong Kong Polytechnic University, Kowloon, Hong Kong; Jockey Club Rehabilitation Engineering Clinic, Department of Biomedical Engineering, The Hong Kong Polytechnic University, Kowloon, Hong Kong

**Kai-Yu Tong** Department of Biomedical Engineering, The Chinese University of Hong Kong, Hong Kong

**Lizhen Wang** Key Laboratory for Biomechanics and Mechanobiology of Ministry of Education, School of Biological Science and Medical Engineering, Beihang University, Beijing, China; Beijing Advanced Innovation Center for Biomedical Engineering, Beihang University, Beijing, China

**Kevin B. Wilkins** Physical Therapy Movement and Human Movement Sciences Department, Northwestern University, USA

**Kedi Xu** Qiushi Academy for Advanced Studies, Zhejiang University, Hangzhou, China; Key Laboratory of Biomedical Engineering of Ministry of Education, Hangzhou, China; Zhejiang Provincial Key Laboratory of Cardio-Cerebral Vascular Detection Technology and Medicinal Effectiveness Appraisal, Hangzhou, China; College of Biomedical Engineering and Instrument Science, Zhejiang University, Hangzhou, China

**Yuan Yang** Department of Physical Therapy and Human Movement Sciences, Northwestern University Feinberg School of Medicine, Chicago, IL, United States

**Jun Yao** Physical Therapy Movement and Human Movement Sciences Department, Northwestern University, USA

**Wei Yao** Beijing Advanced Innovation Center for Biomedical Engineering, Beihang University, Beijing, China; Department of Biomedical Engineering, University of Strathclyde, Glasgow, United Kingdom

**Ling-Fung Yeung** Department of Biomedical Engineering, The Chinese University of Hong Kong, Hong Kong

**Xiaofei Yin** Key Laboratory for Biomechanics and Mechanobiology of Ministry of Education, School of Biological Science and Medical Engineering, Beihang University, Beijing, China

**King-Pong Yu** Department of Biomedical Engineering, The Chinese University of Hong Kong, Hong Kong; Community Rehabilitation Service Support Center, Hospital Authority, Hong Kong

**Shaomin Zhang** Qiushi Academy for Advanced Studies, Zhejiang University, Hangzhou, China; Key Laboratory of Biomedical Engineering of Ministry of Education, Hangzhou, China; Zhejiang Lab, Hangzhou, China

**Xu Zhang** Biomedical Engineering Program, University of Science and Technology of China, Hefei, China

**Ting Zhao** Howard Hughes Medical Institute, Ashburn, VA, United States

**Nenggan Zheng** Qiushi Academy for Advanced Studies, Zhejiang University, Hangzhou, China; Zhejiang Provincial Key Laboratory of Cardio-Cerebral Vascular Detection Technology and Medicinal Effectiveness Appraisal, Hangzhou, China; Innovation Joint Research Center for iCPS, Zhejiang University, Hangzhou, China; College of Biomedical Engineering and Instrument Science, Zhejiang University, Hangzhou, China

**Xiaoxiang Zheng** Qiushi Academy for Advanced Studies, Zhejiang University, Hangzhou, China; Key Laboratory of Biomedical Engineering of Ministry of Education, Hangzhou, China; Zhejiang Provincial Key Laboratory of Cardio-Cerebral Vascular Detection Technology and Medicinal Effectiveness Appraisal, Hangzhou, China; Innovation Joint Research Center for iCPS, Zhejiang University, Hangzhou, China; College of Biomedical Engineering and Instrument Science, Zhejiang University, Hangzhou, China

**Yongping Zheng** Department of Biomedical Engineering, The Hong Kong Polytechnic University, Kowloon, Hong Kong

**Ping Zhou** Department of Physical Medicine and Rehabilitation, University of Texas Health Science Center at Houston, TIRR Memorial Hermann Research Center, Houston, TX, United States

# Preface

With the rapid growth of the aging population, long-term rehabilitation for neurological impairments has become a great challenge that must be faced over the coming decades. In contrast to the increasing aging populations worldwide, traditional healthcare resources (e.g., professional manpower) in the rehabilitation industries are lacking, even in developed countries. It further makes the related services difficult to access or afford based on traditional manual practices. Neurorehabilitation by automation has been a must to release the continuously increasing pressure on the medical and healthcare systems. Intelligent biomechatronics, or bionic robotics, providing a seamless interface and real-time interaction with the biological system and external environment, have been the key in services with automation, whose success depends on the collaboration among neuroscientists, engineers, manufacturers, clinicians, etc., from upstream to downstream in the bioinstrumentation and healthcare industries. The aim of this book is to introduce representative achievements related to the fundamental design, commercialization, and clinical applications of intelligent biomechatronics for neurorehabilitation. Leading research groups, experienced practitioners, and successful entrepreneurs in the related industries contribute from different angles in the book.

In the control design of biomechatronics for neurorehabilitation, interpretation of voluntary intention from the nervous system based on neural coding is the first and most important step to build communication between a biological system and external mechanics. In the first volume of the book, from Chapters 1 to 5, neural coding techniques with detection at different levels in the nervous system are introduced by leading research groups. Neural signals that reflects voluntary intention can be detected by both invasive and noninvasive methods related to brain machine interface technologies (BMIs). In Chapter 1, "Toward bidirectional closed-loop brain–machine interfaces (BMIs): A summary on invasive BME research in China," by Weidong Chen and his colleagues, the neural signal detection and interpretation were introduced by their work on nonprimates, monkeys, and clinical studies on human subjects. The main application of BMIs is in improving the mobility of severely impaired persons. The invasive coding techniques are further illustrated by Shaomin Zhang et al. on electrocorticography (ECoG) for hand gestures decoding in Chapter 2, "Neural decoding by invasive electrocorticography." The voluntary motor intention can also be captured by noninvasive technologies, such as electroencephalography (EEG), as discussed in Chapter 3 by Yuan Yang, electromyography (EMG) in Chapter 4 by Ping Zhou et al., and even by behavioral description as introduced in Chapter 5 by Nenggan Zheng and Ting Zhao's group.

In the second volume of the book, from Chapters 6 to 10, representative biomechatronic robots with sophisticated mechanical designs for neurorehabilitation are introduced when some of the coding techniques in the first volume are applied in the development. In the lower limb robot developed by the team led by Kai-yu Tong, a behavioral sensing system enhanced by 3D printing and special knee locking mechanics was implemented to facilitate persons with gait deficits, as detailed in Chapter 6, "Bionic robotics for post polio walking." A robot for upper limb rehabilitation is introduced in Chapter 6 by Rong Song et al. The robot is a cable-driven system to assist patients with upper limb disabilities to practice in a three-dimensional space, and EMG was integrated into the control strategy in the robot-assisted dynamics. Robotic design for rehabilitation purposes does not only require mechanical support to motor function, but also needs artificial sensory feedback. In Chapter 7, Sliman J. Bensmaia et al. demonstrate the efforts of introducing sensory feedback through electrical stimulation to the nerves for amputees and electrical stimulation to the brain for tetraplegic patients based on a bionic robot hand. Robotic design and electrical stimulation are further integrated to optimize the rehabilitative effects, such as the neuromuscular electrical stimulation (NMES) and robot hybrid systems introduced in Chapter 8, by the research group of Xiaoling Hu. Soft robotics for hand rehabilitation are illustrated in Chapter 9 by Yubo Fan's group. Lightweight pneumatic artificial muscles are applied for mechanical actuation with 3D printed cable guides in the robotic system, providing a comfortable wearing experience during repeated and intensive physical practices in rehabilitation.

Clinical application of biomechatronic automation is covered in the third volume of the book. Robots not only can be applied in rehabilitation training, but also can act as evaluation platforms to save manpower and improve reliability and repeatability through objective measurements. In Chapter 11, Jun Yao and her group discuss how robotic techniques assisted in quantitative evaluations on upper limb functions during rehabilitation. Examples of quantitative and objective evaluation are further introduced in Chapter 12 by Le Li's group, with a focus on muscle functions in different neurological conditions. In Chapter 13, the need for automation in neurorehabilitation is addressed by an experienced rehabilitation professional, Marko Chan, a senior occupational therapist, based on his first-hand experiences in assistive technologies in routine clinical practices. There remains plenty of room for the current automation systems to improve, in order to meet all clinical demands. In Chapter 14, a head-to-head comparison of the robot-assisted rehabilitation effects after stroke is conducted between those achieved in a well-controlled clinical trial and in a real clinical service, based on a self-designed and currently commercialized robotic hand by Kai-yu Tong and Xiaoling Hu's groups. Excitingly, there are no significant differences in rehabilitation effectiveness during the translation from lab to service, which brings further confidence to the application of biomechatronic automations into clinical usage for effective neurorehabilitation. The commercialization of a biomechatronic prototype from the lab into a rehabilitation device in service is a key step to achieve real clinical applications. In Chapter 15, the fourth volume of the book, Michael Tsui shares his valuable and successful experiences in commercialization of the robotic hand used in Chapter 14.

Finally, I am very grateful for the great support from all contributors and for their sharing of their unique expertise and experiences. In addition, I also would like to express my appreciation to the editorial team and publishing office for their help and support throughout the preparation of this book.

**Xiaoling Hu**
Assistant Professor
Department of Biomedical Engineering
The Hong Kong Polytechnic University

PART I

# Neural coding mechanisms

1

# Toward bidirectional closed-loop brain—machine interfaces (BMIs): a summary on invasive BMI research in China

Weidong Chen[1,3,4,5], Shaomin Zhang[1,2,6], Yaoyao Hao[1,2,3,5], Kedi Xu[1,2,3,5], Nenggan Zheng[1,3,4,5], Xiaoxiang Zheng[1,2,3,4,5]

[1]QIUSHI ACADEMY FOR ADVANCED STUDIES, ZHEJIANG UNIVERSITY, HANGZHOU, CHINA; [2]KEY LABORATORY OF BIOMEDICAL ENGINEERING OF MINISTRY OF EDUCATION, HANGZHOU, CHINA; [3]ZHEJIANG PROVINCIAL KEY LABORATORY OF CARDIO-CEREBRAL VASCULAR DETECTION TECHNOLOGY AND MEDICINAL EFFECTIVENESS APPRAISAL, HANGZHOU, CHINA; [4]INNOVATION JOINT RESEARCH CENTER FOR ICPS, ZHEJIANG UNIVERSITY, HANGZHOU, CHINA; [5]COLLEGE OF BIOMEDICAL ENGINEERING AND INSTRUMENT SCIENCE, ZHEJIANG UNIVERSITY, HANGZHOU, CHINA; [6]ZHEJIANG LAB, HANGZHOU, CHINA

## Chapter outline



Intelligent Biomechatronics in Neurorehabilitation. https://doi.org/10.1016/B978-0-12-814942-3.00001-5

# Introduction

Brain–machine interfaces (BMIs) provide a direct communication and control channel between the brain and external devices, independent of the brain's normal output pathways of peripheral nerves and muscles. BMI technology has the potential to assist, augment, or repair human sensorimotor and other cognitive functions [1–4]. It also provides a new method to encode sensory information and train animals to learn the patterns by microstimulating key areas of their brains [5,6]. During the last few decades, significant progress has been made in various BMI systems using noninvasive neural recording technologies, such as electroencephalography (EEG) and magnetoencephalography (MEG). In the past 10 years, invasive BMI has become one of the most enthusiastic research areas along with the quick development of invasive recording technologies, such as intracortical microelectrode arrays that record single/multiunit activity, brain surface electrodes, or electrocorticography (ECoG), etc. As a result, most of the latest reports of elaborate BMI realizations, especially those high dimensional artificial arm control demos are based on invasive recordings. Furthermore, invasive BMI has successfully been carried out in clinical trials, and remarkably progressed toward practical application.

Despite its late start, China has been making rapid progress in invasive BMI research. Multidisciplinary researchers of scientists, engineers, and clinicians have played important roles in the development of BMI technologies and systems on animals or humans, such as robot control [7,8], virtual typewriter [9], animal navigation or behavior control [10–15], seizure detection and inhibition [16], etc., using various invasive technologies including intracortical microelectrode arrays and electrocorticography (ECoG), as well as electrical stimulation and optogenetics, etc. Furthermore, they also used BMI technologies to investigate basic questions in neuroscience or cognitive science. Among these research groups, Zheng and colleagues pioneered a systematic study on invasive BMI in China since 2006, the time at which Qiushi Academy for Advanced Studies (QAAS), an interdisciplinary institute aiming at converging technology research, was founded at Zhejiang University [17]. The BMI group at QAAS consisted of researchers and students with various backgrounds, including biomedical engineering, computer science, neural surgery, neuroscience, materials science, etc. Focusing on brain–machine interfaces, they developed enabling technologies and built platforms to support BMI research on rodents, nonhuman primates, and humans, including system design, hardware, software, coding/decoding algorithms, and control strategies. Just like the international trend, invasive BMI has become the research focus in China. In this chapter, we review recent progress in invasive BMI in China and propose potential development trends and challenges of this technology to translate advances into clinical applications.

# BMIs on nonprimates

## Neural decoding in rodents

In one of the earliest demonstrations of brain control using invasive recording technology in China, Dai et al. [7] designed a "lever-press" system for rat drinking, in which both rat's neural activities and behavior were recorded and analyzed synchronously. Three Sprague–Dawley (SD) rats were trained to perform "lever-press" task for water rewards. When the rat pressed the lever using its forelimb and the pressure value was over a threshold, it was rewarded by a drop of water. After training, a $2 \times 8$-channel microwire electrodes array was implanted into the forepaw region of the primary motor cortex. The routine measurements began after rats' recovery. Sixteen-channel analog signals were recorded with a 30-kHz sample frequency, using Cerebus 128TM (Blackrock Microsystems Inc., USA). The pressure signals were synchronously recorded with a sampling rate of 500 Hz by pressure sensor during the experiment.

Neural spikes were extracted from 15-channel neural signals by thresholding (except the reference electrode in the array). Spikes from each channel were then classified into up to three types by principal components analysis (PCA) and K-means clustering, in which each type represented one neuron physiologically. A range of 22–58 neurons were found in all 15 channels per rat. Neural spiking rate was counted within a 100-ms time bin. Meanwhile, the pressure signal was recorded and synchronized by average filtering within the same bin period [18]. After date preprocessing, two decoding models based on probabilistic neural network (PNN), the PNN and the modified PNN (MPNN) [19], were built to estimate press value from neural ensemble spikes. In the training process, the MPNN decoder used actually recorded pressure value in place of the previous estimated pressure which was used in the PNN decoder. In addition, two commonly used neural decoding algorithms, Wiener filter (WF) and Kalman filter (KF), were compared with PNN and MPNN [20,21]. The waveform MPNN decoder was very smooth. Also, the correlation coefficient (CC) of the MPNN decoder was relatively large, while the mean square error (MSE) of the MPNN decoder was small. The results indicated that MPNN decoder had better performance than traditional algorithms, WF and KF decoders [18].

Additionally, Zhou et al. [22] implemented the decoding algorithm in a field-programmable gate array (FPGA) platform for integrated real-time BMI applications. This paradigm demonstrated that rats could get water reward through intention only, without pressing the lever. It proved that the "lever press" system was a basic BMI system, which was useful for researching many scientific problems, such as coding circumstances of neuronal ensembles.

## Neural coding of sensory information using brain stimulation

Afferent feedback is an essential element in a closed-loop BMI system. Brain stimulation is an important potential approach to encoding sensory information, thus it is necessary for the subjects to learn a certain number of stimulus patterns, through "virtually"

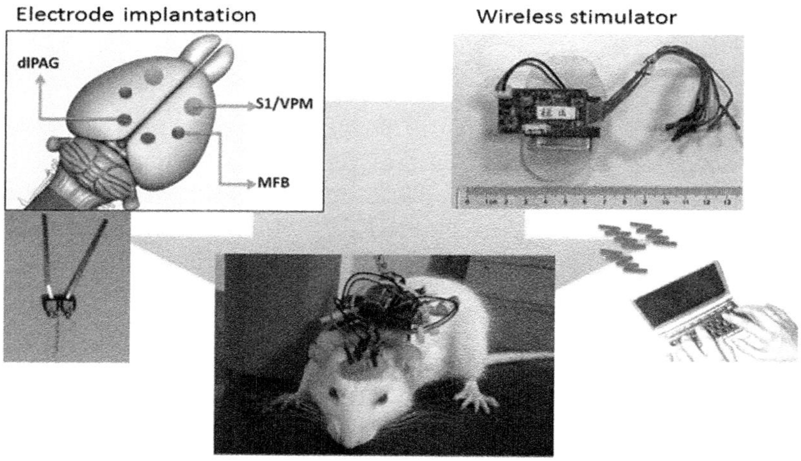


FIGURE 1.1 Rat behavior control system.

training to adjust the microstimulation in different key areas of animal brain [5]. In recent years, researchers have explored different methods of brain stimulation and developed several navigation or behavior control systems in rats [13,23], pigeons [24], geckos [12,25], underwater animals [14], and insects [26]. Among these systems, Feng et al. [10] developed a rat navigation system, which allowed remote control of a rat-robot by a wireless microstimulator in complicated environments (Fig. 1.1). This system consists of the following main components: an integrated PC control program, a transmitter and a receiver based on Bluetooth (BT) modules, a stimulator controlled by C8051 microprocessor, as well as an operant chamber, and an eight-arm radial maze. The microstimulator is featured with its changeable amplitude of pulse output for both constant-voltage and constant-current mode, which provided an easy way to set the proper suitable stimulation intensity for different training. The system had been used in behavior experiments for monitoring and recording bar-pressing in the operant chamber, controlling rat roaming in the eight-arm maze, as well as navigating rats through a 3D obstacle route. The results indicated that the system worked stably and that the stimulation was effective for different types of rat behavior controls. In addition, the results showed that stimulation in the whisker barrel region of rat primary somatosensory cortex (SI) acted like a cue. The animals can be trained to take different desired turns upon the association between the SI cue stimulation and the reward stimulation in the medial forebrain bundle (MFB).

Furthermore, Lin et al. [27] first indicated that the motor status of a rat could be switched between "motion" and "motionless" based on this system. They improved this remote control system and added a "stop" function in the rat robot by applying stimulation on the periaqueductal gray matter (PAG) area. With a new control system, six bipolar stimulating electrodes were separately implanted in the bilateral areas of MFB,

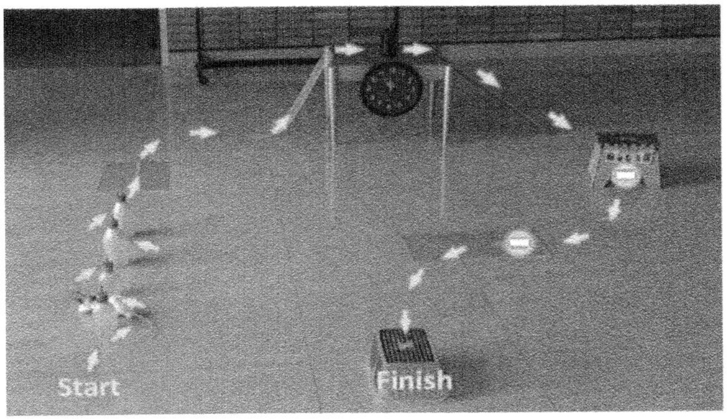


**FIGURE 1.2** An example of controlling rat navigation using brain microstimulation. *Arrows* indicate the movement tracks. *Dashes* denotes the places where the rat was immobile with a planned time span of 3s to 5s.

SI, and PAG. After 1 week of postoperative recovery and 5–7 days of behavior training, they could control the rat to run and freeze in an open field 3D environment by stimulating these electrodes (Fig. 1.2).

Moreover, Sun et al. [28–31] proposed a new method to realize the automatic navigation of rats. A general regression network (GRNN) was adopted to analyze and model the controlling procedure of human operations. Compared to the traditional approaches with explicit controlling rules, this algorithm learned the controlling process and imitated the decision-making of human beings to steer the rat automatically. In real-time navigation experiments, this method successfully controlled rats to follow given paths automatically and precisely. This work would be significant for future applications of animal navigation and provide a new way to realize hybrid intelligent systems with artificial intelligence and natural biological intelligence combined.

As is well known, electrical stimulation can activate all types of neurons unselectively in the brain, which can cause unpredictable side effects on the control of the rat [11,32]. The advent of optogenetics technology provides the possibility to manipulate the activity of specific types of neurons [33]. This novel technology integrates optics and genetics to achieve excitation or inhibition within specific types of cells in vivo. This cell-type-specific optical neural control method opens up the possibility of a more precise manipulation in behavior control, along with the chance for simultaneous electrophysiological recording in awake animals [34], which would be useful in understanding the mechanisms that lie in neural circuit modulations of brain states and behaviors. In Zheng's lab, optogenetics technology was first applied to the control of rat locomotion and the development of a novel rat navigation system [35,36]. Compared with electrical stimulation, optical stimulation on the dPAG area showed much gentler reactions of defensive behavior, indicating a higher precision in stimulation targeting [11]. The study in the ventral tegmental area (VTA) provided an optogenetic approach to regulate the motivation and motor activity of rats [37]. Further studies promise to reveal the rules of the neural circuits in more complex brain areas that guide behaviors.

### Portable system for neural stimulation and recording

Both electrical stimulation and neuronal activity recording are fundamental tools in brain–machine interfaces. Ye et al. [38] developed a portable multichannel telemetry system which could be used concurrently for brain electrical stimulation and neuronal activity recording on free-behaving small animals. This system consists of three major components of headstage, backpack, and portable personal digital assistant (PDA). The headstage contains high-precision instrument amplifiers with high input impedance. The backpack was comprised of two parts: (1) a main board (size: 36 × 22 × 3.5 mm; and weight: 20 g without batteries), with a current/voltage stimulator and special circuit suitable for neuronal activity recording, and (2) a Bluetooth transceiver, with a high data transmission rate of up to 70 kb/s, suitable for downloading stimulation commands and uploading acquired data. The wireless system supported neural signal recording of the primary motor area on a freely behaving rat with 12-bit resolution at 12 k samples/s. The recorded data and analysis results showed that the system was successful by comparing it with the commercial equipment Cerebus(TM) 128-Channel Data Acquisition System (Cyberkinetics Inc.).

A portable telemetry system with brain stimulation and neuronal activity recording in freely behaving small animals has been only rarely reported. This design established a closed-loop system between a specific region of the nervous system and the artificial device, and all the components used in this system are commercially available. For neuroscientists, it provides a method to explore the relationships between electrical stimulation, spiking characteristics, and behavioral response. Moreover, it may be useful as an animal navigation or behavior control system in some emergent situations for specific purposes with the portable PDA.

## BMIs in non-human primates

The monkey is a good subject for BMI research because it has an advantage that it could be trained to do more complex tasks than rats. Moreover, the macaque monkey has been used as a model for human motor behavior research, because its functional anatomical structure is similar to that of humans. Based on their experience on the construction of rat BMI experiment platforms, Zheng et al. [8] built the first nonhuman primate BMI platform in China. They set up an experimental paradigm of invasive BMIs on monkeys, and successfully recorded the brain activities of the primary motor cortex while the subject was performing different tasks. Approval for all experimental procedures was obtained from the Animal Care Committee at Zhejiang University, China, which follows the Guide for the Care and Use of Laboratory Animals (China Ministry of Health).

### Neural data reduction and decoding models

As the dimension of neural recordings in typical nonhuman primate applications is usually high (hundreds of channels), which brings challenges to real-time computation of decoding, it is necessary to ascertain the neuronal importance before the decoding procedure (i.e., neural data reduction). Many techniques have been proposed to identify the important neurons, such as neuron correlation analysis [39], tuning depth [20], mutual information [40], and cross-model validation [41]. However, most of these methods for neuron selection are model-related, and thus can be biased by the type of encoding or decoding methods used. Xu et al. proposed a local learning-based technique to assess the neural importance, without assuming any encoding or decoding method [42]. The high dimensional nonlinear neural data recorded from monkey motor cortex is decomposed into a set of linear ones to capture the local structure of neural activities [43]. Then the neural activities were projected into a feature space by weights, which represent the contributions of neurons. The distance between the local neuronal patterns associated with different grasp movement was maximized according to global costs to obtain the optimal neuronal weights. With a 1-norm regularization added in the cost function, the weight vector is constrained to be sparse for feature selection. The results showed that only 10 out of 70 neurons selected by their method could achieve over 95% of the full recording's performance, no matter which different decoding methods were used (support vector machine and K-nearest neighbor). Moreover, the method provided the most important neuron subset even when simulated irrelevant neurons were added in the neuronal activities and the final decoding accuracy outperformed other approaches, including model sensitivity analysis, single-neuron correlation analysis, support vector machine recursive feature elimination, and mutual information. In addition, the weights of important neurons showed a fast convergence, usually within 10–20 iterations, and thus low demand of computational resources. In total, the identification of the important neurons helps to inspect neural patterns highly associated with the task and greatly reduces the computational burden of systems, which may have an influence on the practical use of BMI.

The design of the decoding algorithm is the core component in BMIs, which determines the general performance of the whole system. Up to now, several signal processing approaches have been applied to extract the functional relationship between the neural activity and the corresponding movement including a bio-inspired population vector, a linear Wiener filter, and state-space model-based Kalman filter, and a point process filter. Recently, sequential Monte Carlo estimation with point process observation (SMCPP) was proposed to decode movement states, in which the spike trains are regarded as point processes and the spiking timing information is exploited by estimating the instantaneous firing rate in a much shorter interval ($\sim$10 ms) [44,45]. The biggest advantage is that there is no restriction on the posterior distribution of the state, which makes the SMCPP more flexible and suitable for the highly nonlinear systems. However, the tuning models used in current SMCPP usually assume the firing rates of neurons are only dependent on the kinematics, which has been shown recently not to be the case [46,47]. The accuracy of spike timings prediction increased by 20% when the spiking of other neurons was taken into account [48]. Recently, Xu et al. implemented a new state-space model in SMCPP by incorporating the recent neural ensemble activity into the tuning functions of neurons and the state transition model [49]. The goodness-of-fit analysis demonstrates that the proposed model can predict the neuronal response more accurately than that only depending on kinematics. A new sequential Monte Carlo algorithm based on the proposed model is constructed to decode the continuous kinematics of a monkey performing a target-pursuit task. The algorithm can significantly reduce the root mean square error of decoding results by an average of 23.6%. In addition, they accelerate the decoding speed by implementing the proposed algorithm in a massive parallel manner on GPU, which results in it being over 10 times faster than the serial implementation. Thus the results confirm neurophysiologic knowledge that cortical neurons are interconnected in a large network and demonstrate a much faster and more accurate high-performance BMI application.

When moving to the clinical BMI applications, these supervised learning approaches mentioned above suffer biological implausibility in, e.g., tetraplegia patients who cannot generate explicit limb movements to train the decoder. Moreover, the neuroplasticity induced by biofeedback could impair the system control performance over time. Reinforcement learning (RL), however, allows the users to learn from the interaction with the environment without the supervising signals, which enables intelligent decoders to learn to interpret the neural modulation for better control on the external devices [50–52]. Wang et al. propose to adopt an efficient learning scheme, attention-gated reinforcement learning (AGREL) [53], in a form of three-layer neural network, to instantaneously interpret the neural states at each time index into seven distinct movements (four directional moves, two holdings, and one resting state) to reconstruct the trajectory of a center-out task [54]. The results showed that AGREL improved the target acquisition rate up to 90.16% on average, even outperforming other reinforcement learning method such as Q-learning techniques. The neural network enabled a soft encoding to find a sequence of the movements according to the probability distribution and defines an intensive function based on a global error to intensify the learning by valuing the unexpected rewards, which contributed to the performance improvement. The method also demonstrated much shorter convergence time and more stable performance to follow neural activity over multiple days, indicating the potential to achieve better online decoding for daily used BMI tasks.

Another issue concerning clinical application is the long-term stability of BMIs. As the spike signals decayed rapidly with time [55], LFP could be a good candidate for long-term usage in BMIs [56]. Many studies have shown LFPs could offer a stable decoding performance [57], however, the decoding stability of LFPs was examined only when high-quality spike signals were recorded, which cannot exclude the influence of spikes on LFPs. We examined the long-term decoding stability of LFPs over a longer time scale when the quality of spike signals was from good to poor or even when no spike was recorded [58]. Neural signals were collected from the motor cortex of three monkeys via silicon arrays over 230, 290, and 690 days postimplantation when they performed 2D center out task. To compare long-term stability between LFPs and spike signals, we examined them in neural signal characteristics, directional tuning properties, and offline decoding performance, respectively. When spike signals' quality decayed over time, we observed much slower decreasing trends of the number of LFP channels and mean LFP power in different frequency bands. Meanwhile, the number of significantly directional tuning LFP channels decreased more slowly than that of tuning single and multiunit activity (SUA and MUA). We also found that LFPs achieved better decoding performance than SUA and MUA, even when the quality of spike signals seriously decayed. Especially, when no spike was recorded in one monkey after 671 days postimplantation, LFPs still provided some kinematic information. The results suggested that LFPs were more durable and could provide better decoding performance when spike signal was quality seriously decayed or even disappeared. This might be due to their resistance to recording degradation and their high redundancy among channels.

## Grasp decoding and neural prosthesis control

The hands are one of the most commonly used human parts, the loss of which will result in a serious reduction of the functional autonomy (e.g., grasp and manipulation) of a person in daily life. BMI researches on nonhuman primates have provided a good platform for hand function reconstruction as the monkey has similar dexterous grasp to humans. Recently, there have been some reports that can decode a large number of joint kinematics from the hand [59,60]. However, none of them can reconstruct the functional grasp gesture, which is crucial for grasping different objects. Actually, it has been suggested that the brain adopts simplifying strategies in order to reduce the complexity of hand movements by controlling the different fingers in a synergic way. In one of the applications, Hao et al. investigated online decoding of grasp movement and real-time neuroprosthetic control [61]. In their study, they obtained neural data from the dorsal premotor cortex (PMd) when a monkey was reaching and grasping one of four differently shaped objects following visual cues. Alternatively, they classified finger configurations into one of the predefined categories based on the kinematic synergy movement in grasping and the neural encoding of grasp postures [62,63], rather than decoding continuous grasp movement. The single unit activity during the grasp movement showed distinct and stable neuron-by-neuron tuning to grasp gestures. In offline decoding, they evaluated the performance of the fuzzy k-nearest neighbor (FKNN) classifier, getting an average decoding accuracy of 97.1% across all sessions. In the online demonstration, a finite state machine (FSM) combined with a two-stage FKNN model was used to classify the motor information (four grip types and one rest state) asynchronously, in which both different movement states and movement onset timing were predicted. The instantaneous status of monkey grasping could be extracted from the firing pattern of the neural ensemble during the entire experiment, with an event accuracy of 85.1%. The monkey enabled real-time control of a prosthetic hand doing the same grasp gesture as itself (Fig. 1.3).

PMd, the area used in a previous study, has been traditionally viewed as part of the reach-related system located in the dorsomedial movement control pathway. Although

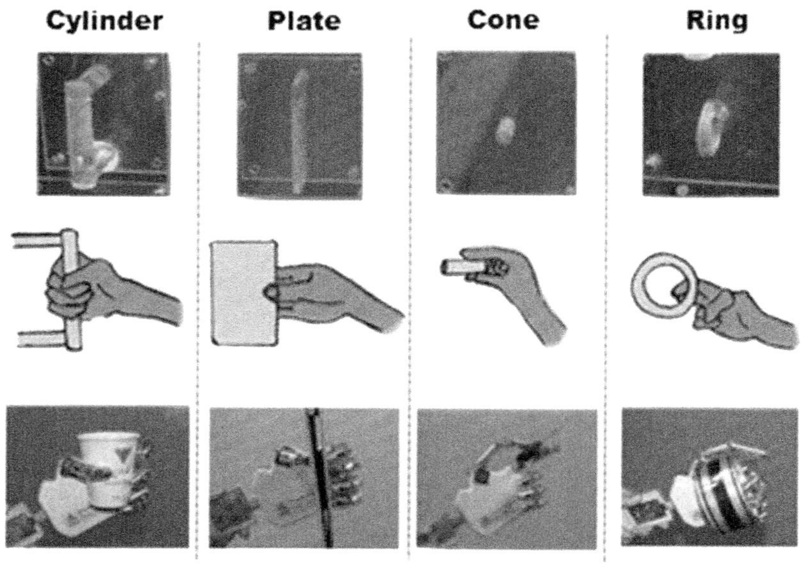


**FIGURE 1.3** Experimental setup and task. Experimental setup showing monkey reaching to grasp an object on a panel. Neural signals are recorded during the grasp task. The subpanel shows the four target objects and corresponding grasp gestures used in this study.

PMd contains neurons highly modulated by parameters of reach in goal-directed movements [64–67], recent developments have shown this area to be also related to distal hand grasp movements [68–71]. In one of these studies, Hao et al. investigated the ensemble firing properties and coding strategy of grasp movement in this area [72]. Single and population neuronal activity showed distinct patterns during execution of different grip types. Cluster analysis of neural ensemble signals indicated that the grasp-related patterns emerged soon (200–300 ms) after the go cue signal, and faded away during the hold period. They found that the grip type discrimination information was encoded only during a short period soon after movement onset rather than along the whole time course of the grasp. Application of a support vector machine model during this period achieved classification accuracies of 94% and 89% for two monkeys. Grasp decoding using local field potential (LFP), especially the high-frequency bands within this period, also produced high decoding accuracies.

In order to investigate the specific contribution of PMd to reaching and grasping respectively, Cao et al. trained the monkeys to reach in four directions to grasp the same object or reach in the same direction to grasp four different objects [73]. In this way, the two factors were separately encoded in the area. Mutual information (MI) is employed to quantitatively evaluate the neurons' tuning property in both tasks. They find that there exist neurons in PMd that are tuned only to reaching, tuned only to grasping, and tuned to both tasks. When applied with a support vector machine (SVM), the movement decoding accuracy by the tuned neuron subset in either task is quite close to the performance by the full ensemble. Furthermore, the decoding performance improves significantly by adding the neurons tuned to both tasks into the neurons tuned to one property only. These findings, combined with previous evidence for reach-related modulation studies, suggest that PMd may play an important role in generation and maintenance of grasp action and may be a suitable locus for brain–machine interface applications.

# Pilot studies in clinic research

## Prosthesis control using human ECoG BMI

Over the past decade, electrocorticographic (ECoG) have become recognized as a promising signal platform for brain–machine interface (BMI) research and clinical application. Compared to signals acquired from the scalp (EEG) and intraparenchymal single neuronal recordings, ECoG recordings have characteristics that make them especially suited for basic neuroscience research and resulting translational opportunities. These characteristics include high spatial resolution and signal fidelity, much less training, resistance to noise, and substantial robustness over long recording periods. ECoG recordings appear to strike an ideal balance between fidelity and clinical practicality.

It has been demonstrated recently that individuated finger and wrist movements, wrist rotations, and grasp patterns of the hand can be decoded from the recorded activity of multiple neurons in the motor cortex. Recent studies have shown the possibility of decoding kinematic parameters of movement during individuated finger movements and simple grasping motion from the ECoG signal in humans [74,75]. However, no one has tried the online control of two more gestures control. Wang et al. [58] investigated the grasp decoding performance of ECoG in human subjects and online controlling of the robotic hand. First, they tried to decode the grasping types which were rock, paper, and scissors offline. Next, they used a proper decoding model to send the command to the robotic arm of the type of gesture in order to control it online. The results of this study provide further evidence that it is possible to employ ECoG signals from sensorimotor cortex to realize grasping movement.

Two subjects who suffered from intractable epilepsy and required surgical treatment in an attempt to control the epileptic seizure were included in this study. All procedures were followed up and followed the guidance and were approved by the Second Affiliated Hospital of Zhejiang University, China. Subjects gave written informed consent after detailed explanations of the potential risks of the research experiment.

Subjects were required to perform a grasping task. A vertical computer screen was placed vertically in front of the subject. On the center of the monitor was a cross mark instructing the subjects to focus on it in the ready period. Then the cue showed, instructing the subject and the subject was required to grasp according to the cue and hold on until the cue disappeared. This period lasted 2 s followed by a release period.

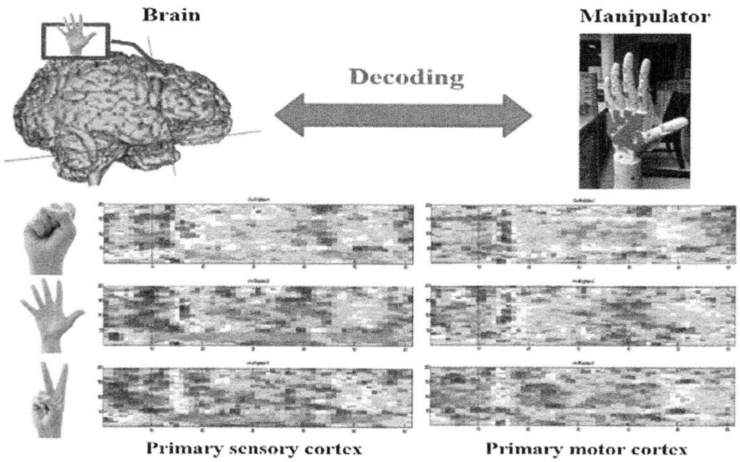


**FIGURE 1.4** Decoding ECoG signals in "Rock-scissor-paper game".

This process was repeated 50 times in one block, with each session containing three blocks. Three different types of gestures, rock, paper, and scissors, were randomly displayed (Fig. 1.4).

Two patients (one male and one female) were included in this study. In the cued grasp movement, the averaged and normalized power spectrum represented a movement-related activation from the channel located both precentrally and postcentrally. With movement onset, the high gamma power increased and the beta power decreased. According to the decoding accuracy after cross-validation, the high gamma frequency (80–120 Hz) of 13 channels in the sensorimotor cortex was included as the decoding input, the outcome reached up to 85%–95% offline.

Besides the routine usage of ECoG signal as a way to localize the foci, these studies provided the possibility that the ECoG signal could also be used to implement the simple hand grasp function. As grasp movement is quite important in daily life, the decoding of the grasp hand could lead to better design of the brain–machine interface. Based on this technique and looking forward into the future, brain–computer interfaces may finally be used to replace or restore useful function for people who have lost natural motor outputs. They also might improve rehabilitation for people with strokes, head trauma, and other disorders.

Real-life application of ECoG BCI demands mitigation of its invasive nature by minimizing both the size of the involved brain regions and the number of implanted electrodes. Zhang et al. [9] proposed a new BMI paradigm that utilizes the attentional modulation of visual motion response. With ECoG data collected from five epilepsy patients, a power increase of the high gamma (60–140 Hz) frequency range was found to be associated with overtly attended movement of visual stimuli in the parietal–temporal–occipital junction and the occipital cortex. Event-related potentials (ERPs) were elicited as well but with broader cortical distribution. They achieved significantly higher BCI classification accuracy by employing both high gamma and ERP responses from a single ECoG electrode than by using ERP responses only ($84.22 \pm 5.54\%$ vs. $75.48 \pm 4.18\%$, $P < .005$, paired $t$-test, three-trial averaging, binary results of attended vs. unattended). More importantly, the high gamma responses were located within brain regions specialized in visual motion processing as mapped by fMRI, suggesting the spatial location for electrode implantation can be determined prior to surgery using noninvasive imaging. Their findings demonstrate the feasibility of implementing a minimally invasive ECoG BCI.

## Closed-loop BMI for seizure detection and inhibition

Bidirectional brain–machine interfaces can provide a direct interaction between the brain and a machine. One application of a bidirectional BMI system is seizure detection and control. Usually, a closed-loop system based on BMI is set up which delivers a therapeutic electrical stimulus only in response to seizure onsets. The side effects of neurostimulation can thus be greatly reduced, as the electricity is not regularly applied to brain tissue. Qi et al. [16] proposed a new BCI-based responsive stimulation system (Fig. 1.5). With an efficient morphology-based seizure detector, seizure events can be identified in the early stage, which could trigger electrical stimulation into the brain. The proposed system was tested on rats with penicillin-induced epileptic seizures. Online experiments show that 83% of seizures could be detected successfully with a short average time delay of 3.11 s. With BCI-based seizure control system therapy, most seizures were suppressed within 10 s. Compared with the control group, the average seizure duration was reduced by 30.7%.

This study demonstrated that with the efficient morphology-based spike identification method, seizures could be recognized accurately within a short time to trigger

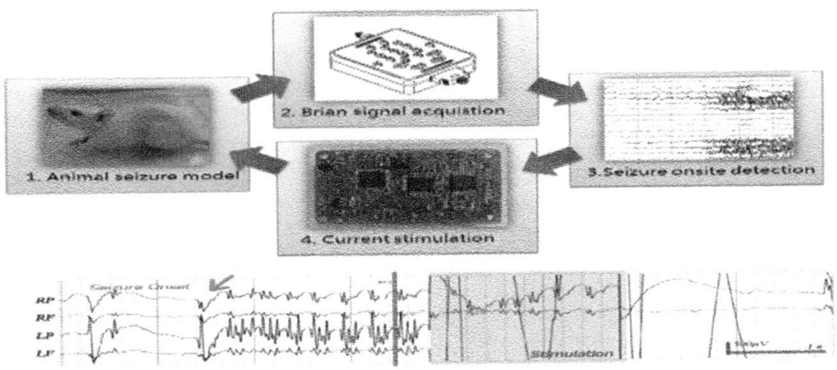


**FIGURE 1.5** Diagram of the BMI-base responsive stimulation system. The BMI system has three main components: a signal acquisition system as the input, a seizure detector as the control, and a current-controlled stimulator as the output. A stimulus is triggered once a seizure is detected.

therapy stimulations in the early stages of onset. Experimental results showed that the responsive seizure suppression system could abort ongoing seizures effectively.

## Conclusion

In recent years China has made encouraging progress in invasive BMI studies, and thus the increasing interest has been attracted by BMI technologies for implementing advanced applications to restore and potentially augment human physical and mental capabilities, and advances in neuroscience and engineering are making this a reality. However, most of the neural prosthetic or human augmentation applications based on BMIs are not yet fully mature technologies, BMIs are still facing the following challenges:

**(1)** Adaptive decoding: how to find nonlinear and dynamic neural decoding methods to precisely interpret tuning properties;
**(2)** Multimodal feedback: how to develop multimodal feedback techniques to the brain, especially the need for tactile and proprioceptive feedback to augment vision;
**(3)** Coding of sensory information: how to encode the sensory information using brain stimulation (intracortical microstimulation or optogenetics), and how to learn the stimulus patterns by the brain if the stimulus is unnatural;
**(4)** Blending of biological and artificial intelligence: how to make the brain and the machine coadaptive and interact in harmony and thus be highly integrated in the BMI system.

Brain science has become a hot issue around the world in recent years, and BMI has already become a powerful tool to address fundamental questions in neuroscience. Through synergistic interactions between traditional engineering and basic science, today in China BMI is growing like a tree that opens its branches into many traditional engineering fields, and also extends deep roots into basic neuroscience.

## Acknowledgments

This work is supported by grants from the National Key Research and Development Program of China (2017YFC1308501), the research fund of PLA of China (BWS17J024), the Fundamental Research Funds for the Central Universities. The authors also thank Mr. Yimin Shen from the Department of BME of Zhejiang University for his assistance and support in their studies.

## References

[1] Schwartz AB, Cui XT, Weber DJ, Moran DW. Brain-controlled interfaces: movement restoration with neural prosthetics. Neuron 2006;52(1):205–20.

[2] Donoghue JP. Bridging the brain to the world: a perspective on neural interface systems. Neuron 2008;60(3):511–21.

[3] Daly J. Brain-computer interfaces in neurological rehabilitation. The Lancet Neurology 2008;5.

[4] Wei W, Collinger JL, Degenhart AD, Tyler-Kabara EC, Schwartz AB, Moran DW, et al. An electrocorticographic brain interface in an individual with tetraplegia. PLoS One 2013;8(2):e55344.

[5] Talwar SK, Xu S, Hawley ES, Weiss SA, Moxon KA, Chapin JK. Rat navigation guided by remote control. Nature 2002;417(6884):37–8.

[6] O'Doherty JE, Lebedev MA, Ifft PJ, Zhuang KZ, Shokur S, Bleuler H, et al. Active tactile exploration using a brain–machine–brain interface. Nature 2011;479(7372):228–31.

[7] Dai J, Zhang H, Zhang S, Li Q, Liu X, Hao Y, et al. Temporal-spatial model decoding for neuron ensemble spikes from rats' cortex. Science in China Series C, Life Sciences 2009;39(8):736–45.

[8] Zhang QS, Zhang SM, Hao YY, Zhang HJ, Zhu JM, Zhao T, et al. Development of an invasive brain–machine interface with a monkey model. Chinese Science Bulletin 2012;57(16):2036–45.

[9] Zhang D, Song H, Xu R, Zhou W, Ling Z, Hong B. Toward a minimally invasive brain–computer interface using a single subdural channel: a visual speller study. NeuroImage 2013;71(Complete): 30–41.

[10] Feng ZY, Chen WD, Ye XS, Zhang SM, Zheng XJ, Wang P, et al. A remote control training system for rat navigation in complicated environment. Journal of Zhejiang University (Science) 2007;8(2): 323–30.

[11] Chen S, Zhou H, Guo S, Zhang J, Qu Y, Feng Z, et al. Optogenetics based rat–robot control: optical stimulation encodes 'stop' and 'escape' commands. Annals of Biomedical Engineering 2015;43: 1851–64.

[12] Wang W, Guo C, Sun J, Dai Z. Gecko's brain stereotaxic method and device. Chinese Science 2007; 52(21):2524–8.

[13] Yang J, Su X, Huai R, Liu X. Animal-robot remote control system based on a new type of multichannel brain stimulation. Progress in Natural Science 2007;17(3):379–84.

[14] Wu Y. Study on the control technique of brain for underwater animal-robot. Yanshan University; 2010.

[15] Zheng NG, Chen WD, HU FL, Li B, Zhao HX, Shen W, et al. Research Progress and Challenges in Cyborg Insects 2011;41(4):259–72.

[16] Qi Y, Ma FQ, Ge TT, et al. Abidirectional brain-computer interface for effective epilepsy control. Journal of Zhejiang University Science 2014;15(10):839–47.

[17] Zhang S, Chen W, Xu K, Hao Y, Zheng X, Zhang J, et al. Brain-machine interfaces: get what you think. IEEE Spectrum 2015;6:98–9.

[18] Yu Y, Zhang S, Zhang H, Liu X, Zhang Q, Zheng X, et al. Neural decoding based on probabilistic neural network. Journal of Zhejiang University Science 2010;11(4):298–306.

[19] Specht DF. Probabilistic neural networks. Neural Networks 1990;3(1):109–18.

[20] Carmena JM, Lebedev MA, Crist RE, O'Doherty JE, Santucci DM, Dimitrov DF. Learning to control a brain–machine interface for reaching and grasping by primates. PLoS Biology 2003;1(2):e2.

[21] Black WWMJ, Gao Y, Serruya M, Shaikhouni A, Bienenstock E, Donoghue JP. Neural decoding of cursor motion using a Kalman filter. Advances in Neural Information 2003;15(1):133–40.

[22] Zhou F, Liu J, Yu Y, Tian X, Liu H, Hao YY, et al. Field-programmable gate array implementation of a probabilistic neural network for motor cortical decoding in rats. Journal of Neuroscience Methods 2010;185(2):299–306.

[23] Zhang S, Wang P, Jiang J, Liu F, Ye X, Chen W, et al. Research of control system for rat remote navigation and behavior training. Chinese Journal of Biomedical Engineering 2007;26(6):830–6.

[24] Su X, Huai R, Yang J, Wang H, Lv C. Brain mechanism and methods for robo-animal motor behavior control. Scientia Sinica (Informationis) 2012;42(9):1130–2012.

[25] Li H, Dai Z, Tan H, Guo C, Sun J. A remote system for gecko animal-robot. Computer Technology 2008;18(8):16–9.

[26] Bao L, Zheng N, Zhao H, Hao Y, Zheng H, Hu F, et al. Flight control of tethered honeybees using neural electrical stimulation. In: International IEEE/EMBS conference on neural engineering. IEEE; 2011.

[27] Lin J, Yu C, Jia J, Zhang S, Wang Y, Chen W. Using dlPAG-evoked immobile behavior in animal-robotics navigation. In: International conference on computer science & education. IEEE; 2010.

[28] Sun C, Zheng N, Zhang X, Chen W, Zheng X. Automatic navigation for rat-robots with modeling of the human guidance. Journal of Bionic Engineering 2013;10(1):46–56.

[29] Sun C, Zheng N, Zhang X, Chen W, Zheng X. An automatic control model for rat-robot. In: Annual International conference of the IEEE engineering in medicine and biology society. IEEE; 2011.

[30] Zhang Y, Sun C, Zheng N, Zhang S, Lin J, Chen W, et al. An automatic control system for ratbot navigation. In: IEEE/ACM international conference on green computing and communications & 2010 IEEE/ACM international conference on cyber, physical and social computing. IEEE Computer Society; 2010.

[31] Sun C, Zhang X, Zheng N, Zheng X. Bio-robots automatic navigation with electrical reward stimulation. In: International conference of the IEEE engineering in medicine & biology society. Conference proceedings: annual international conference of the IEEE Engineering In Medicine And Biology Society; 2012.

[32] Mcintyre CC, Mori S, Sherman DL, Thakor NV. Electric field and stimulating influence generated by deep brain stimulation of the subthalamic nucleus. Clinical Neurophysiology 2004;115(3):589–95.

[33] Deisseroth K, Feng G, Majewska AK, Miesenböck G, Ting A, Schnitzer MJ. Next-generation optical technologies for illuminating genetically targeted brain circuits. Journal of Neuroscience 2006; 26(41):10380–6.

[34] Guo S, Zhou H, Zhang J, Xu K, Zheng X. A multi-electrode array coupled with fiberoptic for deep-brain optical neuromodulation and electrical recording. Conference Proceedings: Annual International Conference of the IEEE Engineering in Medicine and Biology Society 2013;2013(2013): 2752–5.

[35] Chen S, Qu Y, Guo S, Xu K, Zheng X. Encode the "STOP" command by photo-stimulation for precise control of rat-robot. Conference Proceedings: Annual International Conference of the IEEE Engineering in Medicine and Biology Society 2013;2013:5.

[36] Guo SC, Zhou H, Wang YM, Xu K, Zheng X. A rat-robot control system based on optogenetics. Applied Mechanics and Materials 2013;461:5.

[37] Guo S, Chen S, Zhang Q, Wang YM, Xu K, Zheng X. Optogenetic activation of the excitatory neurons expressing CaMKIIα in the ventral tegmental area upregulates the locomotor activity of free behaving rats. BioMed Research International 2014;2014(1–6):687469.

[38] Ye XS, Wang P, Liu J, Zhang SM, Jiang J, Wang QB, et al. A portable telemetry system for brain stimulation and neuronal activity recording in freely behaving small animals. Journal of Neuroscience Methods 2008;174(2):186–93.

[39] Sanchez JC, Carmena JM, Lebedev MA, Nicolelis MAL, Harris JG, Principe JC. Ascertaining the importance of neurons to develop better brain-machine interfaces. IEEE Transactions on Biomedical Engineering 2004;51(6):943–53.

[40] Wang Y, Principe JC, Sanchez JC. Ascertaining neuron importance by information theoretical analysis in motor Brain–Machine Interfaces. Neural Networks 2009;22(5–6):781–90.

[41] Kahn K, Sheiber M, Thakor N. Neuron selection for decoding dexterous finger movements. In: International conference of the IEEE engineering in medicine & biology society. IEEE; 2011.

[42] Xu K, Wang Y, Wang Y, Wang F, Hao Y, Zhang S, et al. Local-learning-based neuron selection for grasping gesture prediction in motor brain machine interfaces. Journal of Neural Engineering 2013; 10(2):026008.

[43] Sun Y, Todorovic S, Goodison S. Local-learning-based feature selection for high-dimensional data analysis. IEEE Transactions on Pattern Analysis and Machine Intelligence 2010;32(9):1610–26.

[44] Wang Y, Sanchez JC. Sequential Monte Carlo point-process estimation of kinematics from neural spiking activity for brain-machine interfaces. Neural Computation 2009;21(10):2894–930.

[45] Ergun A, Barbieri R, Eden UT, Wilson MA, Brown EN. Construction of point process adaptive filter algorithms for neural systems using sequential Monte Carlo methods. IEEE Transactions on Biomedical Engineering 2007;54(3):419–28.

[46] Truccolo W, Hochberg LR, Donoghue JP. Collective dynamics in human and monkey sensorimotor cortex: predicting single neuron spikes. Nature Neuroscience 2010;13(1):105–11.

[47] Truccolo W, Eden UT, Fellows MR, Donoghue JP, Brown EN. A point process framework for relating neural spiking activity to spiking history, neural ensemble, and extrinsic covariate effects. Journal of Neurophysiology 2004;93(2):1074–89.

[48] Pillow JW, Shlens J, Paninski L, Sher A, Litke AM, Chichilnisky EJ, et al. Spatio-temporal correlations and visual signalling in a complete neuronal population. Nature 2008;454(7207):995–9.

[49] Xu K, Wang Y, Wang F, Liao Y, Zhang Q. Neural decoding using a parallel sequential Monte Carlo method on point processes with ensemble effect. BioMed Research International 2014;2014:1–11.

[50] Digiovanna J, Mahmoudi B, Fortes J, Principe JC, Sanchez JC. Coadaptive brain-machine interface via reinforcement learning. IEEE Transactions on Biomedical Engineering 2009;56(1):54–64.

[51] Mahmoudi B, Sanchez JC. A symbiotic brain-machine interface through value-based decision making. PLoS One 2011;6.

[52] Chavarriaga R, Millan JDel R. Learning from EEG error-related potentials in noninvasive brain-computer interfaces. IEEE Transactions on Neural Systems and Rehabilitation Engineering: A Publication of the IEEE Engineering in Medicine and Biology Society 2010;18(4):381–8.

[53] Roelfsema PR, Ooyen AV. Attention-gated reinforcement learning of internal representations for classification. Neural Computation 2005;17(10):2176–214.

[54] Wang Y, Wang F, Xu K, Zhang Q, Zhang S, Zheng X. Neural control of a tracking task via attention-gated reinforcement learning for brain-machine interfaces. IEEE Transactions on Neural Systems and Rehabilitation Engineering 2015;23(3):458–67.

[55] Chestek CA, Gilja V, Nuyujukian P, Foster JD, Fan JM, Kaufman MT, et al. Long-term stability of neural prosthetic control signals from silicon cortical arrays in rhesus macaque motor cortex. Journal of Neural Engineering 2011;8(4):045005.

[56] Mehring C, Rickert J, Vaadia E, Cardosa de Oliveira S, Aertsen A, Rotter S. Inference of hand movements from local field potentials in monkey motor cortex. Nature Neuroscience 2003;6(12): 1253–4.

[57] Flint RD, Wright ZA, Scheid MR, Slutzky MW. Long term, stable brain machine interface performance using local field potentials and multiunit spikes. Journal of Neural Engineering 2013;10(5): 056005.

[58] Wang D, Zhang Q, Li Y, Wang Y, Zhu J, Zhang S, et al. Long-term decoding stability of local field potentials from silicon arrays in primate motor cortex during a 2D center out task. Journal of Neural Engineering 2014;11(3):036009.

[59] Acharya S, Tenore F, Aggarwal V, Etienne-Cummings R, Schieber MH, Thakor NV. Decoding individuated finger movements using volume-constrained neuronal ensembles in the M1 hand area. IEEE Transactions on Neural Systems and Rehabilitation Engineering 2008;16(1):15–23.

[60] Vargas-Irwin CE, Shakhnarovich G, Yadollahpour P, Mislow JMK, Black MJ, Donoghue JP. Decoding complete reach and grasp actions from local primary motor cortex populations. Journal of Neuroscience 2010;30(29):9659–69.

[61] Hao YY, Zhang QS, Zhang SM, Zhao T, Wang Y, Chen W, et al. Decoding grasp movement from monkey premotor cortex for real-time prosthetic hand control. Chinese Science Bulletin 2013; 58(20):2512–20.

[62] Santello M, Flanders M, Soechting JF. Postural hand synergies for tool use. The Journal of Neuroscience 1998;18(23):10105–15.

[63] Santello SM. Invited Review Hand function: peripheral and central constraints on performance. Journal of Applied Physiology 2004;96(6):2293–300.

[64] Wise SP, Boussaoud D, Johnson PB, Caminiti R. Premotor and parietal cortex: corticocortical connectivity and combinatorial computations. Annual Review of Neuroscience 1997;20(20):25.

[65] Caminiti R, Johnson PB, Galli C, Ferraina S, Burnod Y. Making arm movements within different parts of space: the premotor and motor cortical representation of a coordinate system for reaching to visual targets. Journal of Neuroscience 1991;11(5):1182–97.

[66] Fu QG, Suarez JI, Ebner TJ. Neuronal specification of direction and distance during reaching movements in the superior precentral premotor area and primary motor cortex of monkeys. Journal of Neurophysiology 1993;70(5):2097–116.

[67] Messier J, Kalaska JF. Covariation of primate dorsal premotor cell activity with direction and amplitude during a memorized-delay reaching task. Journal of Neurophysiology 2000;84(1):152.

[68] Raos V, Umiltá MA, Murata A, Fogassi L, Gallese V. Functional properties of grasping-related neurons in the ventral premotor area F5 of the macaque monkey. Journal of Neurophysiology 2005;95(2):709–29.

[69] Raos V, Umiltá R, Murata A, Fogassi L, Gallese V, Umiltáet R, et al. Functional properties of grasping-related neurons in the dorsal premotor area F2 of the macaque monkey. Journal of Neurophysiology 2004;92(4):1990.

[70] Fogassi L. Somatotopic organization of the lateral part of area F2 (dorsal premotor cortex) of the macaque monkey. Journal of Neurophysiology 2003;89(3):1503–18.

[71] Fogassi L, Gallese V, Raos V, Gallese V, Raos V, Rizzolatti G. Object representation in the ventral premotor cortex (area F5) of the monkey. Journal of Neurophysiology 1997;78(4):2226.

[72] Hao Y, Zhang Q, Controzzi M, Cipriani C, Li Y, Li J, et al. Distinct neural patterns enable grasp types decoding in monkey dorsal premotor cortex. Journal of Neural Engineering 2014;11(6):066011.

[73] Cao Y, Hao Y, Liao Y, Xu K, Wang Y, Zhang S, et al. Information analysis on neural tuning in dorsal premotor cortex for reaching and grasping. Computational & Mathematical Methods in Medicine 2013;2013(7398):730374.

[74] Acharya S, Fifer MS, Benz HL, Crone NE, Thakor NV. Electrocorticographic amplitude predicts finger positions during slow grasping motions of the hand. Journal of Neural Engineering 2010;7(4). 046002-13.

[75] Kubánek J, Miller KJ, Ojemann JG, Wolpaw JR, Schalk G. Decoding flexion of individual fingers using electrocorticographic signals in humans. Journal of Neural Engineering 2009;6(6):066001.

2

# Neural decoding by invasive electrocorticography

Yue Li[1,2,4], Yu Qi[2,3,4], Gang Pan[2,3,4], Shaomin Zhang[1,2,4]

[1]ZHEJIANG LAB, HANGZHOU, CHINA; [2]QIUSHI ACADEMY FOR ADVANCED STUDIES, ZHEJIANG UNIVERSITY, HANGZHOU, CHINA; [3]COLLEGE OF COMPUTER SCIENCE AND TECHNOLOGY, ZHEJIANG UNIVERSITY, HANGZHOU, CHINA; [4]KEY LABORATORY OF BIOMEDICAL ENGINEERING OF MINISTRY OF EDUCATION, HANGZHOU, CHINA

## Chapter outline



**Intelligent Biomechatronics in Neurorehabilitation. https://doi.org/10.1016/B978-0-12-814942-3.00002-7**

## Introduction

Brain–computer interfaces (BCIs) have great potential to provide new rehabilitation technologies to people with motor disabilities because of their capabilities of bypassing the interrupted motor pathways caused by neurological disorders and building direct communication between the brain and external devices by decoding brain signals [1]. In the past several decades, electrocorticography (ECoG) has been widely applied to localize the seizure foci of patients with intractable epilepsy. Since ECoG electrodes were directly placed on the surface of the cortex, ECoG signals provide higher spatial resolution and signal quality than noninvasive neural signals, such as EEG (electroencephalography) signals [2]. Compared with neural signals obtained from the invasive microelectrode array (MEA), ECoG recording is less invasive and reduces damage to the cerebral cortex [3]. Therefore, ECoG not only contains richer information than EEG widely used in noninvasive BCIs but also provides better long-term stability than neural ensemble recording used in invasive BCIs. Because of its good trade-off between performance and long-term stability, an ECoG-based BCI system has attracted considerable and extensive interest as a semiinvasive BCI.

In the early studies, ECoG signals were used to decode high-dimensional arm movement and predict movement directions [4–6]. Because of the importance of hand grasping movement in daily living and communication, many researchers have been trying to decode the movement of fingers or hands with ECoG signals in recent years. Most existing approaches for hand movement decoding fall into two categories: finger movement regression and hand gesture classification. Finger movement regression aims to predict the flexion trajectories of individual fingers [7–10]. However, in most cases only a single finger moved when the flexion trajectories of individual fingers were recorded. Very few studies have tried to decode flexion trajectories of fingers when multiple fingers move simultaneously [11]. When multiple fingers move simultaneously, although the mixed signals of multiple finger movements could be recorded by ECoG electrodes, the temporal overlapping and spatially sparse sampling make it difficult.

Instead of decoding the flexion trajectories of multiple fingers, hand gesture classification directly regards hand posture decoding as a classification problem, which is a more practical solution for prosthesis control. Yanagisawa et al. proposed a real-time decoding system to classify three hand gestures with a linear classifier [12]. Chestek et al. proposed using a naive Bayes decoder to effectively classify five hand postures from the ECoG signals [13].

Although ECoG has been proven to be a good candidate for BCI signal sources for prosthetic hand control, there are still several issues such as surgery risks and decoding performance that need to be carefully investigated before being applied into clinical practice. One of the main challenges in ECoG-based BCIs is how to minimize the cortical area required for the ECoG recording. Conventional intracranial ECoG electrodes are arranged in grids or strips covering a large area of the cortex surface, which aims to monitor brain activities in large regions simultaneously and localize the seizure onset zone. Some studies have mentioned that the number and locations of electrodes which were chosen exclusively for clinical purposes did not precisely agree with the requirements for the BCI usage. The large extent of conventional implantation of ECoG grids exerts a serious effect on the brain, which significantly increases the risk in surgery and postoperative recovery [14]. For these reasons, BCI users prefer a smaller but less invasive configuration of electrode grid without sacrificing decoding performance. Decreasing the number of channels on the premise of certain decoding accuracy could be an alternative for reducing the area for grid covering. Different strategies have been proposed to optimize the number of ECoG electrodes by reducing the number of input features of the decoder and improve both decoding accuracy and computational speed. Among them, Milekovic et al. restricted electrode channels in a limited cortical region using neighboring channels to decode arm movement [15]. Moreover, Zhang et al. decoded visual stimuli using a single ECoG electrode [16]. However, to our knowledge, few studies have investigated the optimal number of electrodes in hand gesture classification and studied the corresponding anatomical distribution of these working channels.

Another big issue in ECoG-based BCIs is how to decode the hand gesture with less delay when applying it on real-time prosthetic control. Current ECoG-based BCIs have demonstrated the strength of ECoG signals in hand gesture classification, however, they extracted features using statistics over a long time window which ignored the dynamics of ECoG signals. Since the hand grasping movement is a process, temporal patterns in ECoG signals contain potential information for decoding. To capture the temporal patterns, Bleichner et al. and Branco et al. proposed a temporal template-matching method to decode four gestures from ECoG signals [17,18]. With short-term time windows, the temporal patterns of different gestures can be characterized, which provides useful information to improve the accuracy of hand gesture decoding. However, the sequential relationship among these windows was not explicitly modeled for accurate decoding. There is still much room to improve hand gesture decoding by further exploiting the underlying temporal patterns and structures in ECoG signals.

In this chapter, we will introduce our trials in improving the safety and performance of ECoG-based BCIs for hand gesture decoding through one of our clinical trials, which has been reported [14,19]. In this study, we tried to minimize the area for electrode grid covering by analyzing the ECoG representation in the sensorimotor cortex during the execution of hand gestures. We further proposed an RNN-based decoder combined with a greedy strategy for rapid and accurate decoding hand gestures from ECoG signals. Its performance was compared with some typical methods reported in other studies.

# Experimental paradigm and data collection

## Participants and implantation

Two participants in this study were suffering from intractable epilepsy and required surgical treatment for epileptic seizure control. The clinical subdural electrodes were surgically implanted in the sensorimotor cortex for clinical monitoring and localization of the seizure foci. The configuration and location of the electrodes, as well as the duration of the implantation, were determined by clinical requirements. The clinical electrodes were platinum electrodes with a diameter of 4 mm (2.3 mm exposed), spacing at 10 mm, and generally implanted only for a period ranging from several days up to 2 weeks.

The key information on the participants and their implantation sites are shown in Table 2.1. All procedures were followed from the guide and approved by the Second Affiliated Hospital of Zhejiang University, China. Participants gave written informed consent after a detailed explanation of the potential risks of the research experiment.

## Cortical mapping

The location of the electrodes was confirmed by postoperative computed tomography (CT) scans. Both participants went through the clinical examination routine of the motor, sensory, language function, and so on with functional stimulation mapping (FSM), which helped us to further localize the electrodes functionally. None of the hand motor areas of the two participants was the seizure onset zone in our studies.

## Behavioral tasks

Participants were instructed to perform one of three hand gestures ("scissors," "rock," and "paper") or relax their hands in a rest position according to visual cues presented on the screen in front of them. In the rest position, participants were asked to relax their task hands and flex the fingers slightly with palms facing up. A trial began with a verbal cue "ready" meanwhile a cross displayed on the center of the screen, indicating participants to keep task hands in the rest position and be prepared. This was the "baseline period" (2–2.5 s randomly). After the baseline period, the cross was replaced by a gesture picture ["go" cue (GC)], which randomly displayed one of the three gestures. Participants

**Table 2.1** Information on the participants and electrode locations.

| Participants | Gender | Age | Handedness (task hand) | Implanted ECoG grids | Seizure focus |
|---|---|---|---|---|---|
| P1 | Female | 28 | Right (right) | Left hemisphere: temporal, parietal, occipital lobe | Temporal lobe |
| P2 | Male | 22 | Right (left) | Right hemisphere: frontal medial, dorsal surface, parietal lobe | Frontal lobe |

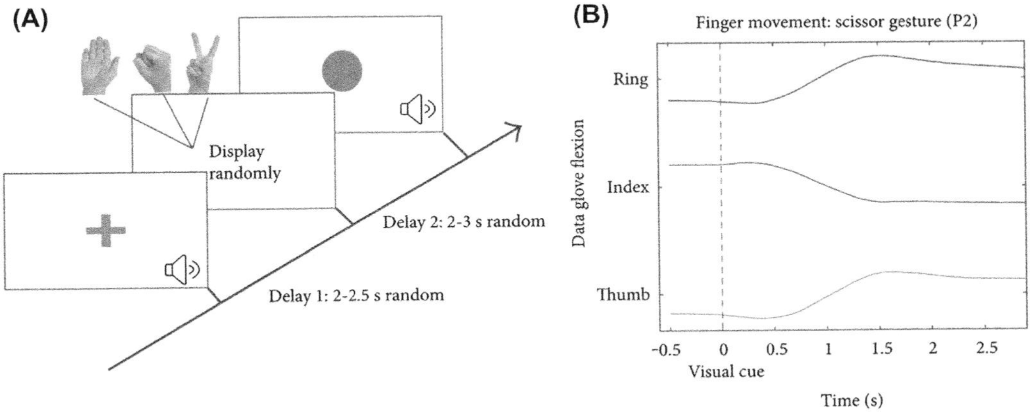


**FIGURE 2.1** The behavioral task and recordings. (A) Hand grasp experimental paradigm. The trial was initiated by a *red cross* [gray in print version] displayed on the center of the screen with a verbal cue "ready." After a random delay ranging from 2 to 2.5 s, the *red cross* [gray in print version] disappeared and the gesture cue appeared on the screen, indicating the participant to replicate the gesture shown and maintain it until the *red dot* [gray in print version] came out 2–3 s later. The correction of the trial was fed back by another verbal cue at the end of the trial. Three types of gestures were displayed randomly and equally. (B) The flexion of the data glove sensors on the ring finger, index finger, and thumb. The *purple dashed line* [light gray in print version] represents the timing when the visual cue was displayed.

were informed to perform the gesture instantly and hold on it until a red circle ["stop" cue (SC)] appeared. The gesture displayed 2–3.5 s randomly. After SC, participants could release the gesture and return to the rest position. At the end of each trial, a verbal feedback, that is *Correct* or *Wrong*, was given by the experimenter to inform the participant whether the trial was successful or not. The entire course of a task is illustrated in Fig. 2.1A. The trials were failed and excluded from the final dataset if participants were not able to maintain the gestures until SC appeared or forgot to release the gestures. Before the ECoG electrode implantation, participants were trained to acquaint themselves with the task until they fully understood the processes and requirements.

Each session was composed of three blocks, and each block was composed of 50 trials (five sessions for P1 and P2). Participants had a short break between the blocks. In practice, the number of trials and the duration of each break depended on the medical condition and the willingness of the participants.

## Neural signals and behavioral data recording

Clinical ECoG signals collected through subdural ECoG grids were recorded by a 128-channel NeuroPort system (Blackrock Microsystems, US). The ECoG signals were firstly low filtered with a cutoff frequency of 500 Hz and stored continuously during the whole task at the sampling rate of 2 kHz. The channels which contained a high level of noise were excluded by visual inspection. The timestamps of external events, such as "go" cues and "stop" cues, were synchronized with recorded ECoG signals by acquiring timestamps from the NeuroPort system. The behavioral data were collected by a 5DT data glove with 14 sensors (5DT Inc., US). Each sensor simultaneously yielded flexion values for posture detection. Fig. 2.1B shows the flexion values of three sensors on the ring, index and thumb, respectively. The data were collected when P2 performed the scissor gesture, and the curves were smoothed by a Savitzky–Golay filter (three orders, 101 points). We defined the occurrence of a movement onset when consecutively five first derivative of the flexion values exceeded a specific threshold. In the experiment, we rejected trials with movement artifacts or electrode failures by visual inspection. After removing invalid trials, the dataset included 243 samples for P1, and 394 samples for P2.

# Hand gesture encoding within ECoG

## Data analysis and channel selection

### *Feature extraction*

Offline data processing was performed on a MATLAB® platform. First, a spatial filter, which is a common average reference, was applied to all the remaining channels after visual inspection to remove common noise.

The ECoG feature was the power spectrum captured in different frequency components. By analyzing its dynamical spatiotemporal pattern, we could characterize the neural features associated with different movement status and hand gesture types. To obtain the time-resolved power spectrum of ECoG signals, we segmented the ECoG time series of an entire session into 300 ms width windows with an overlap of 200 ms. Then a three-order multitaper spectral estimation was employed to calculate the power spectrum in each window of all the selected channels. Multitaper spectral estimations apply a set of orthogonal tapers to the time series and minimize variance by averaging all the tapered and independent spectra.

Whereas due to the fact that the power spectrum of brain signals decreases with increasing frequency, known as a "power-law," the changes in low frequency dominate the entire frequency range. A normalization in each separated frequency band of the spectrum is very necessary to eliminate this phenomenon for the observation of the power spectrum variation in high frequency. Therefore, the baseline power spectrum was calculated by averaging the power spectrum obtained during the baseline period before the visual cue across trials. Then the frequency-resolved power spectrum was normalized by dividing the averaged baseline power spectrum for each trial in each frequency band.

### *Decoding performance evaluation*

We used the MATLAB® Libsvm package to build a multiclass support vector machine (SVM) classification model and realize the gesture type classification [20]. The input features of this classifier were the frequency-resolved power spectrum in different frequency bands across specific time bins and channels, while the corresponding target outputs were the labels of gestures ("Scissors" = 1, "Rock" = 2, "Paper" = 3). Here, the feature of each trial was a 1 * $n$ vector, in which $n$ equaled the number of channels * -frequency bins * 10-time bins after the movement onset.

In offline decoding, we pooled all the trials of each participant and applied threefold cross-validation to the dataset. The decoding performance was the average of the percentage of correct predictions using testing data 50 times. In addition, the chance level was the result of the 95th percentile of the decoding accuracy distribution which contained 10,000 results generated from the testing data with randomly shuffled labels. We used t-test as the significance test and calculated the $P$-values.

### *Channel selection strategies*

We used single-channel selection and greedy selection to progressively select the $n$-best channels offline in our study. In the single-channel selection, we first calculated the decoding performance of each channel. The $n$-best channels were the channels achieving the $n$ highest decoding performances on the single-channel level, constituting the input vector of SVM for training and testing. In the greedy selection, we first picked out the channel which yielded the best single-channel performance from the total number of channels. Then, in the next round, the second best channel was selected which could improve the decoding performance most when paired with the first channel. The $n$ best channels were selected successively by repeating the process and then constituted the input vector of SVM for training and prediction. The decoding performance was saturated with $n$ channels when there was no significant increase in decoding performance using $n + 1$ best channels. In this study, we also compared the decoding performance using most neighboring 4 and 9 channels.

## Results

### *Time–frequency analysis and decoding performance*

Fig. 2.2A shows the time–frequency plots illustrating the normalized and averaged power spectrum of three different gestures recorded by one representative channel of P1. The time–frequency plots were aligned with the "Go" cue. The power spectrum of ECoG signals showed a movement-related modulation during hand movement. The power increased from the start of the movement to the end in both the high gamma frequency band (>70–135 Hz) and the low-frequency band including the theta (4–8 Hz) and alpha frequency bands (8–12 Hz). This task-related modulation existed in similar frequency bands (high gamma and low-frequency bands) across participants. However, the modulation patterns of the power spectrum in these frequency bands varied with different gesture types, indicating that the modulation patterns might contain exclusive information on each gesture type. Therefore, it was highly possible to distinguish the gesture types using power modulation patterns in specified high gamma and low-frequency bands.

The decoding performances of both participants are shown in Fig. 2.2B. The ECoG features were extracted from low, high gamma, and hybrid frequency bands of all

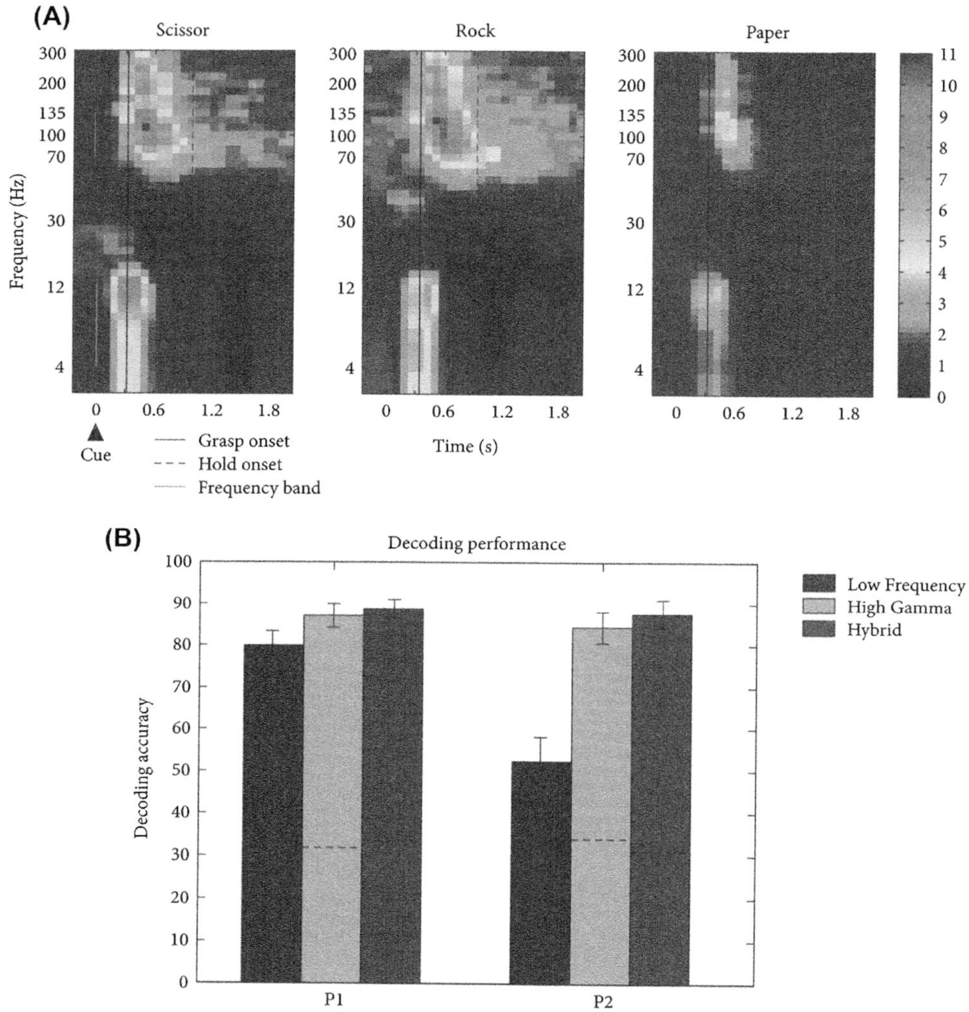


**FIGURE 2.2** Time–frequency plots of the ECoG signals and decoding performance obtained from all channels and selected channels. (A) Normalized power spectrum of ECoG signal from a representative channel 35 over sensorimotor cortex of P1 which was averaged and aligned with visual cue (time = 0). The frequency increases in a log scale, and the color bar gives the scale of the spectrum amplitude values. The gestures from left to right are scissor, rock, and paper, respectively. *Vertical magenta bars* [dark gray in print version] in the subplot indicate the frequency bands used in decoding [high gamma frequency band (70–135 Hz) and low-frequency band (4–12 Hz)]. *Black solid lines* represent the averaged grasping onset across trials, and the *gray dashed lines* represent the averaged end timing of grasping. (B) The decoding performance of different frequency bands (low-, high-, and hybrid frequency bands, from left to right) using all channels (left column). The performance of the hybrid frequency band was the highest and the performance of the low-frequency band was the lowest in both participants (*P* value < .01). The *black dashed lines* represent the chance level.

channels, respectively. The offline decoding performances of both participants in different frequency bands were significantly above the chance level and the performance reached almost 90%. The hybrid decoding results significantly outperformed the other two frequency bands and the high-frequency decoding results significantly outperformed the low-frequency performance in all conditions (t-test, $P < .01$).

### *Channel selections and anatomical patterns*

In this study, the single-channel selection was first employed to intuitively select the channels with the highest decoding accuracy. Fig. 2.3 plots the decoding performances which varied with channel numbers based on single-channel selection, and they are saturated with 9 or 10 channels in each participant. We found that many plateau points occur before the decoding performance reached a saturation point in the single-channel selection strategy. This indicates that there is some movement-related information redundancy among these selected channels. Fig. 2.4 maps these selected channels to the implanted electrode grids. Except for a few channels, the general distributions of these channels present a pattern which can be categorized as clustered.

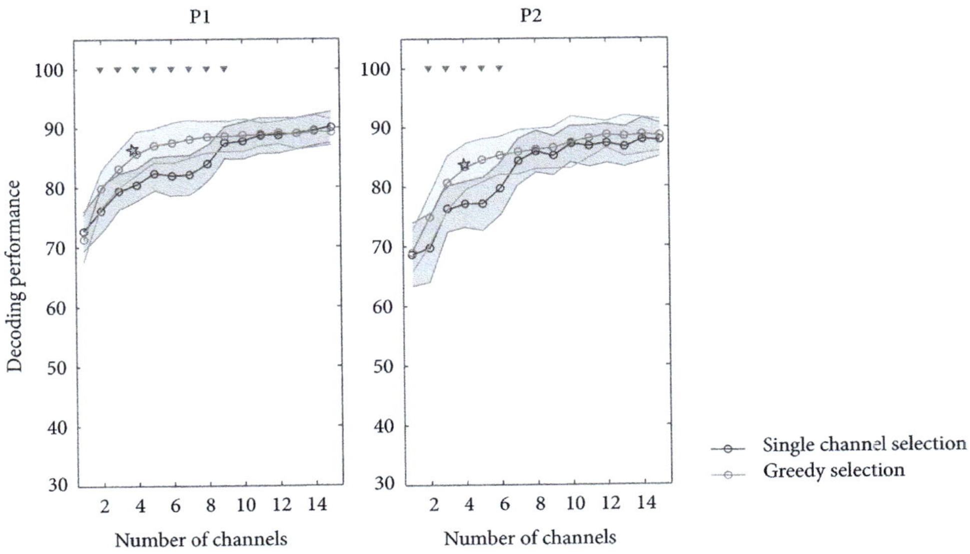


**FIGURE 2.3** The decoding performance varies as a function of the number of channels. The *green curves* [light gray in print version] represent the decoding performance using the channels selected by greedy selection, and the *red curves* [gray in print version] represent the result of the channels selected by single-channel selection. The performance of greedy selection yields a significantly higher performance than that of single-channel selection when the same number of channels was used. The *upside-down triangles* indicate that the greedy selection performance is significantly higher than the single-channel selection performance. The *yellow stars* [light gray in print version] indicate that the decoding performance obtained from the channel subset reaches the saturated point in the greedy selection. No significant improvement of decoding performance was found after adding more channels into the subset.

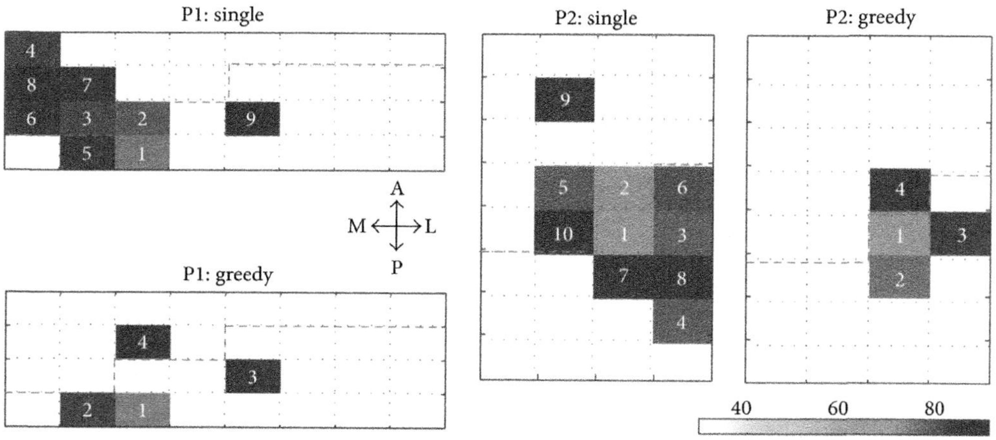


**FIGURE 2.4** The spatial distributions of the selected channels using single-channel selection and greedy selection of both participants. The *color bars* give the scale for the decoding performance. The numbers on the grid indicate the order of the selection. The performance yielded when the current channel was added onto the selected decoding channel subset. All the ECoG grids are in 4 * 8 configuration. Blue dashed lines mark the central sulcus.

To reduce the redundancy brought about by single-channel selection and select the most contributive channels, we further tried the greedy selection. The decoding performance of both participants displayed a trend of a fast increase at the beginning and then saturation after a certain number of channels were added, as shown in Fig. 2.4 using greedy selection. Notably, the decoding performance using the four best channels selected by the greedy algorithm reached their saturated points in both participants, at 85.7% and 84.5%, respectively. Although their performances were slightly lower than those obtained from all channels, the decoding performances were no longer significantly improved when much more additional channels were added into the subsets. Furthermore, the greedy selection showed significantly higher and stable performance than the single-channel selection algorithm did when the same number of channels was used for neural decoding. This implied that the greedy selection algorithm was much more appropriate and effective for selecting the minimum number of channels than was the single-channel selection. Furthermore, we found that most of these selected channels are clustered together and distributed along the central sulcus (shown in Fig. 2.4). For both participants, the selected channels were situated mainly in the precentral gyrus except one in the postcentral gyrus. All of the channels were next to or very close to the central sulcus at a maximum distance of two electrodes spacing.

*Decoding performance using nearest-neighboring channels*

In the above results from both participants, we found that the saturated decoding performance could be achieved by only a few number of the best channels selected by the greedy algorithm. This saturated performance is very close to that obtained with the neural signals from all channels. It is worth noting that the best three or four channels selected by the greedy algorithm are spatially close to each other and most were located near to the central sulcus. This suggests that the ECoG signals from these neighboring channels along the central sulcus might contain the most information about the difference between three types of gestures. Therefore we tried to investigate whether such a subset of neighboring electrodes from a small subregion of ECoG grid could provide enough neural signals for distinguishing three gestures by evaluating the decoding performance of the nearest neighboring 4 and 9 channels. Fig. 2.5A illustrates the highest decoding performances with different channel selection strategies (neighboring 4, neighboring 9, and greedy algorithm). All three strategies for channel selecting achieved a promising performance, which was significantly higher than the chance level. In P1 and P2, the decoding performances of three strategies were all above 82% and very close to each other. No significant difference was found among three strategies in P2. Although the decoding performance of the greedy algorithms is significantly higher than that of the neighboring 4 selection in P1, the difference between them was less than 4.0%. A general spatial pattern emerges, as Fig. 2.5B shows, where most of the channels selected by the greedy algorithm and by the neighboring 4 channels are also included in the square of optimal neighboring 9 channels and the channels are distributed along the central sulcus.

The results of both greedy and neighboring channel strategies consistently demonstrate here that the channels for hand gesture decoding could be restricted to the area of $3 \times 3$ square (approximately $4 \times 4$ cm) when using clinical subdural electrodes with a high decoding accuracy.

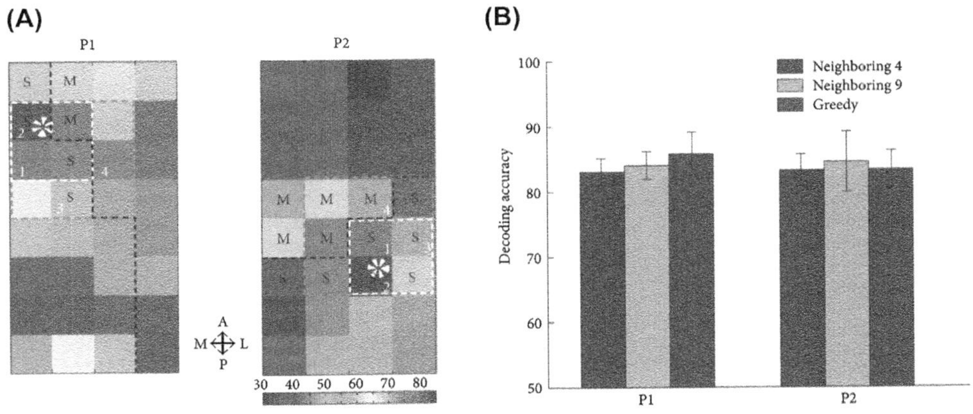


**FIGURE 2.5** Topographies of decoding performance using neighboring 4 channels and the decoding performance comparison of different channel selection strategies (neighboring 4, neighboring 9, and greedy) of both participants. (A) The color of each cell indicates the decoding performance using the recordings from neighboring 4 channels. The *color bars* give the scale of decoding performance. The *white dashed outline* marks out the channels with the highest decoding performance (marked with a *white dashed circle*), and the *red dashed outline* [gray in print version] marks out the neighboring 9 channels. *Black dashed line* represents the central sulcus (*M*, motor; *S*, sensory). (B) The decoding performance with optimal channel subsets using different channel selection strategies of both participants.

# Rapid decoding of hand gestures with recurrent neural networks

## RNN-based hand gesture recognition

As shown in the section *Participants and Implantation*, the placement of clinical subdural electrodes was determined by surgery requirements, and most channels are unrelated to hand grasping movements. The unrelated signals can bring the noise in hand gesture decoding and lead to unnecessary computational costs. However, the neural signals collected through the clinical ECoG grid are still the most important among a few available signal sources from the motor cortex which could be applied in BCI studies in clinical trials. Therefore, the neural signals obtained through the clinical ECoG grid were first divided into sequential short-time segments, and power spectrum features were extracted from each segment. Then we selected the most informative signal channels along with the frequency bands using a greedy strategy, to compose compact features for decoding. Finally, the features of the segments were sequentially put into an RNN-based decoder to improve hand gesture recognition using the sequential relationship among the segments.

### *Feature extraction*

In this study, we first evaluated several feature selection strategies and examined their influence on gesture recognition performance. Then, experiments were designed to find the appropriate number of features for gesture recognition. After that, the channels and frequency bands selected were presented and analyzed.

We evaluated the gesture recognition performance by using all the channels and frequency bands by the SVM classifier to serve as the baseline in the experiment. A single feature selection strategy, which independently selects the top $n$ features with the best decoding performance, was further implemented and compared. Then we applied a greedy strategy-based method for selecting the most informative channels along with the frequency bands. The greedy strategy performs in an iterative manner. In brief, we chose the feature with the highest decoding performance using an SVM classifier, and put it into the selected set at the first step. Then, at each following step, we iteratively chose one candidate feature that improved accuracy the most when combined with the selected features, and added it into the selected set. The iteration stopped when the request feature number was reached or there was no improvement in decoding performance after adding the newly selected feature. The performance of the greedy-based feature selection was further evaluated in comparison with other methods.

In this experiment, we segmented the signals into 300-ms sliding windows with a stride of 100 ms, and a total of 10 temporal segments following the movement onsets were used. We examined and analyzed the decoding performance of the features selected by three types of strategies. The performance was presented in the average accuracy of threefold cross-validation. In hand gesture classification evaluation, we applied 10-fold cross-validation, and the average accuracy over 10-fold was used as the criterion. For each fold in cross-validation, we randomly selected 20% of the data as the validation dataset to select the hyperparameters.

*Recurrent neural network-based gesture recognition*

In this study, we proposed an RNN-based decoder to recognize hand gestures with ECoG signals. To capture the underlying temporal information in ECoG signals, we proposed using a gated RNN model, i.e., long short-term memory (LSTM) models, to learn the temporal patterns of different gestures. The LSTM model can sequentially update the gates in memory cells to determine which features in the preceding windows should be considered for gesture decoding. To benefit temporal pattern learning, our method selected the most temporally informative features as inputs to the LSTM decoder. After feature selection, the feature representation of a task trial can be denoted as $\{x_1, x_2, \ldots, x_t\}$, where $x_i$ is the feature vector at the $i$th temporal segments. The feature representation took rich information in both spectrum and temporal dynamics for hand gesture classification. Since most classifiers require inputs in the form of vectors, the decoders based on such classifiers need to concatenate the temporal features into a vector. This procedure loses the temporal structure of data, which limits the improvement of neural decoding. The RNN-based method overcomes the above problem by inputting data in a recurrent way. As shown in Fig. 2.6, the feature vectors are sequentially put into the model, and the temporal information could be well preserved by the temporal connections. In our approach, the LSTM model is adopted [15]:

$$i(t) = \sigma(W_i x(t) + U_i h(t-1) + b_i),$$

$$f(t) = \sigma(W_f x(t) + U_f h(t-1) b_f),$$

$$o(t) = \sigma(W_o x(t) + U_o h(t-1) + b_o),$$

$$c(t) = i(t)\tanh(W_c x(t) + U_c h(t-1) + bc) + f(t)c(t-1),$$

$$h(t) = o(t)\tanh(c(t)),$$

where $x(t)$ is the feature vector at the $t$-th time window, $o(t)$ is the recognition result output from the model after the last time window, $\sigma(x)$ is the sigmoid function, $c(t)$ is the memory cell, $h(t)$ is the hidden layer units, and $i(t)$, $f(t)$, $o(t)$ are the input gate, forget gate, and output gate, respectively. The memory cell can remember useful information through time, and the gates control how many time windows should be used for the current gesture recognition task. Therefore, in the LSTM model, temporal information can be well preserved for accurate gesture decoding.

We tested the RNN model with different settings to select the optimal parameters for gesture decoding. After that, the RNN-based decoder was compared with four other competitors to demonstrate the advantages of our method. In addition, we investigated the decoding performance in a time interval as short as possible after motion onsets for rapid gesture recognition. The RNN model was implemented with Keras on the top of TensorFlow.

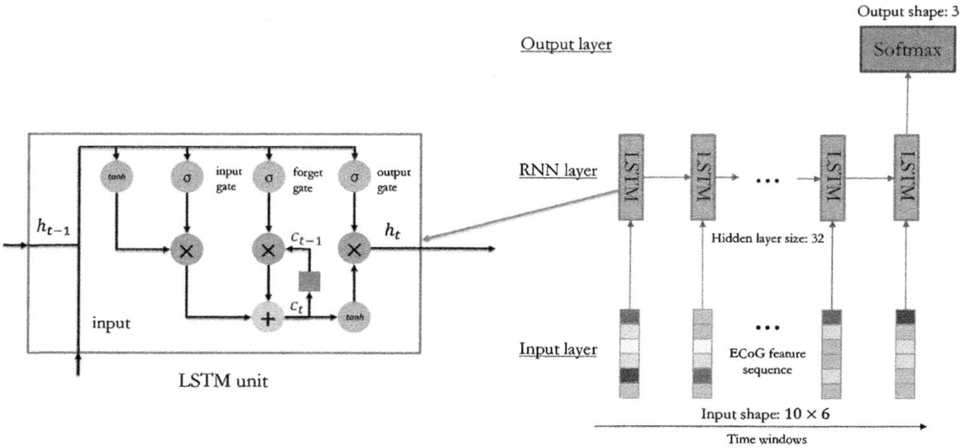


**FIGURE 2.6** Architecture of an RNN model for gesture classification from ECoG.

## Results

### *Feature analysis*

As shown in Fig. 2.7, we compared three feature selection strategies with their performance in the gesture classification. Our results show that the baseline method using all the features reached high performance. With the feature selection strategies, performance close to the baseline can be achieved using only a small set of features. We further compared the decoding performance with a different number of features. The greedy strategy achieves better performance than single-feature selection does on both participants. Because the candidate feature is evaluated together with the selected features in the greedy selection, redundant features are not likely to be selected and the information on hand gesture can be efficiently collected.

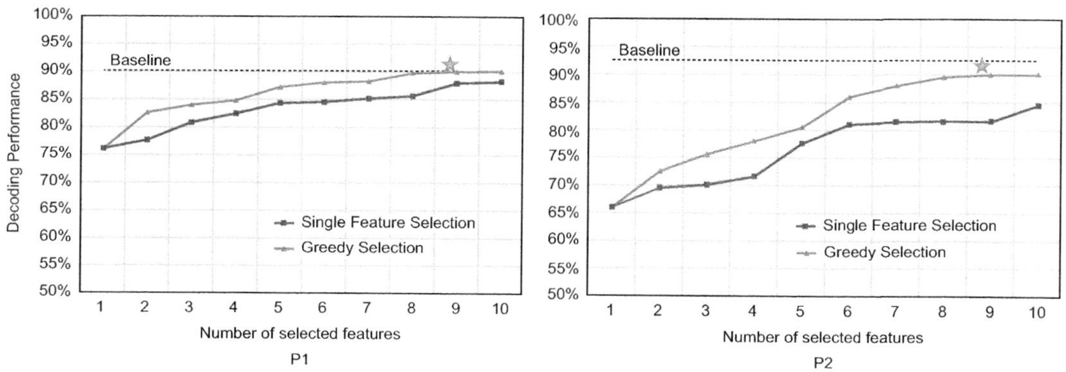


**FIGURE 2.7** Decoding performance of feature selection strategies using a different number of features. The *yellow stars* [light gray in print version] indicate the points that the greedy algorithm stops, and the performance converges after the points.

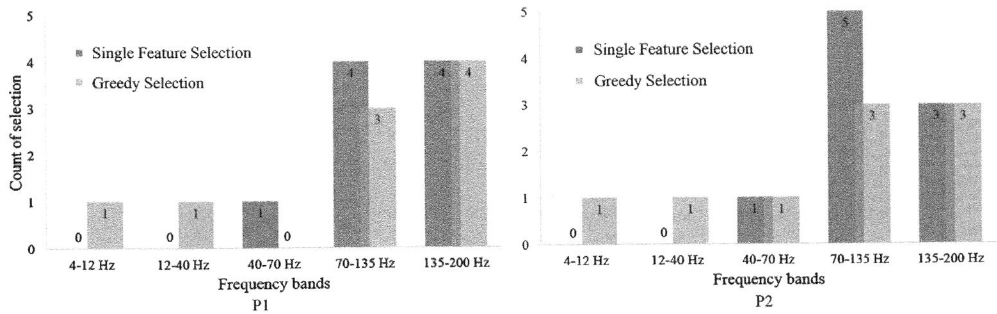


**FIGURE 2.8** Distribution of selected frequency bands in both participants. The first nine features are illustrated and analyzed. The vertical axis presents the number of selections in each frequency band.

Here we present some statistical analysis of the channels and the frequency bands selected by our method. The feature distribution of frequency bands is shown in Fig. 2.8, which shows that most information is from the bands 70–135 Hz and 135–200 Hz. The results suggested that high-frequency bands in ECoG are highly correlated to hand motions, which is in agreement with previous studies [17,18].

For the number of features, we only used the first six features selected by the greedy algorithm. Using more features might bring further improvement in decoding performance as in Fig. 2.7, but the improvements brought by the later selected electrodes would not be great. In addition, since our dataset was small, a slight improvement could result from overfitting rather than useful information. Most of the selected electrodes were close to the central sulcus and within the sensorimotor region, as shown in the section *Channel Selections and Anatomical Patterns.*

### *Performance of gesture recognition*

To investigate the decoding performance of our method, we first carried out experiments on the validation dataset to select the optimal settings and maximize the performance of the classifiers. Then, we compared our method with other classical decoders to demonstrate the effectiveness of temporal information and the ability of RNN in ECoG time series decoding.

#### Model selection

Experiments were carried out to select the optimal setting for the LSTM RNN model. For the LSTM model, one important setting is how many hidden units are used. Models with a small set of hidden units may not provide sufficient information, while models with large sets of hidden units are prone to overfitting. In this study, we tried the number of hidden units from 8 to 128 to test the performance of the LSTM model.

We used the top six features selected by the greedy strategy, and the settings of temporal segments were the same as in the section Feature Extraction. As shown in Fig. 2.9, the LSTM model with 32 hidden units got the best performance (90.56% for P1 and 88.18% for P2) for both participants on validation dataset. Therefore, we used 32

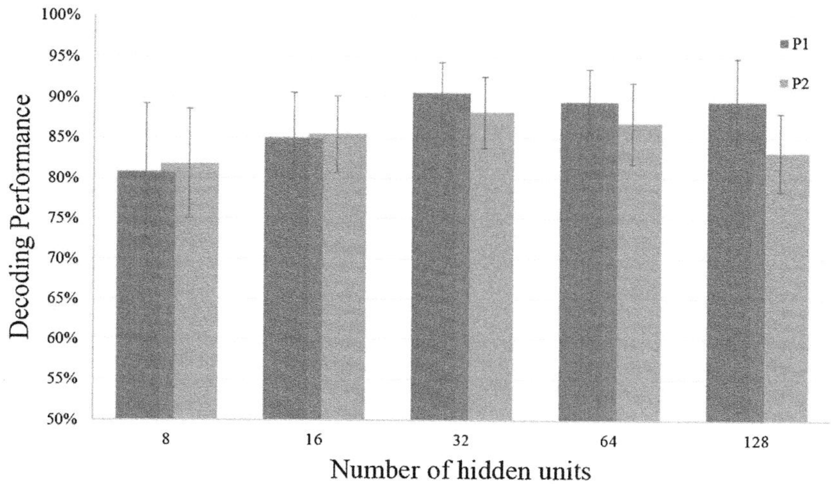


**FIGURE 2.9** Performance of a different number of hidden units in RNN. The LSTM model with 32 hidden units got the best performance (90.56% for P1 and 88.18% for P2) for both participants on validation dataset. The *black dashed line* represents the standard deviation.

hidden units for gesture recognition in our decoder. In model training, we used Adam optimization algorithm, the learning rate was set at 0.001 with a decay rate of 0.0005 for each epoch. An early stop was applied by selecting the epoch with the best performance on the validation set.

**Comparison with other methods**

To evaluate the performance of our proposed RNN model, we compared it with other decoders. Our RNN model was first compared with decoders using long time windows to evaluate the effectiveness of temporal information. Then the RNN decoder was compared with other classifiers to demonstrate the strength of RNN models in sequential modeling. In this study, ECoG signals were divided into temporal segments using a 300-ms sliding window with a stride of 100 ms. A total of 10 temporal segments were used. We evaluated our method in comparison with other methods using 10-fold cross-validation. The significance test was performed using a paired t-test.

The implementation and settings of the competitors in this experiment are as follows:

- SVM-Global: an SVM-based decoder using features calculated over long time windows. For a fair comparison, the length of the ECoG signal used is the same as the following competitors. RBF kernel is used in the SVM model, and the parameters of C and gamma are selected by cross-validation. The parameter C is selected from 0.1, 1, 10, 100, and 1000, and gamma was selected from 0.01, 0.001, and 0.0001.
- SVM-Segments: an SVM-based decoder using features in temporal segments as used in the previous section. The segment settings are the same as the RNN method. The features in sequence are reshaped into a single vector to input to the SVM classifier. RBF kernel is used in the SVM model, and the parameters of C and

gamma are selected by cross-validation. The parameter C is selected from 0.1, 1, 10, 100, and 1000, and gamma was selected from 0.01, 0.001, and 0.0001.

- MLP-Segments: a multilayer perception-based decoder from a previous work [16]. The segment settings are the same as the RNN method. The features in sequence are reshaped into a single vector to input to the MLP classifier.
- LR-Segments: a logistic regression-based decoder from a previous work [21]. The segment settings are the same as the RNN method. The features in sequence are reshaped into a single vector to input to the LR classifier.

The results are shown in Table 2.2. Overall, the RNN-based decoder obtains the highest accuracies for both participants. For participant P1 the gesture recognition accuracy is 92.43%, and for participant P2 the gesture recognition accuracy is 92.92%. Among the competitors, the SVM-Global gives the worst performance. It is reasonable since it calculates the features using the whole time window and ignores the information in time. The SVM-Segments method improves the accuracy by 7.3% and 7.5% for P1 and P2, respectively, by using the temporal segments. The results demonstrate the importance of considering the temporal information in ECoG decoding. The results show that our method statistically significantly outperforms other approaches under a significance of 0.05 (paired $t$-test).

### *Rapid recognition*

In this study, we examined the possibility of recognizing the gestures in a time interval as short as possible after movement onsets. In the experiments, we changed the time interval from 100 ms to 1200 ms after motion onsets. For each time interval, the ECoG signals were segmented using a 300-ms sliding window with a stride of $(t-w)/9$ ms, where $t$ is the time interval and $w=300$ ms is the length of the sliding window. If the time interval is less than 300 ms, we use a $w=t/2$ ms sliding window with a stride of $w/9$ ms.

Fig. 2.10 shows the results. As the time interval becomes longer, we could achieve better gesture decoding performance. The results of these experiments also demonstrated the possibility of rapid recognition. Fig. 2.10 shows that recognition accuracies of over 75% could be obtained at the 0.3-s interval for both participants. If we used a 500-ms time interval, the gesture recognition accuracy was above 80%. Our results indicated that the temporal dynamic was especially informative for quick decoding within short time intervals.

**Table 2.2** Gesture recognition comparison of different decoders.

| **Decoder** | **P1** | **P2** |
|---|---|---|
| SVM-global | 79.03% ± 6.50 | 78.94% ± 7.60 |
| SVM-segments | 86.51% ± 5.23 | 87.66% ± 6.19 |
| MLP-segments | 84.35% ± 5.76 | 87.11% ± 7.13 |
| LR-segments | 83.82% ± 5.69 | 85.76% ± 6.77 |
| RNN | 89.34% ± 4.67 | 90.83% ± 5.94 |

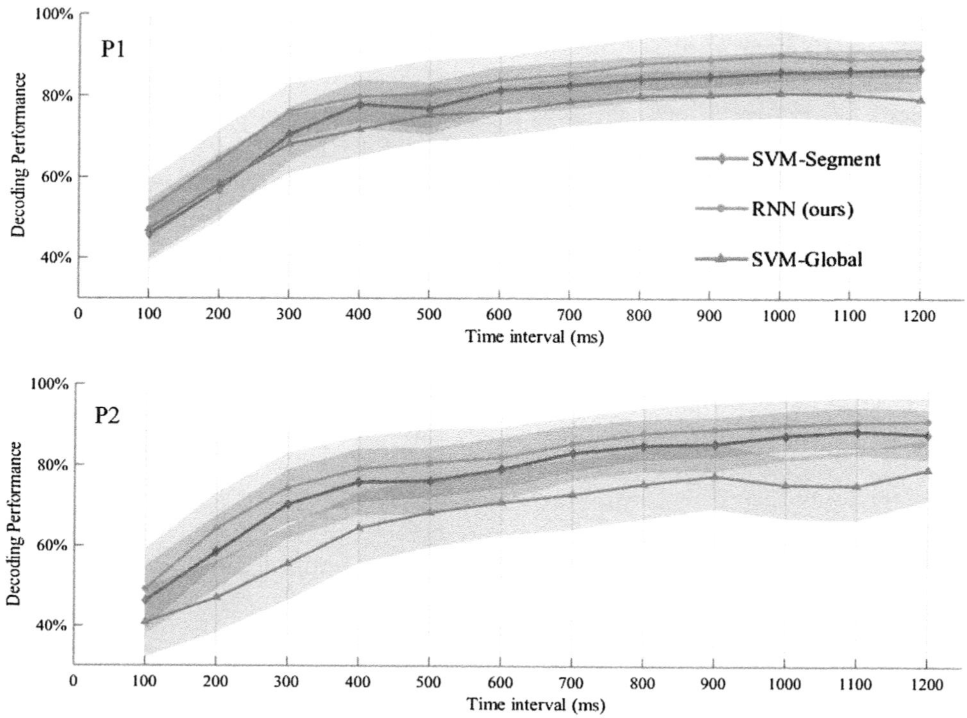


**FIGURE 2.10** Gesture recognition performance within different time intervals after motion onsets are the decoding performance of different methods on both participants. Recognition accuracies of over 75% could be obtained at the 0.3-s interval for both participants. If we use a 0.5-s time interval, the gesture recognition accuracy is over 80%. The results also indicate that the temporal dynamic is especially informative for quick decoding within short time intervals.

# Conclusion

Although ECoG has been proven to be a good candidate of signal sources for BCI control, its safety issues must be carefully considered before practical application. One of the main remaining challenges in this field is related to minimization of the cortical area. A typical intracranial ECoG electrode grid covers a large area of the cortex surface, which aims to simultaneously monitor brain regions and localize the seizure onset zone. However, the large extent of the conventional implantation of ECoG grids might lead to risks during surgery and the postsurgery period if the ECoG grid were used for chronic recording in BCI application [22]. On the other hand, some studies have found that the number and locations of electrodes chosen exclusively for clinical purposes did not exactly agree with the requirements in BCI application.

In recent years, many works have attempted to identify a smaller but less invasive configuration of the electrode grid without sacrificing decoding performance. Some studies customized micro-ECoG (4 mm center-to-center spacing) by decreasing both the electrode diameter and the interelectrode distance to realize a less invasive procedure and detect movement intent with very local cortical activity. Meanwhile, decreasing the number of channels could be another way to reduce the area for grid covering. Different strategies have been proposed to optimize the number of ECoG electrode channels for reducing the number of input features of the decoder and to improve both decoding accuracy and computational speed. Milekovic et al. restricted electrode channels in a limited cortical region using neighboring channels to decode arm movement [23]. Zhang et al. decoded visual stimuli using a single ECoG channel [24]. In this work, we first investigated the optimal number of electrode channels in hand gesture discrimination and the corresponding anatomical distribution of these functional channels. The electrodes chosen by greedy selection were found to be located around the central sulcus and clustered together as shown in Fig. 2.4. These selected electrodes were informative in distinguishing hand gesture types with high decoding performance with an SVM decoder.

Many studies have demonstrated that ECoG signals could provide useful information about upper-limb movement, such as arm reaching and hand grasping movements. In most existing approaches, the ECoG signals were segmented into short-term time windows and the temporal patterns of different gestures were extracted. However, the sequential relationship among these windows was ignored in most cases and the features in short-term windows were reshaped into a vector for classification, which broke the temporal structure of these features. Compared with the existing approaches using linear methods or SVM classifiers, the RNN model can better preserve the structure in feature sequence and is capable of learning from nonlinear relationships. Elango et al. proposed to use RNN-based models to classify individual finger movements [25]. In contrast to their approach which manually selected the ECoG channels and frequencies from empirical observations, we proposed an RNN-based method combined with a greedy feature extraction to exploit the temporal information in ECoG signals for rapid and robust gesture recognition. Our decoder recognized three hand gestures with a high accuracy of 92%, and a quick response was achieved within 0.5 s with an accuracy of 80%. The results showed that our RNN-based model could extract sufficient information from ECoG signals for effective hand gesture classification, and demonstrated the feasibility of robust and practical ECoG-based system for real-time prosthetic use.

## References


[1] Wolpaw JR, et al. Brain–computer interfaces for communication and control. Clinical Neurophysiology 2002;113(6):767–91.

[2] Gharabaghi A, et al. From assistance towards restoration with epidural brain–computer interfacing. Restorative Neurology and Neuroscience 2014;32(4):517–25.

[3] Chao ZC, Nagasaka Y, Fujii N. Long-term asynchronous decoding of arm motion using electrocorticographic signals in monkeys. Frontiers in Neuroengineering 2010;3:3.

[4] Leuthardt EC, et al. A brain–computer interface using electrocorticographic signals in humans. Journal of Neural Engineering 2004;1(2):63–71.

[5] Schalk G, et al. Decoding two-dimensional movement trajectories using electrocorticographic signals in humans. Journal of Neural Engineering 2007;4(3):264–75.

[6] Schalk G, et al. Two-dimensional movement control using electrocorticographic signals in humans. Journal of Neural Engineering 2008;5(1):75–84.

[7] Kubanek J, et al. Decoding flexion of individual fingers using electrocorticographic signals in humans. Journal of Neural Engineering 2009;6(6).

[8] Miller KJ, et al. Human motor cortical activity is selectively phase-entrained on underlying rhythms. PLoS Computational Biology 2012;8(9).

[9] Miller KJ, et al. Broadband changes in the cortical surface potential track activation of functionally diverse neuronal populations. NeuroImage 2014;85:711–20.

[10] Xie Z, Schwartz O, Prasad A. Decoding of finger trajectory from ECoG using deep learning. Journal of Neural Engineering 2018;15(3).

[11] Acharya S, et al. Electrocorticographic amplitude predicts finger positions during slow grasping motions of the hand. Journal of Neural Engineering 2010;7(4).

[12] Yanagisawa T, et al. Real-time control of a prosthetic hand using human electrocorticography signals. Journal of Neurosurgery 2011;114(6):1715–22.

[13] Chestek CA, et al. Hand posture classification using electrocorticography signals in the gamma band over human sensorimotor brain areas. Journal of Neural Engineering 2013;10(2).

[14] Pan G, et al. Rapid decoding of hand gestures in electrocorticography using recurrent neural networks. Frontiers in Neuroscience 2018;12.

[15] Hochreiter S, Schmidhuber J. LSTM can solve hard long time lag problems.

[16] Chatterjee R, Bandyopadhyay T. EEG based motor imagery classification using SVM and MLP. In: 2016 2nd international conference on computational intelligence and networks (CINE); 2016.

[17] Bleichner MG, et al. Give me a sign: decoding four complex hand gestures based on high-density ECoG. Brain Structure and Function 2016;221(1):203–16.

[18] Branco MP, et al. Decoding hand gestures from primary somatosensory cortex using high-density ECoG. NeuroImage 2017;147:130–42.

[19] Li Y, et al. Gesture decoding using ECoG signals from human sensorimotor cortex: a pilot study. Behavioural Neurology 2017. 3435686.

[20] Chang CC, Lin CJ. LIBSVM: a library for support vector machines. ACM Transactions on Intelligent Systems and Technology 2011;2(3).

[21] Subasi A, Ercelebi E. Classification of EEG signals using neural network and logistic regression. Computer Methods and Programs in Biomedicine 2005;78(2):87–99.

[22] Lesser RP, Crone NE, Webber WR. Subdural electrodes. Clinical Neurophysiology 2010;121(9):1376–92.

[23] Milekovic T, et al. An online brain–machine interface using decoding of movement direction from the human electrocorticogram. Journal of Neural Engineering 2012;9(4).

[24] Zhang D, et al. Toward a minimally invasive brain–computer interface using a single subdural channel: a visual speller study. NeuroImage 2013;71:30–41.

[25] Elango V, et al. Sequence transfer learning for neural decoding. 2018.

3

# Neural coding by electroencephalography (EEG)

Yuan Yang

DEPARTMENT OF PHYSICAL THERAPY AND HUMAN MOVEMENT SCIENCES, NORTHWESTERN UNIVERSITY FEINBERG SCHOOL OF MEDICINE, CHICAGO, IL, UNITED STATES

**Chapter outline**



## Introduction

The brain is the commander of voluntary movement control. It generates the oscillatory neural activity at specific frequency bands traveling through the corticospinal tract to activate the musculoskeletal system for movement execution [1]. Neural oscillations at mu (8–13 Hz) and beta (15–35 Hz) bands measured around the sensorimotor cortex, known as cortical sensory–motor rhythms (CSMRs), are thought to be associated with voluntary control of movements [2]. The coupling between CSMR and muscle activities has been previously reported at these frequency bands, confirming the key functional role of CSMR in movement control [3,4]. Damage to the corticospinal tract following a brain or spinal injury can result in a decrease in the coupling between CSMR and muscle activities, and associated motor impairments, such as muscle weakness and loss of independent movement control [5,6]. However, the CSMR may be preserved at the sensorimotor cortex that allows the identification of motor intentions via measuring and decoding the CSMR [7].

Intelligent Biomechatronics in Neurorehabilitation. https://doi.org/10.1016/B978-0-12-814942-3.00003-9

Electroencephalography (EEG) is an electrophysiological monitoring technique that records the oscillatory cortical activity including the CSMR. By placing the electrodes on the scalp, the EEG measures cortical activity without surgery. Compared to other brain signal recording methods (e.g., functional MRI, electrocorticography, positron emission tomography), the advantages of EEG are that it is inexpensive, low-risk, and portable [8]. These advantages allow EEG to be an online monitoring method for daily use. However, due to the volume conduction through the scalp, skull, and other layers of the brain, EEGs recorded by a scalp sensor are a "blurred" copy of multisource activities, which increases the difficulty of EEG signal decoding. Advanced signal processing methods are required to address this challenge. Traditional signal processing methods such as independent component analysis [9] and common spatial pattern (CSP) filter [10,11] need a large number of EEG electrodes covering the whole scalp during the measurement for disentangling the mixed multisource signals. This whole-scalp recording reduces the feasibility of EEG in neuro-rehabilitation for daily use. Several novel signal processing methods have been recently proposed to improve EEG data analysis for accurate identification of motor intentions using only a few electrodes. These novel methods can be combined with various new EEG devices with very few electrodes, such as the Emotiv Epoc headset and LooxidVR package for daily use.

## Novel signal processing methods for few EEG electrode-based neural decoding

### Spatial filter for improving signal-to-noise ratio

EEG signals recorded at each time point can be considered as a spatial matrix corresponding to the spatial distribution of the electrodes. Spatial filters are typically required to improve the signal-to-noise ratio (SNR) of EEG by using either the neighborhood [12] or global information [11]. Spatial filters using global information, such as the CSP algorithm [13] and the common average reference (CAR) [14], typically require a whole-scalp recording with a large number of EEG electrodes. To reduce the number of electrodes, we recommend the use of local spatial filters such as bipolar derivation [15] and Laplacian derivation [16]. Recent studies have demonstrated that the combination of local spatial filters and time-frequency analysis algorithm for extracting CSMR patterns can yield better results and use less electrodes than using global spatial filters [16–19].

#### *Bipolar derivation*

Bipolar derivation can increase the SNR by reducing the common noise of both electrodes. Let $D_1$ and $D_2$ be the discriminative patterns (e.g., CSMR) of EEG signals at the two recording electrodes. Let N represent the mixture of common additive background noise of these two electrodes. The signals $X_1$ and $X_2$ recorded at these two EEG electrodes can be expressed by the following additive model:

$$X_1 = D_1 + N, \; X_2 = D_2 + N$$

This additive model is based on a basic assumption that a scalp EEG is a linear combination of source components. Although this assumption may not exactly reflect the complex composition of neural responses (which is still under exploration) in the scalp EEG, it is typically used as an approximate model in practice for artifact reduction and source analysis. Based on this model, the spatially filtered EEG signal derived from these two electrodes is:

$$X_{BI} = X_1 - X_2 = D_1 - D_2$$

Thus, the bipolar derivation can remove the shared additive background noise between two "active" electrodes.

*Laplacian derivation*

Laplacian derivation is also a widely used local spatial filtering algorithm in EEG signal processing. The Laplacian value of each electrode is obtained by subtracting the mean activity at surrounding electrodes from the electrode of interest. Denote by $X_{LAP}$ the Laplacian filtered EEG signal at electrode i, and $S_i$ an index set of the four electrodes surrounding the i-th electrode. The Laplacian derivation is computed according to the following formula:

$$X_{LAP} = X_i - \frac{1}{4}\sum_j X_j \quad (j \in S_j)$$

Two different sets of four surrounding electrodes are commonly used: nearest-neighbor electrodes (the distance from each surrounding electrode j to the center i is $d_{ij} = 3$ cm) and next-nearest-neighbor electrodes ($d_{ij} = 6$ cm). The Laplacian derivation employing nearest-neighbor electrodes is called a small Laplacian, while the one using next-nearest-neighbor electrodes is named a large Laplacian. The characteristics of the Laplacian derivation are highly dependent on $d_{ij}$. Experiments show that the Laplacian derivation becomes more sensitive to the components with high spatial frequencies when $d_{ij}$ increases [20]. Thus, the small Laplacian is often used to improve the SNR when the signal is highly localized and stable over time, while the large Laplacian is more suitable when the signal is more distributed and nonstationary. Compared to the CAR, the Laplacian derivation uses local instead of global information to obtain a reference-free EEG. This method can be used where there are limited electrodes available to record EEG data.

## Subject-specific channel selection for individualized recording setup

Previous studies show that actual or imaginary movements of different body parts can cause a decrease of signal power in the CSMR, called event-related desynchronization

(ERD), at corresponding "active" cortical regions; meanwhile, a power increase in the CSMR, known as event-related synchronization (ERS), might be observed at other "idling" areas [21,22]. In practice, many researchers simply place electrodes at three key positions (C3, Cz, and C4 of 10–20 EEG recording system [23]) over the sensorimotor areas, which generally covers the "active" regions controlling the upper and lower limbs. However, the precise functional regions may vary from subject to subject. Especially after a brain injury, the functional regions in the brain may be reorganized [24]. To address this problem, Yang and colleagues recently proposed a method based on the discriminative powers of different regions of interest (ROIs) to identify the optimal ROIs in the scalp for placing EEG electrodes [25]. The ROIs can be defined using EEG electrode positions according to an extended 10–20 recording system [23]. Assuming that the task is to identify the movement intention of the upper limb ($U$) versus the lower limb ($L$) as a two-class problem, the discriminative power of each ROI can be estimated using the Fisher's type $F$-score [26]:

$$F = \frac{\|\vec{\mu}_U - \vec{\mu}_L\|_2}{tr(C_U) + tr(C_L)}$$

where $C$ denotes the covariance matrix of the feature vector extracted at this ROI, $\vec{\mu}$ denotes the mean of the feature vector, $\|\cdot\|_2$ denotes the $L2$-norm (Euclidean norm), and tr($\cdot$) the trace of a matrix.

Given an n-dimensional feature vector, $v(k) = [v_1(k), v_2(k), \ldots, v_n(k)]$, $k = 1, \ldots, K$, where $K$ is the number of samples (trials) for one class ($U$ or $L$). Thus, the mean of the feature vector for the class is $\vec{\mu} = [\mu_1, \mu_2, \ldots, \mu_n]$, where $\mu_1, \mu_2, \ldots, \mu_n$ are the mean values of $v_1(k), v_2(k), \ldots, v_n(k)$, respectively. We denote by ${\sigma_1}^2, {\sigma_2}^2, \ldots, {\sigma_n}^2$ the variances of $v_1(i)$, $v_2(i), \ldots, v_n(i)$, respectively. Then the trace of the covariance matrix for each class can be computed as:

$$tr(C) = \sum_{i=1}^{n} \sigma_i^2 = \sum_{i=1}^{n}\left(\frac{1}{K-1}\sum_{k=1}^{K}(v_i(k) - \mu_i)^2\right) = \frac{1}{K-1}\sum_{k=1}^{K}\left(\sum_{i=1}^{n}(v_i(k) - \mu_i)^2\right)$$
$$= \frac{1}{K-1}\sum_{k=1}^{K}\|\vec{v}(k) - \vec{\mu}\|_2^2$$

Thus, the trace of the covariance matrix for each class is the mean Euclidean distance between samples to the class center, which reflects intraclass spread.

The $F$-score uses the Euclidean distance between class centers, $\|\vec{\mu}_U - \vec{\mu}_L\|_2$ to estimate the difference between classes and employs the trace of the covariance matrix to evaluate the variance within each class. The $F$-score can be extended to multiclass cases using either a one-versus-rest (OVR) strategy [16] or pairwise strategy [13]. The optimal ROIs should be selected with a large $F$-score to maximize the difference between classes and minimize the variance within each class. We can normalize the $F$-score by its maximum value over all ROI to obtain a topography of the discriminative power for selecting the key ROIs. Fig. 3.1 show an example of the selected ROI in different

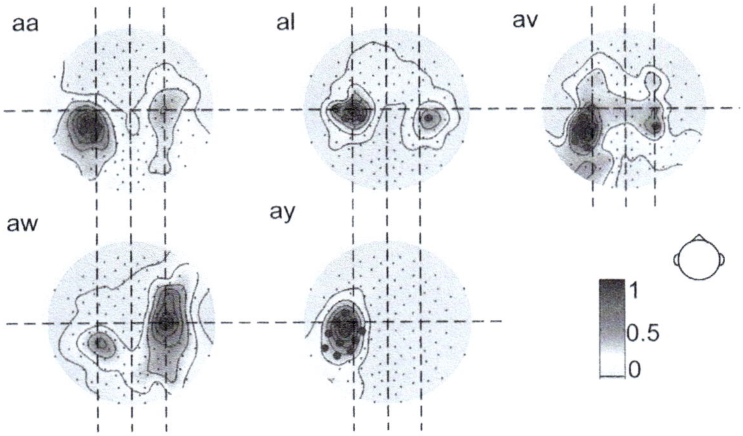


**FIGURE 3.1** Selected ROIs (*blue dots* [black in print version]) in five different individuals for the identification of upper limb versus lower limb movement intentions. The *color bar* indicates the normalized *F*-score (range of value is between 0 and 1). The *dashed line* indicates the grid for defining the positions of Cz (middle cross), C3 (left cross), and C4 (right cross) in the 10–20 system. *Reproduced from YangY, Bloch I, Chevallier S, Wiart J. Subject-specific channel selection using time information for motor imagery brain–computer interfaces. Cognitive Computation 2016;8(3):505–518.*

individuals for the identification of upper limb versus lower limb movement intentions [26]. We can see that the selected regions could be different from subject to subject, indicating that a personalized setup is necessary for EEG recording based on few electrodes.

## Time–frequency analysis for extracting CSMR

After placing the EEG electrodes on the optimal ROIs, the time–frequency analysis is required to extract the most discriminative CSMR patterns, since they are typically short-lasting (e.g., half to a few seconds depending on the movement duration) with the frequency range varying between subjects. Thus, a subject-specific time–frequency parameterization is required for CSMR pattern extraction. Various approaches have been previously proposed for this purpose [16,17,26–32]. Among the existing methods, the approach developed by Yang and colleagues yielded the best performance [16,17]. This time–frequency analysis approach is also based on the *F*-score and can be applied to both two-class and multiclass cases.

In this approach, the EEG signal at each electrode is decomposed into components in a series of overlapping time–frequency bins $(\omega_m, \tau_n)$, $m \in \{1, 2, \ldots, \mathrm{M}\}$, $n \in \{1, 2, \ldots, \mathrm{N}\}$ with different frequency bands $\omega_m = [f_m, f_m + F - 1]$, $f_{m+1} = f_m + F_s$ ($F$ is the bandwidth, $F_s$ is the frequency step), and time intervals $\tau_n = [t_n, t_n + T{-}1]$, $t_{m+1} = t_m + T_s$ ($T$ is the interval width, $T_s$ is the time step). The goal is to find a time–frequency bin that contains the most discriminative CSMR patterns for identifying the movement intentions.

**Table 3.1** Comparison of using Yang's method and other methods on BCI competition IV dataset IIb for the identification of right versus left hand.

| | Subject ID | | | | | | | | | |
|---|---|---|---|---|---|---|---|---|---|---|
| | **1** | **2** | **3** | **4** | **5** | **6** | **7** | **8** | **9** | **Mean** |
| Yang's method | 0.39 | **0.25** | 0.13 | 0.93 | **0.88** | **0.63** | 0.55 | **0.88** | **0.78** | **0.60** |
| FBCSP | 0.40 | 0.21 | **0.22** | **0.95** | 0.86 | 0.61 | 0.56 | 0.85 | 0.74 | **0.60** |
| CSSD | **0.43** | 0.21 | 0.14 | 0.94 | 0.71 | 0.62 | **0.61** | 0.84 | **0.78** | 0.58 |
| NTSPP | 0.19 | 0.12 | 0.12 | 0.77 | 0.57 | 0.49 | 0.38 | 0.85 | 0.61 | 0.46 |

*FBCSP*, filter band CSP algorithm [34]; *CSSD*, common spatial subspace decomposition [35]; *NTSPP*, neural time series prediction preprocessing [36]. Yang's method yielded the best results on more subjects than the FBCSP, though their mean performances are the same. The best results are highlighted in bold.
Reproduced from Yang Y, Chevallier S, Wiart J, Bloch I. Time-frequency optimization for discrimination between imagination of right and left hand movements based on two bipolar electroencephalography channels. EURASIP Journal on Advances in Signal Processing 2014;2014(1):38.

We used the F-score to estimate the discriminative power of each time–frequency bin. The optimal time–frequency bin ($\omega^* \times \tau^*$) is found by exhaustively searching the largest $F$-score value among all regions:

$$F(\omega^*, \tau^*) = \max\{F(\omega_m, \tau_n)|m \in \{1,\ 2, \ldots, \mathrm{M}\}, n \in \{1,\ 2, \ldots, \mathrm{N}\}\}$$

Then, the CSMR patterns can be extracted from the optimal time–frequency bin by computing the variance of the $\omega^*$ band-pass filtered signal at the time segment defined by $\tau_n$ [17].

Tables 3.1 and 3.2 present a comparison using Yang's approach and other methods on two open-access EEG datasets for identifying movement intentions. The results are given in kappa coefficient [33]:

$$\kappa = (\mathrm{Acc} - \mathrm{Pe})/(1 - \mathrm{Pe})$$

where Pe is the chance level for agreement (i.e., Pe $= 1/n$ for n-class problems). Thus, a larger $\kappa$ value indicates a better identification performance.

**Table 3.2** Comparison of using Yang's method and other methods on BCI competition III dataset IIIa [37] for the identification of intentions of right hand, left hand, both feet, and tongue movement.

| | Subject ID | | | |
|---|---|---|---|---|
| | **1** | **2** | **3** | **Mean** |
| Yang's method | 0.64 | **0.71** | **0.72** | **0.69** |
| AAR | 0.70 | 0.37 | 0.39 | 0.49 |
| ICA + PCA | **0.95** | 0.41 | 0.52 | 0.63 |
| JAD-CSP | 0.76 | 0.41 | 0.53 | 0.57 |

*AAR*, adaptive autoregressive method [37]; *ICA + PCA*, combined method based on ICA and principal component analysis [38]; *JAD-CSP*, joint approximate diagonalization-based CSP [39]. The best results are highlighted in bold.
Reproduced from Yang Y, Chevallier S, Wiart J, Bloch I, Subject-specific time-frequency selection for multi-class motor imagery-based BCIs using few Laplacian EEG channels. Biomedical Signal Processing and Control 2017;38:302–311.

## Remaining challenges and future directions

The state-of-the-art EEG signal processing methods allow the use of a small number of electrodes to identify movement intentions. However, the identification accuracy is still a challenge for reliable control of assistive devices (e.g., an electrical wheelchair) and neuro-prostheses. More advanced methods such as deep learning may be required to improve the identification accuracy in neural decoding using EEG. Meanwhile, other physiological signals such as electromyography and electro-oculography can also be used to realize a hybrid control [40–42], since these signals generally have a better SNR than EEG.

Traditional EEG sensors are "wet" electrodes that need a conductive gel applied to reduce the skin–electrode impedance. The user experience can be further improved by using a dry electrode. Although dry electrodes are now produced by several hardware producers, the quality of the signal might not be as good as with traditional "wet" electrodes. Most proposed EEG signal processing methods were tested on data recorded with "wet" electrodes. In the future, they should be further tested on data recorded using dry electrodes.

Despite the remaining challenges, EEG is still a promising technique for decoding movement intentions for helping individuals suffering from movement disabilities. EEG-based human–machine interfaces have the potential to become a next-generation intelligent biomechatronic technique for neurorehabilitation.

## References


[1] Negro F, Farina D. Linear transmission of cortical oscillations to the neural drive to muscles is mediated by common projections to populations of motoneurons in humans. The Journal of Physiology 2011;589(3):629–37.

[2] McFarland DJ, Miner LA, Vaughan TM, Wolpaw JR. Mu and beta rhythm topographies during motor imagery and actual movements. Brain Topography 2000;12(3):177–86.

[3] Mehrkanoon S, Breakspear M, Boonstra TW. The reorganization of corticomuscular coherence during a transition between sensorimotor states. NeuroImage 2014;100:692–702.

[4] Yang Y, Dewald JP, van der Helm FC, Schouten AC. Unveiling neural coupling within the sensorimotor system: directionality and nonlinearity. European Journal of Neuroscience 2018;48(7): 2407–15.

[5] Fang Y, et al. Functional corticomuscular connection during reaching is weakened following stroke. Clinical Neurophysiology 2009;120(5):994–1002.

[6] Rossiter HE, et al. Changes in the location of cortico-muscular coherence following stroke. NeuroImage: Clinical 2013;2:50–5.

[7] Shahid S, Sinha RK, Prasad G. Mu and beta rhythm modulations in motor imagery related post-stroke EEG: a study under BCI framework for post-stroke rehabilitation. BMC Neuroscience 2010; 11(1):P127.

[8] Yang Y, Wiart J, Bloch I. Towards next generation human–computer interaction—brain–computer interfaces: applications and challenges. In: The first international symposium of Chinese CHI (Chinese CHI 2013); 2013. p. 1.

[9] Kachenoura A, Albera L, Senhadji L, Comon P. ICA: a potential tool for BCI systems. IEEE Signal Processing Magazine 2008;25(1):57–68.

[10] Ramoser H, Muller-Gerking J, Pfurtscheller G. Optimal spatial filtering of single trial EEG during imagined hand movement. IEEE Transactions on Rehabilitation Engineering 2000;8(4):441–6.

[11] Müller-Gerking J, Pfurtscheller G, Flyvbjerg H. Designing optimal spatial filters for single-trial EEG classification in a movement task. Clinical Neurophysiology 1999;110(5):787–98.

[12] Lou B, Hong B, Gao X, Gao S. Bipolar electrode selection for a motor imagery based brain–computer interface. Journal of Neural Engineering 2008;5(3):342.

[13] Yang Y, Chevallier S, Wiart J, Bloch I. Automatic selection of the number of spatial filters for motor-imagery BCI. In: The proceedings of the 20th European symposium on artificial Neural networks, computational Intelligence and Machine Learning (ESANN 2012); 2012. p. 109–14.

[14] Kübler A, et al. Patients with ALS can use sensorimotor rhythms to operate a brain–computer interface. Neurology 2005;64(10):1775–7.

[15] Yang Y, Chevallier S, Wiart J, Bloch I. Time-frequency selection in two bipolar channels for improving the classification of motor imagery EEG. In: The proceedings of the 34th annual international conference of the IEEE Engineering in medicine and biology society (EMBC'12); 2012. p. 2744–7.

[16] Yang Y, Chevallier S, Wiart J, Bloch I. Subject-specific time-frequency selection for multi-class motor imagery-based BCIs using few Laplacian EEG channels. Biomedical Signal Processing and Control 2017;38:302–11.

[17] Yang Y, Chevallier S, Wiart J, Bloch I. Time-frequency optimization for discrimination between imagination of right and left hand movements based on two bipolar electroencephalography channels. EURASIP Journal on Applied Signal Processing 2014;2014(1):38.

[18] Pfurtscheller G, Müller-Putz GR, Pfurtscheller J, Rupp R. EEG-based asynchronous BCI controls functional electrical stimulation in a tetraplegic patient. EURASIP Journal on Applied Signal Processing 2005;2005:3152–5.

[19] Kyrgyzov O, Bloch I, Yang Y, Wiart J, Souloumiac A. Data ranking and clustering via normalized graph cut based on asymmetric affinity. In: International conference on Image Analysis and Processing. Springer; 2013. p. 562–71.

[20] McFarland DJ, McCane LM, David SV, Wolpaw JR. Spatial filter selection for EEG-based communication. Electroencephalography and Clinical Neurophysiology 1997;103(3):386–94.

[21] Pfurtscheller G, Brunner C, Schlögl A, Da Silva FL. Mu rhythm (de)synchronization and EEG single-trial classification of different motor imagery tasks. NeuroImage 2006;31(1):153–9.

[22] Neuper C, Wörtz M, Pfurtscheller G. ERD/ERS patterns reflecting sensorimotor activation and deactivation. Progress in Brain Research 2006;159:211–22.

[23] Homan RW, Herman J, Purdy P. Cerebral location of international 10–20 system electrode placement. Electroencephalography and Clinical Neurophysiology 1987;66(4):376–82.

[24] Filatova OG, et al. Dynamic information flow based on EEG and diffusion MRI in stroke: a proof-of-principle study. Frontiers in Neural Circuits 2018;12.

[25] Yang Y, Kyrgyzov O, Wiart J, Bloch I. Subject-specific channel selection for classification of motor imagery electroencephalographic data. In: Acoustics, speech and signal processing (ICASSP), 2013 IEEE international conference on. IEEE; 2013. p. 1277–80.

[26] Yang Y, Bloch I, Chevallier S, Wiart J. Subject-specific channel selection using time information for motor imagery brain–computer interfaces. Cognitive Computation 2016;8(3):505–18.

[27] Bostanov V. BCI competition 2003-data sets Ib and IIb: feature extraction from event-related brain potentials with the continuous wavelet transform and the t-value scalogram. IEEE Transactions on Biomedical Engineering 2004;51(6):1057–61.

[28] Deng J, Yao J, Dewald JP. Classification of the intention to generate a shoulder versus elbow torque by means of a time–frequency synthesized spatial patterns BCI algorithm. Journal of Neural Engineering 2005;2(4):131.

[29] Luo T-j, Chao F. Exploring spatial-frequency-sequential relationships for motor imagery classification with recurrent neural network. BMC Bioinformatics 2018;19(1):344.

[30] Wang T, Deng J, He B. Classifying EEG-based motor imagery tasks by means of time–frequency synthesized spatial patterns. Clinical Neurophysiology 2004;115(12):2744–53.

[31] Yamawaki N, Wilke C, Liu Z, He B. An enhanced time-frequency-spatial approach for motor imagery classification. IEEE Transactions on Neural Systems and Rehabilitation Engineering 2006; 14(2):250–4.

[32] Gursel Ozmen N, Gumusel L, Yang Y. A biologically inspired approach to frequency domain feature extraction for EEG classification. Computational and Mathematical Methods in Medicine 2018; 2018:9890132.

[33] Carletta J. Assessing agreement on classification tasks: the kappa statistic. Computational Linguistics 1996;22(2):249–54.

[34] Ang KK, Chin ZY, Wang C, Guan C, Zhang H. Filter bank common spatial pattern algorithm on BCI competition IV datasets 2a and 2b. Frontiers in Neuroscience 2012;6:39.

[35] Wang Y, Zhang Z, Li Y, Gao X, Gao S, Yang F. BCI competition 2003-data set IV: an algorithm based on CSSD and FDA for classifying single-trial EEG. IEEE Transactions on Biomedical Engineering 2004;51(6):1081–6.

[36] Coyle D, McGinnity TM, Prasad G. Improving the separability of multiple EEG features for a BCI by neural-time-series-prediction-preprocessing. Biomedical Signal Processing and Control 2010;5(3): 196–204.

[37] Schlögl A, Lee F, Bischof H, Pfurtscheller G. Characterization of four-class motor imagery EEG data for the BCI-competition 2005. Journal of Neural Engineering 2005;2(4):L14.

[38] Blankertz B, et al. The BCI competition III: validating alternative approaches to actual BCI problems. IEEE Transactions on Neural Systems and Rehabilitation Engineering 2006;14(2):153–9.

[39] Grosse-Wentrup M, Buss M. Multiclass common spatial patterns and information theoretic feature extraction. IEEE Transactions on Biomedical Engineering 2008;55(8):1991–2000.

[40] Witkowski M, Cortese M, Cempini M, Mellinger J, Vitiello N, Soekadar SR. Enhancing brain-machine interface (BMI) control of a hand exoskeleton using electrooculography (EOG). Journal of Neuroengineering and Rehabilitation 2014;11(1):165.

[41] Yang Y, Chevallier S, Wiart J, Bloch I. A self-paced hybrid BCI based on EEG and EOG. In: 3rd workshop of tools for Brain-computer interaction (TOBI 2012); 2012. p. 42–3.

[42] Leeb R, Sagha H, Chavarriaga R, Millán J del R. A hybrid brain–computer interface based on the fusion of electroencephalographic and electromyographic activities. Journal of Neural Engineering 2011;8(2):025011.

4

# Electromyography (EMG) examination on motor unit alterations after stroke

Ping Zhou[1], Cliff Klein[2], Xu Zhang[3], Xiaoyan Li[1], Sheng Li[1]

[1]DEPARTMENT OF PHYSICAL MEDICINE AND REHABILITATION, UNIVERSITY OF TEXAS HEALTH SCIENCE CENTER AT HOUSTON, TIRR MEMORIAL HERMANN RESEARCH CENTER, HOUSTON, TX, UNITED STATES; [2]GUANGDONG WORK INJURY REHABILITATION CENTER, GUANGZHOU, CHINA; [3]BIOMEDICAL ENGINEERING PROGRAM, UNIVERSITY OF SCIENCE AND TECHNOLOGY OF CHINA, HEFEI, CHINA

**Chapter outline**



## Introduction

Stroke is one of the leading causes of adult disability and death worldwide. Studies have shown that stroke has a detrimental effect on health-related quality of life [1]. Following a hemispheric stroke, motor control of the extremities on one side of the body is usually impaired. Patients suffer disabling physical symptoms on the contralesional side such as weakness, spasticity, and muscle incoordination that contribute to motor impairment. In general, these symptoms likely reflect altered neural (i.e., muscle activation, denervation, trophic support) and muscular (i.e., atrophy, contracture) related processes, but the details and relative contributions of each are equivocal.

Since the motor unit is the final common pathway for the control of muscle force, it provides a structure–function framework for the examination of neural and muscular

Intelligent Biomechatronics in Neurorehabilitation. https://doi.org/10.1016/B978-0-12-814942-3.00004-0

disorders. Understanding the changes in the different components of a motor unit (e.g., spinal motor neuron, axon, and muscle fibers) can help identify specific mechanisms contributing to poststroke motor impairment, and thereby help guide the development of new treatments.

To assess paretic muscle and motor unit changes after stroke, different electromyography (EMG) techniques have been applied. With technical advances in both electrode design and signal processing, EMG has broad applicability for the assessment of motor impairment. EMG has been used to assess properties of single muscle fibers, motor units, gross muscle, or muscle groups in stroke, and a number of abnormalities have been identified (Table 4.1). In this chapter, we provide a summary of findings using different EMG techniques for examination of paretic muscles of stroke patients, with a view toward better understanding motor unit alterations underlying muscle weakness and other associated intrinsic property changes.

**Table 4.1** EMG examination of stroke subjects.

| EMG recording methods | EMG processing methods | Major findings from paretic muscles (compared with contralateral muscles) |
|---|---|---|
| Surface EMG | EMG–force relation<br>Power spectrum analysis<br>Spike distribution analysis<br>Clustering index analysis | Increased/decreased slope of the regression of EMG–force relation [3–8]<br>Increased/decreased mean or median frequency [9,10]<br>Shifting of spike distribution toward smaller/larger amplitude [12]<br>Increased/decreased clustering index [13]<br>In summary, rather mixed findings from interference surface EMG analysis |
| Concentric needle EMG, fine wire EMG | Spontaneous motor activity detection<br>EMG decomposition<br>MUAP analysis | Evidence of denervation and reinnervation: fibrillation and positive sharp waves, enlarged and complex MUAPs [48–55]<br>Spontaneous motor unit activity [56,57]<br>Reduction of motor unit firing rate and compression of motor unit recruitment range [4,37–39]<br>Altered motor unit synchronization [36]<br>Ordered motor unit firing behavior still maintained [40] |
| Single-fiber EMG | Fiber density analysis<br>Jitter analysis | Significantly larger fiber density [15,60]<br>Abnormal jitters [61] |
| Macro EMG | Macro-MUAP analysis | Similar Macro-MUAP amplitudes at high and low muscle forces [63] |
| Motor unit number estimation (MUNE) | Multipoint stimulation MUNE<br>F wave-based MUNE<br>MUNIX | Significantly lower M-wave amplitude [20–26,30,31,34]<br>Decreased number of functional motor units [20–26]<br>Reduced MUNIX [30,31,34] |
| Electrode array surface EMG | EMG decomposition<br>MUAP analysis | Reduced muscle fiber conduction velocity [10,11]<br>Reduction of motor unit firing rate and compression of motor unit recruitment range [44–46]<br>Reduced MUAP mean frequency [58,59] |

## Complex neuromuscular changes demonstrated by interference surface EMG analysis

Conventional surface EMG, using relatively large (centimeter) electrodes, is commonly used to assess muscle activity in research laboratories. Measures of EMG amplitude and power spectral analysis provide a global measure of muscle activity. The relation between isometric muscle force and surface EMG amplitude, a useful marker of changes in neural and/or muscular components of the motor units, was profoundly affected in some stroke patients [2]. In about half of the tested stroke subjects, the slope of biceps (brachialis) EMG–force relation was significantly larger in the paretic side compared with contralateral or neurologically intact muscles, whereas the other half had smaller or unchanged slopes [3,4]. The findings were further confirmed in distal muscles. There were diverse changes in the slope of the paretic first dorsal interosseous (FDI) EMG–force relation, with different subjects showing significant increases or decreases relative to the contralateral side [5,6]. It is noteworthy that a recent study of brachii biceps EMG–force relation using a linear surface electrode array revealed the importance of electrode placement for reliable measurement and comparison between paretic and contralateral muscles [7,8]. For example, as Fig. 4.1 shows, the EMG amplitude varies with different linear electrode array channels (smallest near the innervation zone or tendons), resulting in an "M"-shaped distribution of slopes for the regression lines between EMG amplitude and torque [7].

Power spectral analysis has also been performed to characterize surface EMG features in stroke. Power spectrum analysis of 14 stroke subjects indicated a significant reduction in mean power frequency in the paretic FDI muscle compared with the contralateral muscle at a given percentage of the maximal voluntary isometric force [9]. More specifically, nine subjects showed a smaller mean power frequency, three showed no difference, and only two had larger mean power frequency in the paretic than the contralateral muscle. Similar findings were also observed in a different study based on frequency analysis of linear electrode array recordings [10]; the median frequency and the mean power frequency, averaged from different channels, were significantly smaller in the paretic biceps muscle compared to the contralateral muscle at matched relative forces (Fig. 4.2A and B). The muscle fiber conduction velocity, calculated from the linear electrode array, was also found to be significantly slower in the paretic muscle than in the contralateral muscle (Fig. 4.2C) [10,11].

When analyzing spike amplitude distribution of surface EMG signals post stroke, different changes were observed in the paretic muscle compared with the contralateral one at the matched contraction levels [12]. In 14 chronic stroke subjects with mild to severe muscle weakness, nine showed a distribution with a broader range and a shift toward larger amplitude for the paretic FDI muscle when compared with the contralateral muscle. In contrast, the remaining five subjects displayed a different histogram

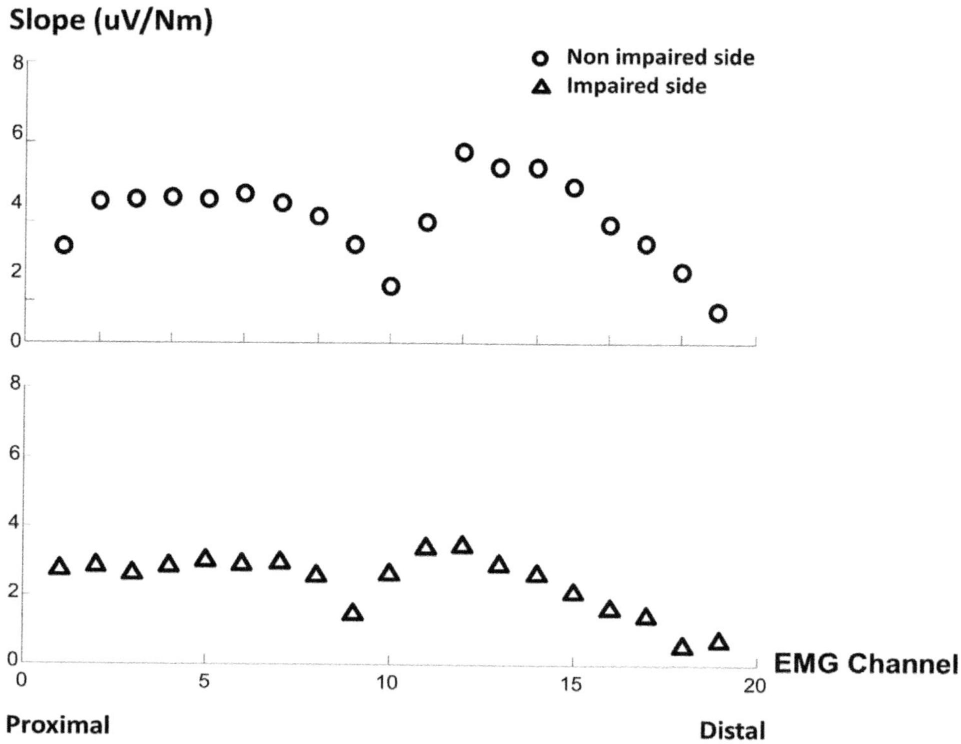


**FIGURE 4.1** The "M"-shaped distribution of slopes with the lowest slope over the innervation zone channel. Slopes are obtained from linear regressions between individual channel EMG and torque from a stroke subject. *Reproduced from Fig. 4 of Bhadane M, Liu J, Rymer WZ, Zhou P, Li S. Re-evaluation of EMG-torque relation in chronic stroke using linear electrode array EMG recordings. Scientific Reports 2016;6:28957, Open Access.*

distribution with narrower range of peak amplitudes and a shift toward smaller amplitudes for the paretic muscle.

Recently, a novel clustering index (CI) analysis of conventional surface EMG was also applied to characterize complex neuromuscular changes induced by a hemisphere stroke [13]. The CI is a marker derived from surface EMG morphology, which can be used to discriminate neurogenic and myopathic changes [14]. The CI analysis was performed bilaterally for the thenar muscles of 17 stroke subjects and 12 matched healthy controls during varying levels of isometric force. Compared with the neurologically intact or contralateral muscles, mixed (increased, decreased, or unaltered) CI patterns were observed in the paretic muscles. Two stroke subjects showed significantly increased CI, implying dominant neurogenic changes, whereas three stroke subjects had significantly reduced CI, indicating dominant myopathic changes. The remaining stroke subjects did not demonstrate a significant CI alternation, likely due to a combined effect of neuromuscular factors, or a deficit of descending central drive (which might not have a significant effect on EMG morphology or CI) [13].

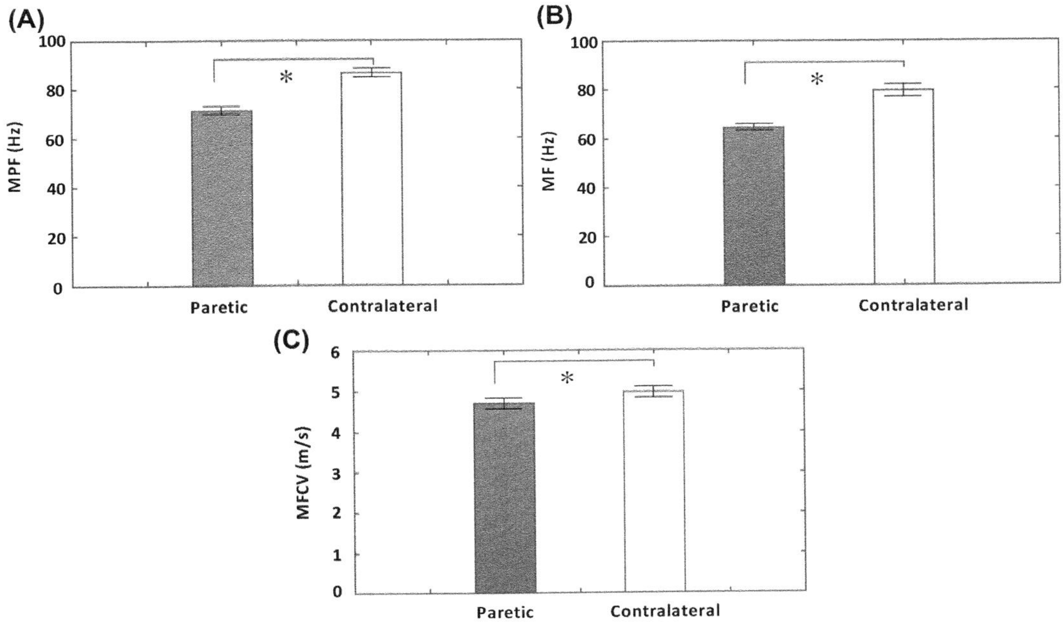


**FIGURE 4.2** A comparison of global averages of surface EMG measurements between the paretic and contralateral sides from all subjects. * indicates statistical significance, $P < .05$. (A) Mean power frequency (MPF); (B) median frequency (MF); (C) muscle fiber conduction velocity (MFCV). *Reproduced from Fig. 4 of Yao B, Zhang X, Li S, Li X, Chen X, Klein CS, Zhou P. Analysis of linear electrode array EMG for assessment of hemiparetic biceps brachii muscles. Frontiers in Human Neuroscience 2015;9:569, Open Access.*

In summary, the findings demonstrated a rather inconsistent picture of interference surface EMG patterns after stroke, with some patients showing dramatic but different changes and others showing little change. This may be related to patient-specific differences in the extent of paralysis (i.e., reduced muscle activation), motor unit loss, reinnervation, and atrophy. For example, patients that have a greater proportion of large-amplitude spikes [12] and larger EMG–force slopes [5,6] in paretic limb muscles may indicate more effective reinnervation and/or less atrophy. Some of the inconsistent findings may also relate to the duration and severity of stroke. For example, jitter was larger in patients with longer stroke duration or greater weakness [15] (see section: Remodeling of surviving motor units after stroke). Besides true physiological differences, inconsistent findings may also result from technical factors of surface EMG, such as electrode placement, size, and configurations. Hence, the recorded surface EMG may be dominated by the largest motor units closest to the recording electrode, which may not be representative of the rest of the muscle. The significance of global surface EMG parameters remains limited because of their ambiguous character for examining specific motor unit properties.

## Motor unit loss after stroke

Various forms of motor unit number estimation (MUNE) have been developed since its first introduction in 1973 [16]. Different procedures have been applied to improve the MUNE estimation including multipoint stimulation, EMG decomposition or spike triggered averaging, high density surface EMG, statistical or Bayesian analysis, F-waves, and compound muscle action potential scans [17–19]. Despite the differences in technique, MUNE relies on the same premises in terms of the CMAP measure and estimation of the mean motor unit action potential (MUAP). There are various factors that can influence MUNE performance. For example, sample motor units used for calculation of mean MUAP are generally obtained from relatively low-intensity electrical stimulation or low voluntary contraction efforts. It is unclear whether the sample motor units can truly represent the whole motor neuron pool. In addition, alternation can occur when two or more motor axons have overlapping thresholds and uncertain firing probability. There is lack of a gold-standard MUNE method due to the underlying theoretical and practical limitations of different MUNE techniques.

Several different MUNE techniques, such as multipoint stimulation and F wave recordings, have been applied to examine motor unit loss after stroke [20–26]. The examined muscles included FDI, thenar, and hypothenar muscles. The estimated MUNE was significantly smaller in paretic muscles compared with contralateral muscles (Fig. 4.3) [25], suggesting that degeneration of spinal motor neurons may arise after a hemispheric brain lesion [20–26]. Conversely, others reported no significant difference in the number of anterior horn cells between the affected and unaffected sides (although their trans-sectional areas associated with the hemispheric lesion side were significantly decreased), suggesting that upper motor neuron lesions in stroke patients do not induce anterograde trans-synaptic degeneration of spinal cord horn cells [27,28]. Because of the conflicting results of MUNE [20–26] versus anatomical-based [27,28] studies, the question of motor neuron loss after stroke remains somewhat controversial. It may be that in humans there is a reduction in the number of functional motor units (i.e., axons do not respond to electrical stimulation) without loss of spinal motoneurons [23].

Toward estimating motor unit number in a clinical environment, a convenient method called motor unit number index (MUNIX) has been developed [29] and recently applied in stroke patients [30,31]. It was found that the MUNIX was reduced in paretic muscles compared with contralateral muscles, but there was no significant change in motor unit size index (MUSIX). Simulation analysis indicated that with the lack of significant MUSIX changes, it is difficult to obtain definite information as the MUNIX drop could be from actual motor unit loss, muscle fiber atrophy, or their combination [32]. Although a modified MUNIX method has been developed recently [33], the impact of fiber atrophy on the estimated motor unit number remains problematic. This remains a dilemma for its application in stroke [34].

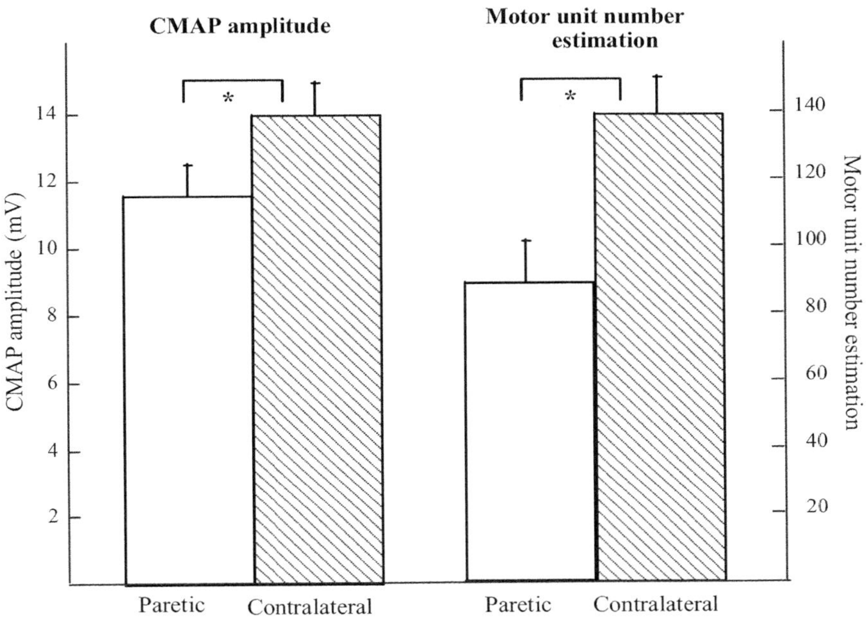


**FIGURE 4.3** Comparisons of the CMAP amplitude and motor unit number estimation (MUNE) between the paretic and contralateral hands of 12 stroke subjects. CMAP amplitude (mean ± standard error): 11.55 ± 0.94 mV (paretic), 13.94 ± 1.00 mV (contralateral), $P < .01$. MUNE: 89 ± 13 (paretic), 139 ± 11 (contralateral), $P < .01$. *From Li X, Fisher M, Rymer WZ, Zhou P (2016) Application of the F-response for estimating motor unit number and amplitude distribution in hand muscles of stroke survivors. IEEE Trans Neural Syst Rehabil Eng. 24(6):674–681 © 2016 IEEE.*

## Motor unit control property alterations after stroke

EMG decomposition is a procedure by which the EMG signal can be separated into its constituent MUAP trains [35]. It provides a unique approach to observing human motor neuron behavior by monitoring the recruitment and firing rate of many concurrently active motor units. Altered motor unit control properties post stroke have been reported using EMG decomposition [4,36–40]. For example, an earlier study using needle EMG decomposition revealed that tested stroke subjects showed compression of motor unit recruitment range and reduction of motor unit firing rates [4].

Needle EMG is limited by its sampling volume. The invasive character of needle EMG also prevents its wider use in testing of stroke survivors. Recent advances in surface EMG recording and processing have made it feasible to extract single motor unit activity (mainly for those superficial large motor units) from surface EMG, thus providing a noninvasive approach for examination of motor unit control properties [41–43]. Using an advanced surface sensor array to record EMG over a range of isometric contractions, the control properties of a substantial population of motor units were characterized in both paretic and contralateral FDI muscles of stroke subjects [44,45]. A significant disruption of motor unit control properties was reported in paretic muscles, demonstrated as a weak correlation between motor unit recruitment threshold force and firing rate, compressed motor unit recruitment range, and a deteriorated firing rate modulation with the contraction level, which correlated well with the impairment of voluntary force generation.

High-density surface EMG recording and decomposition techniques were used to assess motor unit firing behavior alterations post stroke [46]. By decomposing surface EMG signals collected using a 64-channel two-dimensional electrode array at different isometric contraction levels, it was found that paretic FDI muscles showed decreased motor unit firing rates compared with contralateral muscles. Regression analysis indicated a linear relation between the mean motor unit firing rate and the muscle contraction level for both paretic and contralateral muscles (Fig. 4.4A), with the former demonstrating a smaller increment in firing rate compared with the latter (Fig. 4.4B) [46].

In a recent study [47], the motor unit activity was discriminated from EMG signals recorded by a double differential surface electrode in stroke patients during low-level sustained voluntary contractions in different tasks (ankle dorsiflexion, wrist flexion, and elbow flexion), and compared with healthy control subjects. It was reported that mean motor unit firing rates after stroke were significantly lower on the paretic than the contralateral side. Interestingly, firing rates on the contralateral side of stroke subjects were found to be significantly higher than those of healthy control subjects, suggesting that data from matched healthy subjects should be recorded to provide a valid control for both sides of stroke patients.

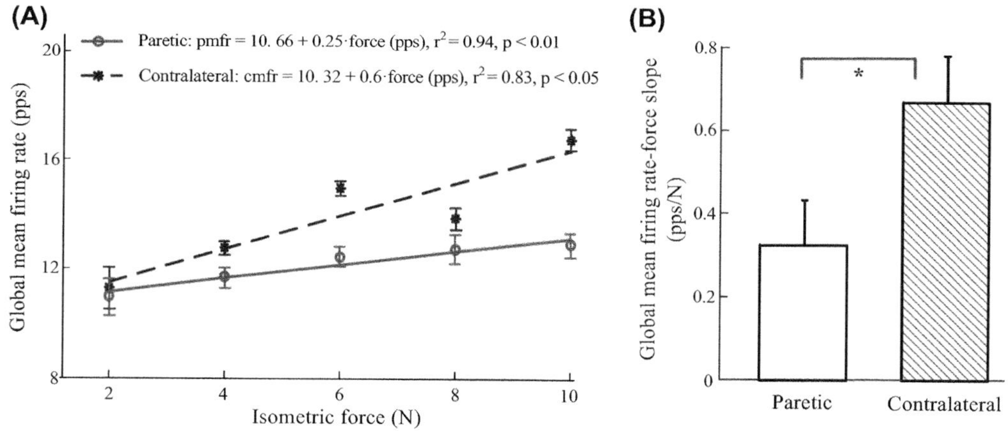


**FIGURE 4.4** (A) A representative stroke subject's global mean firing rate (and standard error) at each contraction level. A linear relation was found in both paretic and contralateral sides. Paretic: global mean firing rate (pmfr) = 10.66 + 0.25·f ($r^2$ = 0.94, $P < .01$); contralateral: global mean firing rate (cmfr) = 10.32 + 0.6·f ($r^2$ = 0.83, $P < .05$). (B) Slope comparison of global mean motor unit firing rate–force relation for nine tested subjects. Paretic, mean ± standard error: 0.32 ± 0.11 pulses per second (pps)/N; contralateral: 0.67 ± 0.11 pps/N. The asterisk (*) represents statistical significance ($P < .05$). *From Li X, Holobar A, Gazzoni M, Merletti R, Rymer WZ, Zhou P (2015) Examination of poststroke alteration in motor unit firing behavior using high-density surface EMG decomposition. IEEE Trans Biomed Eng. 62(5):1242–1252 © 2015 IEEE.*

## Remodeling of surviving motor units after stroke

Motor unit structural changes after stroke have also been examined, primarily using intramuscular EMG. Fibrillation activity and positive sharp waves, indicative of denervation, were reported to occur in proximal and distal paretic muscles. The spontaneous activity appeared during the acute phase following stroke and then tended to disappear over a period of months [48–55]. In addition to fibrillations and positive sharp waves, spontaneous motor unit activity has also been shown to be prevalent in some spastic-paretic muscles, due to hyperexcitability of spinal motor neurons [56,57].

Quantitative MUAP analysis relying on needle EMG is a routine approach for diagnosis of neuromuscular diseases, and has been used to assess muscles after stroke [50,52]. Similar to that occurring in neuropathies, the duration of MUAPs, the number of MUAP phases and turns, and the outliers above maximum for MUAP duration and amplitude were significantly higher on the paretic side than the contralateral side. MUAP analysis was also performed using a surface electrode array [58,59], and one finding was that the MUAP mean frequency was smaller in the paretic than the contralateral biceps, indicating an increased contribution of low-threshold motor units. However, in a recent study of the abductor digiti minimi (ADM) muscles in 46 stroke patients, although increased surface MUAP area and reduced motor unit number were observed in the paretic side compared with the contralateral side, there were no remarkable differences in needle EMG parameters between the sides [26]. In this case, an increase in MUAP size may have been masked by muscle fiber atrophy, although direct measures of fiber size in this muscle are not available.

Single-fiber EMG provides another approach to examining muscle fiber denervation and reinnervation after stroke [15,60,61]. The fiber density in the ADM muscle was determined by single-fiber EMG in 59 stroke patients with the duration of hemiparesis ranging from 2 weeks to 48 months. The fiber density was increased on the paretic side relative to the contralateral side and the control group, and this change correlated with clinical impairment [15]. The fiber density increased during the first 10 months following stroke and thereafter remained stable, indicating that reinnervation took place in the acute phase. The functional integrity of neuromuscular transmission after stroke was also assessed using single-fiber EMG along with axonal microstimulation, by measuring neuromuscular jitter [61]. For patients with a disease duration of 2 weeks–48 months, mean jitters of the extensor digitorum communis (EDC) and anterior tibial muscles were found to be significantly larger than normal controls. These findings indicate dysfunction of neuromuscular transmission in paretic muscles and suggest an ongoing muscle fiber reinnervation process. Increased jitter seemed to correlate positively with disease duration and severity. However, a recent FDI study of chronic stroke indicated little change in jitter although the fiber density was significantly higher in the paretic muscle than the contralateral muscle [62].

Macro-EMG, triggered by single-fiber EMG, was also used to assess paretic muscle changes after stroke [63]. Two (low and high) levels of ADM muscle contraction were used in the testing of stroke and control subjects. It was found that for the contralateral or control muscles, the macro-MUAPs were significantly larger at the high force output than at the low one. However, in the paretic muscles, the macro-MUAPs were of the same amplitude at the high and low forces, suggesting a selective loss of the large high-threshold motor units following stroke.

## Significance and future perspectives

Although stroke-related motor impairment is often thought to arise from dysfunction (loss) of upper motor neurons, there is growing evidence of long-term changes in motor unit properties. However, whether altered motor unit properties reflect mostly neural or muscular changes, and their relative contributions to motor impairment, is unclear. A better understanding of motor unit alterations is essential for development of patient-specific treatment and rehabilitation. For example, weakness due to primarily spinal motor neuron degeneration may require a different treatment strategy than weakness that arises mostly from atrophy. In the case of spinal motor neuron degeneration and associated weakness, treatment strategies can be focused on preventing motor unit loss, including drug interventions, during the acute phase of stroke. However, given that some motor unit loss may be unavoidable, enhancing the natural reinnervation process through exercise, electrical stimulation and other treatments may be necessary. If muscle fiber atrophy (or other pathological changes) is the major factor contributing to muscle weakness rather than loss of motor units, active therapy or exercise is required to prevent further atrophy and regain muscle strength. If muscle weakness is primarily induced by drop of motor unit firing rates, it may be treatable through appropriate training and/or drug interventions. The monitoring of motor unit activity in paretic muscles can help assess not only the nature of the neuromuscular deficits post stroke but also the response of these deficits to medication and physical therapies.

We would like to indicate that changes in different motor unit components may contribute to motor impairment, but the relative contribution of each is uncertain, and likely varies between patients. Simultaneous examination of the neural and muscular motor unit components in the same stroke patient, and their association with motor impairment, is still lacking. We advocate application of a range of EMG techniques in the same stroke patients to obtain a more comprehensive picture of motor unit alterations. This can also be supplemented by other approaches such as medical imaging, electrical impedance myography [64,65], and twitch characterization for quantifying paretic muscle changes. In addition, most of the previous studies have focused on a cross-sectional examination and the statistical analysis of the patient EMG data with different time durations between stroke onset and data collection. Longitudinal studies are needed to track EMG parameters in paretic muscles early after stroke and into the chronic phase of recovery in order to provide data on the time course of motor unit alterations, and their association with motor recovery.

## Funding statement

This work was supported by the National Institutes of Health of the U.S. Department of Health and Human Services under Grant R01NS080839.

## Authors contributions

PZ wrote the first draft of the manuscript. CK, XZ, XL, and SL contributed to developing, writing, revising, and finalizing the manuscript.

## Conflict of interest

The authors declare that the research was conducted in the absence of any commercial or financial relationships that could be construed as a potential conflict of interest.

## References

[1] Xie J, Wu EQ, Zheng ZJ, Croft JB, Greenlund KJ, Mensah GA, Labarthe DR. Impact of stroke on health-related quality of life in the noninstitutionalized population in the United States. Stroke 2006;37:2567–72.

[2] Zhou P, Suresh NL, Rymer WZ. Model based sensitivity analysis of EMG-force relation with respect to motor unit properties: applications to muscle paresis in stroke. Annals of Biomedical Engineering 2007;35:1521–31.

[3] Tang A, Rymer WZ. Abnormal force–EMG relations in paretic limbs of hemiparetic human subjects. Journal of Neurology, Neurosurgery, and Psychiatry 1981;44:690–8.

[4] Gemperline JJ, Allen S, Walk D, Rymer WZ. Characteristics of motor unit discharge in subjects with hemiparesis. Muscle & Nerve 1995;18:1101–14.

[5] Zhou P, Li X, Rymer WZ. EMG–force relations during isometric contractions of the first dorsal interosseous muscle after stroke. Topics in Stroke Rehabilitation 2013;20(6):537–44.

[6] Suresh NL, Concepcion NS, Madoff J, Rymer WZ. Anomalous EMG–force relations during low-force isometric tasks in hemiparetic stroke survivors. Experimental Brain Research 2015;233(1):15–25.

[7] Bhadane M, Liu J, Rymer WZ, Zhou P, Li S. Re-evaluation of EMG–torque relation in chronic stroke using linear electrode array EMG recordings. Scientific Reports 2016;6:28957.

[8] Zhang X, Wang D, Yu Z, Chen X, Li S, Zhou P. EMG–torque relation in chronic stroke: a novel EMG complexity representation with a linear electrode array. IEEE Journal of Biomedical and Health Informatics 2017. 2016 Nov 8. [Epub ahead of print] PMID: 27845680.

[9] Li X, Shin H, Zhou P, Niu X, Liu J, Rymer WZ. Power spectral analysis of surface electromyography (EMG) at matched contraction levels of the first dorsal interosseous muscle in stroke survivors. Clinical Neurophysiology 2014;125(5):988–94.

[10] Yao B, Zhang X, Li S, Li X, Chen X, Klein CS, Zhou P. Analysis of linear electrode array EMG for assessment of hemiparetic biceps brachii muscles. Frontiers in Human Neuroscience 2015;9:569.

[11] Conrad MO, Qiu D, Hoffmann G, Zhou P, Kamper DG. Analysis of muscle fiber conduction velocity during finger flexion and extension after stroke. Topics in Stroke Rehabilitation 2017;24(4):262–8.

[12] Li X, Suresh A, Zhou P, Rymer W. Alternations in the spike amplitude distribution of the surface electromyogram post-stroke. IEEE Transactions on Biomedical Engineering 2013;60(3): 845–52.

[13] Zhang X, Wei Z, Ren X, Gao X, Chen X, Zhou P. Complex neuromuscular changes post-stroke revealed by clustering index analysis of surface electromyogram. IEEE Transactions on Neural Systems and Rehabilitation Engineering May 24, 2017. https://doi.org/10.1109/TNSRE.2017.2707582. [Epub ahead of print] PMID: 28541902.

[14] Uesugi H, Sonoo M, Stålberg E, Matsumoto K, Higashihara M, Murashima H, Ugawa Y, Nagashima Y, Shimizu T, Saito H, Kanazawa I. Clustering Index method: a new technique for differentiation between neurogenic and myopathic changes using surface EMG. Clinical Neurophysiology 2011;122(5):1032–41.

[15] Lukacs M, Vecsei L, Beniczky S. Changes in muscle fiber density following a stroke. Clinical Neurophysiology 2009;120:1539–42.

[16] McComas AJ, Fawcett PR, Campbell MJ, Sica RE. Electrophysiological estimation of the number of motor units within a human muscle. Journal of Neurology, Neurosurgery, and Psychiatry 1971; 34(2):121–31.

[17] Bromberg MB. Updating motor unit number estimation (MUNE). Clinical Neurophysiology 2006; 118:1–8.

[18] Gooch CL, Doherty TJ, Chan KM, Bromberg MB, Lewis RA, Stashuk DW, Berger MJ, Andary MT, Daube JR. Motor unit number estimation: a technology and literature review. Muscle & Nerve 2014; 50(6):884–93.

[19] Bostock H. Estimating motor unit numbers from a CMAP scan. Muscle & Nerve 2016;53(6):889–96.

[20] McComas AJ, Sica REP, Aguilera N. Functional changes in motoneurones of hemiparetic patients. Journal of Neurology, Neurosurgery, and Psychiatry 1973;36:183–93.

[21] Arasaki K, Igarashi O, Ichikawa Y, Machida T, Shirozu I, Hyodo A, Ushijima R. Reduction in the motor unit number estimate (MUNE) after cerebral infarction. Journal of Neurological Sciences 2006;250:27–32.

[22] Hara Y, Akaboshi K, Masakado Y, Chino N. Physiologic decrease of single thenar motor units in the F-response in stroke patients. Archives of Physical Medicine and Rehabilitation 2000;81:418–23.

[23] Hara Y, Masakado Y, Chino N. The physiological functional loss of single thenar motor units in the stroke patients: when does it occur? Does it progress? Clinical Neurophysiology 2004;115:97–103.

[24] Choi I, Kim J, Han J, Lee S. The correlation between F-wave motor unit number estimation (F-MUNE) and functional recovery in stroke patients. Journal of Korean Medical Science 2007;22: 1002–6.

[25] Li X, Fisher M, Rymer WZ, Zhou P. Application of the F-response for estimating motor unit number and amplitude distribution in hand muscles of stroke survivors. IEEE Transactions on Neural Systems and Rehabilitation Engineering 2016;24(6):674–81.

[26] Kouzi I, Trachani E, Anagnostou E, Rapidi CA, Ellul J, Sakellaropoulos GC, Chroni E. Motor unit number estimation and quantitative needle electromyography in stroke patients. Journal of Electromyography and Kinesiology 2014;24(6):910–6.

[27] Qiu Y, Wada Y, Otomo E, Tsukagoshi H. Morphometric study of cervical anterior horn cells and pyramidal tracts in medulla oblongata and the spinal cord in patients with cerebrovascular diseases. Journal of Neurological Sciences 1991;102:137–43.

[28] Terao S, Li M, Hashizume Y, Osano Y, Mitsuma T, Sobue G. Upper motor neuron lesions in stroke patients do not induce anterograde transneuronal degeneration in spinal anterior horn cells. Stroke 1997;28:2553–6.

[29] Nandedkar SD, Barkhaus PE, Stålberg EV. Motor unit number index (MUNIX): principle, method, and findings in healthy subjects and in patients with motor neuron disease. Muscle & Nerve 2010; 42(5):798–807.

[30] Li X, Wang YC, Suresh N, Rymer WZ, Zhou P. Motor unit number reductions in paretic muscles of stroke survivors. IEEE Transactions on Information Technology in Biomedicine 2011;15(4):505–12.

[31] Li X, Liu J, Li S, Wang YC, Zhou P. Examination of hand muscle activation and motor unit indices derived from surface EMG in chronic stroke. IEEE Transactions on Biomedical Engineering 2014; 61(12):2891–8.

[32] Li X, Rymer WZ, Zhou P. A simulation-based analysis of motor unit number index (MUNIX) technique using motoneuron pool and surface electromyogram models. IEEE Transactions on Neural Systems and Rehabilitation Engineering 2012;20(3):297–304.

[33] Li X, Nandedkar SD, Zhou P. Modified motor unit number index: a simulation study of the first dorsal interosseous muscle. Medical Engineering & Physics 2016;38(2):115–20.

[34] Zhou P, Li X, Li S, Nandedkar SD. A dilemma in stroke application: standard or modified motor unit number index? Clinical Neurophysiology 2016;127(8):2756–9.

[35] Parsaei H, Stashuk DW, Rasheed S, Farkas C, Hamilton-Wright A. Intramuscular EMG signal decomposition. Critical Reviews in Biomedical Engineering 2010;38(5):435–65.

[36] Farmer SF, Swash M, Ingram DA, Stephens JA. Changes in motor unit synchronization following central nervous lesions in man. Journal of Physiology 1993;463:83–105.

[37] Mottram CJ, Heckman CJ, Powers RK, Rymer WZ, Suresh NL. Disturbances of motor unit rate modulation are prevalent in muscles of spastic-paretic stroke survivors. Journal of Neurophysiology 2014;111(10):2017–28.

[38] Hu XL, Tong KY, Hung LK. Firing properties of motor units during fatigue in subjects after stroke. Journal of Electromyography and Kinesiology 2006;16(5):469–76.

[39] Chou LW, Palmer JA, Binder-Macleod S, Knight CA. Motor unit rate coding is severely impaired during forceful and fast muscular contractions in individuals post stroke. Journal of Neurophysiology 2013;109(12):2947–54.

[40] Sauvage C, Manto M, Adam A, Roark R, Jissendi P, De Luca CJ. Ordered motor-unit firing behavior in acute cerebellar stroke. Journal of Neurophysiology 2006;96(5):2769–74.

[41] Holobar A, Farina D, Gazzoni M, Merletti R, Zazula D. Estimating motor unit discharge patterns from high-density surface electromyogram. Clinical Neurophysiology 2009;120(3):551–62.

[42] Chen M, Zhou P. A novel framework based on FastICA for high density surface EMG decomposition. IEEE Transactions on Neural Systems and Rehabilitation Engineering 2016;24(1):117–27.

[43] Nawab SH, Chang SS, De Luca CJ. High-yield decomposition of surface EMG signals. Clinical Neurophysiology 2010;121(10):1602–15.

[44] Hu X, Suresh AK, Rymer WZ, Suresh NL. Assessing altered motor unit recruitment patterns in paretic muscles of stroke survivors using surface electromyography. Journal of Neural Engineering 2015;12(6):066001.

[45] Hu X, Suresh AK, Rymer WZ, Suresh NL. Altered motor unit discharge patterns in paretic muscles of stroke survivors assessed using surface electromyography. Journal of Neural Engineering 2016; 13(4):046025.

[46] Li X, Holobar A, Gazzoni M, Merletti R, Rymer WZ, Zhou P. Examination of poststroke alteration in motor unit firing behavior using high-density surface EMG decomposition. IEEE Transactions on Biomedical Engineering 2015;62(5):1242–52.

[47] McNulty PA, Lin G, Doust CG. Single motor unit firing rate after stroke is higher on the less-affected side during stable low-level voluntary contractions. Frontiers in Human Neuroscience 2014;8:518.

[48] Benecke R, Berthold A, Conrad B. Denervation activity in the EMG of patients with upper motor neuron lesions: time course, local distribution and pathogenetic aspects. Journal of Neurology 1983; 230:143–51.

[49] Brown WF, Snow R. Denervation in hemiplegic muscles. Stroke 1990;21:1700–4.

[50] Dattola R, Girlanda P, Vita G, Santoro M, Roberto ML, Toscano A. Muscle rearrangement in patients with hemi paresis after stroke: an electrophysiological and morphological study. European Neurology 1993;33:109–11.

[51] Johnson EW, Denny ST, Kelley JP. Sequence of electromyographic abnormalities in stroke syndrome. Archives of Physical Medicine and Rehabilitation 1975;56:468–73.

[52] Lukacs M. Electrophysiological signs of changes in motor units after ischemic stroke. Clinical Neurophysiology 2005;116:1566–70.

[53] Segura RP, Sahgal V. Hemiplegic atrophy: electrophysiological and morphological studies. Muscle & Nerve 1981;4:246–8.

[54] Spaans F, Wilts G. Denervation due to lesions of the central nervous system. Journal of Neurological Sciences 1982;57:291–305.

[55] Zalis AW, Lafratta CW, Fauls LB, Oester YT. Electrophysiological studies in hemiplegia: lower motor neurone findings and correlates. Electromyography & Clinical Neurophysiology 1976;16:151–62.

[56] Chang SH, Francisco GE, Zhou P, Rymer WZ, Li S. Spasticity, weakness, force variability, and sustained spontaneous motor unit discharges of resting spastic-paretic biceps brachii muscles in chronic stroke. Muscle & Nerve 2013;48(1):85–92.

[57] Mottram CJ, Suresh NL, Heckman CJ, Gorassini MA, Rymer WZ. Origins of abnormal excitability in biceps brachii motoneurons of spastic-paretic stroke survivors. Journal of Neurophysiology 2009; 102:2026–38.

[58] Kallenberg LA, Hermens HJ. Motor unit properties of biceps brachii in chronic stroke patients assessed with high-density surface EMG. Muscle & Nerve 2009;39:177–85.

[59] Kallenberg LA, Hermens HJ. Motor unit properties of biceps brachii during dynamic contractions in chronic stroke patients. Muscle & Nerve 2011;43:112–9.

[60] Cruz Martinez A, Del Campo F, Mingo P, Ferrer MT, Perez Conde MC. Altered motor unit architecture in hemiparetic patients. A single fiber EMG study. Journal of Neurology, Neurosurgery, and Psychiatry 1982;45:756–7.

[61] Chang CW. Evident trans-synaptic degeneration of motor neurons after stroke: a study of neuromuscular jitter by axonal microstimulation. Electroencephalography and Clinical Neurophysiology 1998;109:199–202.

[62] Yao B, Klein CS, Hu H, Li S, Zhou P. Motor unit properties of the first dorsal interosseous in chronic stroke subjects: concentric needle and single fiber EMG analysis. Frontiers in Physiology 2018;9: 1587.

[63] Lukacs M, Vécsei L, Beniczky S. Large motor units are selectively affected following a stroke. Clinical Neurophysiology 2008;119:2555–8.

[64] Rutkove SB. Electrical impedance myography: background, current state, and future directions. Muscle & Nerve 2009;40(6):936–46.

[65] Li X, Li L, Shin H, Li S, Zhou P. Electrical impedance myography for evaluating paretic muscle changes after stroke. IEEE Transactions on Neural Systems and Rehabilitation Engineering May 26, 2017. https://doi.org/10.1109/TNSRE.2017.2707403. [Epub ahead of print] PMID: 28574361.

# 5

# Automatic analysis of segmentwise locomotion details of *Drosophila* larva

Nenggan Zheng[1,5,6,7], Xin Chu[1,2], Zhenhuan Ouyang[1,2], Ting Zhao[3], Zhefeng Gong[4]

[1]QIUSHI ACADEMY FOR ADVANCED STUDIES, ZHEJIANG UNIVERSITY, HANGZHOU, CHINA; [2]COLLEGE OF COMPUTER SCIENCE AND TECHNOLOGY, ZHEJIANG UNIVERSITY, CHINA; [3]HOWARD HUGHES MEDICAL INSTITUTE, ASHBURN, VA, UNITED STATES; [4]DEPARTMENT OF NEUROBIOLOGY, ZHEJIANG UNIVERSITY SCHOOL OF MEDICINE, HANGZHOU, ZHEJIANG, CHINA; [5]ZHEJIANG PROVINCIAL KEY LABORATORY OF CARDIO-CEREBRAL VASCULAR DETECTION TECHNOLOGY AND MEDICINAL EFFECTIVENESS APPRAISAL, HANGZHOU, CHINA; [6]INNOVATION JOINT RESEARCH CENTER FOR ICPS, ZHEJIANG UNIVERSITY, HANGZHOU, CHINA; [7]COLLEGE OF BIOMEDICAL ENGINEERING AND INSTRUMENT SCIENCE, ZHEJIANG UNIVERSITY, HANGZHOU, CHINA

## Chapter outline



**Intelligent Biomechatronics in Neurorehabilitation. https://doi.org/10.1016/B978-0-12-814942-3.00005-2**

## Introduction

Neuroscience research often uses model organisms that have relatively simple nervous systems but rich behavior reservoirs. For example, the *Drosophila* larva is an ideal model to study sensorimotor functionalities [1,2], since it is practically possible to identify all the neural pathways in the larva and further discover the neural mechanisms underlying their complex movements in an affordable way. These scientific results are believed to be significantly helpful in finding the conserved principles of sensorimotor functionalities in the human brain. And it can also inspire fundamental studies in the labs as well as real clinical applications for neural rehabilitations.

As in humans, muscles fibers and other motor actuators of *Drosophila* larvae are also commanded by the neural circuits that compose the integral functionalities of the sensory input, integration, decision-making, and motor control, exhibiting flexible and sophisticated behaviors in various environments. It is required to quantify and analyze the behaviors in an unbiased manner in order to discriminate the intricate motion patterns, which can provide detailed locomotion evidence of high temporal-spatial resolution. In order to achieve this goal, automatic behavior analysis systems are expected to replace the traditional research paradigm based on visual inspection and manual operations, which is tedious, subjective, and inefficient.

Modern computer vision methods [3,4] have been adopted to automatically assess the behavior videos, providing efficient and convenient approaches, which are becoming popular in the neuroscience research community. Besides the global spatiotemporal parameters such as the trajectory, current researchers want to estimate the detailed pose as the fundamental data snapshot to build a hierarchical behavior structure, aiming to compare the subtle behavior changes carefully and reveal the novel underlying neuroscience mechanisms. This thus requires us to develop a robust and practical method to locate key body points and infer two-dimensional (2D) animal poses from biological images and videos.

In previous work on automatic animal behavior analysis, several methods [3–7] were considered to infer coarse pose information of specific animal species, such as curve segments for *Drosophila* larvae, geometric shapes for bees or rats, or even the center of mass representation to form a motion trajectory. These representations perform well for capturing overall movements, but are not detailed enough to reflect subtle variations in a complicated pose, which is critical for sensorimotor neural circuits research scenarios. Inspired by landmark regression methods [8,9] in vision, we instead present a landmark-based segmentation detection method to extract pose series data for the larva movements and thus present a more informative behavior descriptor for scientists than ever before.

To detect the larval segments automatically in a behavior video, we face the following open challenges. First, the larval body is nonrigid, which makes it difficult to establish a

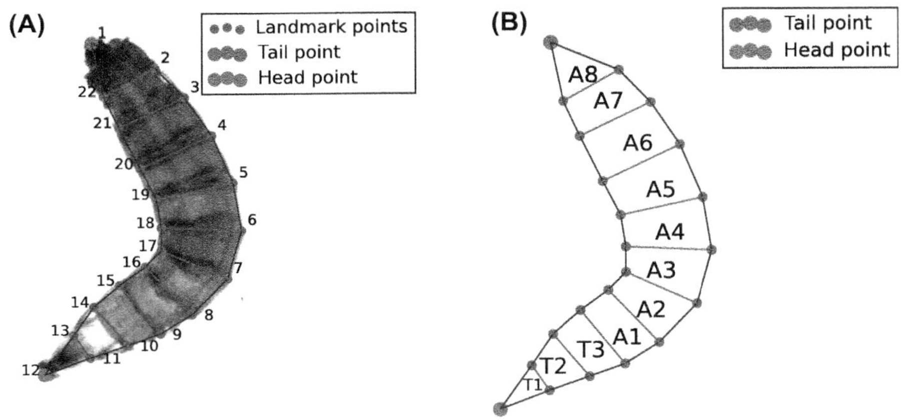


**FIGURE 5.1** The larva segment landmark model: (A) 20-segment landmark points; (B) 11 segments of a larva where T1–T3 are for thoracic segments and A1–A8 for abdominal segments.

rigid geometric model. The peristalsis results in largely deformed body segments and complicated motion shapes. Second, a larva can move freely in all directions, causing a scattered distribution on similarity transform. We formalize the problem of segment recognition as the landmark regression for the bilateral segment endpoints. Given a larva image, we use the proposed method to find 22 landmarks which partition the larva body (see Fig. 5.1A) and define a more informative pose. To locate the segment endpoints precisely, we first construct the dataset of body segment landmarks of *Drosophila* larvae for training and testing our regression model. Then, a cascaded pose estimation model is designed to solve the larval segment detection problem. Extensive experiments show that our model can capture the subtle somatic movements precisely and succeed in modeling the free crawling motion of *Drosophila* larvae.

# Related work

## Automatic behavior analysis

Animal behavior is a macroscopic manifestation of neural activity, which is produced by the contraction of muscles and glands acting on the body, and forming a structured and regular time series [10]. To analyze the detailed behavior patterns and discover the neural mechanisms of sensorimotor circuits, we should construct precise and consistent methods to automatically abstract the discriminative features from the behavior raw data. Videos recorded in behavior experiments are dominantly ubiquitous for convenience and high-performance cost ratio and novel automated computation tools for behavior feature extraction and analysis are proposed [1,2]. To represent the instant locomotion state of an animal, a pose is defined in these works at considerably coarse levels [1,3] based on the geometric model or the pictorial structure [11–13]. For the model organisms of *Drosophila* larva, body model parameters can be calculated by the pose estimation methods and animal locomotion parameters such as movement speed, head turn angular velocity can be achieved straightforwardly for further behavior analysis. A single larva in the research arena is segmented from the clear background and the head or tail points are located by the curvature maximum [3]. Multiple larvae are detected using the predefined gray threshold and high-throughput behavior data of the time series of the moving larvae are processed in real-time [14].

### Pose estimation

Pose estimation is a widely studied problem in computer vision. It defines an appearance model and learns a set of parameters of the model, which is known as pose [11,15,16]. It is a common way to establish a parametric geometric model [11,12,17] for the target in video frames. Another way is to adopt the pictorial structures [13,18,19] framework, which represents a target in terms of a series of components and their positional relationships to each other. In addition, we can also directly use the key point with semantic information to represent the pose. The face alignment methods [20–23] can be migrated to the animal key point detection. At present, regression-based methods have been widely used on the facial point location. The regression-based method learns a regression function, which takes the characters around the current estimated points as input and the real location of the key points as the target of regression. The performance of a single regressor might be weak, similar to boosting [24] methods, Dollar et al. proposed a cascade pose regression framework (CPR) [8] for pose estimation. In CPR, pose can be a set of model parameters or key points directly. It learns a series of cascading regressors and regresses iteratively. Cao et al. proposed the explicit regression shape algorithm (ESR) [9] on the basis of CPR which directly regresses coordinates. ESR uses two layer regressors to improve the speed of training and testing. It also uses a local coordinate system to locate feature points, which reduces the influence of facial differences and small deformations on the result. Meanwhile, they proposed a fast way to calculate feature correlation, which can quickly select features with low correlation from a large number of candidate features for training of the weak regressor. Robust cascade pose regression (RCPR) [25] mainly dealt with the problem of occlusion. It also puts forward an interpolation shape-index feature. RCPR chose a linear interpolation between key points as a feature point, which improves the accuracy and speeds up the feature calculation.

## Method

In this section, we present the cascaded method to automatically locate the body segments of a larva in each frame of the behavior videos. Unlike previous experiments, to make the larval segments more visible, we first dyed larvae with black ink. In the arena back-lighted by an LED surface source, we record videos with high temporal and spatial resolutions for analysis. These acquisition steps made larval segments easier to identify, as shown in Fig. 5.1A.

## Problem formulation

Given a larval movement image, we need to extract $N$ segment endpoints to represent the pose or shape $\boldsymbol{P} = [x_1, y_1, \cdots, x_N, y_N]$. Each $(x_i, y_i)$ pair is the coordinates of a landmark point (i.e., a segment endpoint). Referring to previous work [5], there are three thoracic body segments ($T_1$, $T_2$, $T_3$) and eight abdominal segments ($A_1$, ..., $A_8$) for a larva. We set the tail as the first point. Fig. 5.1B shows the clockwise order of all 22 endpoints that form the 11 segments. Thus pose estimation is formulated as the problem of estimating the larval body segments by regression.

For the input image $\boldsymbol{I}$ the output pose $\boldsymbol{P}$, we should construct a mapping function $\mathcal{R}$:

$$\boldsymbol{P} = \mathcal{R}(\boldsymbol{I}) \tag{5.1}$$

where $\mathcal{R}$ is called a regressor, which predicts the 22 segment endpoints. We combine weak cascaded regressors to achieve an efficient function $\mathcal{R}$, as described in the following subsection.

## Cascaded regression model for larval segment endpoint localization

### *Explicit shape regression-based endpoint location*

We apply the explicit shape regression (ESR) [9] method to calculate the segment endpoint coordinates directly in an iterative way, which uses multiple weak regressors to shorten the shape residual. The $T$ cascaded weak regressors ($\boldsymbol{R}^1$, ..., $\boldsymbol{R}^T$) gradually approximate the unknown function $\mathcal{R}$. Each cascaded weak regressor re-extracts the pose index feature at each step based on the image appearance and the current estimated pose.

Given the input picture $\boldsymbol{I}$ and an initialization posture $\boldsymbol{S}^0$, first we calculate the residual posture $\Delta\boldsymbol{S}^t$ at each step, which is the output of the $t$-th regressor:

$$\Delta\boldsymbol{S}^t = \boldsymbol{R}^t\left(\boldsymbol{\Phi}^t\left(\boldsymbol{I}, \boldsymbol{S}^{t-1}\right)\right), \quad t = 1, \ldots, T \tag{5.2}$$

where $\boldsymbol{S}^{t-1}$ represents the landmark shape of the $(t-1)$-th stage estimation, and $\boldsymbol{\Phi}^t$ is the shape index feature extractor of the $t$-th regressor. Note that this feature extractor is based on the image appearance and the result from the previous step to estimate the current shape residual. We add the residual regressed to the previous estimation as the current estimated shape:

$$\boldsymbol{S}^t = \boldsymbol{S}^{t-1} + \Delta\boldsymbol{S}^t, \quad t = 1, \cdots, T \tag{5.3}$$

As shown in Fig. 5.2, starting from the initial shape $\boldsymbol{S}^0$, the estimated posture is completely different from the ground-truth shape with a large deviation. After $T$ steps, the estimated posture $\boldsymbol{S}^T$ is already very close to the ground-truth shape and the segment endpoints precisely outline the larval contour. The whole iterative regression is a process of gradually approaching the ground-truth. Fig. 5.3 shows the superposition results, forming a process from rough estimation to refinement.

**FIGURE 5.2** Example of landmark regression process with *T* stages.

**FIGURE 5.3** Example of landmark regression process with *T* stages.

*Learning framework*

Given the training data $\left\{\boldsymbol{I}_i, \widehat{\boldsymbol{S}}_i\right\}_{i=1}^{N}$, the regressor $(\boldsymbol{R}^1,\cdots,\boldsymbol{R}^T)$ needs to be learned from the training data. We first define the loss function and initialize a pose $\{S_i^0\}_{i=1}^N$ for each image in the training set. There are multiple initialization strategies, such as calculating the average of all poses as the initial pose, or randomly selecting one from the training samples, then we optimize the following functions:

$$\boldsymbol{S}_i^t = \boldsymbol{R}^t(\boldsymbol{\Phi}^t(\boldsymbol{I}_i, \boldsymbol{S}_i^{t-1})) + \boldsymbol{S}_i^{t-1}, t \in 1, \cdots, T \tag{5.4}$$

$$R^1, \Phi^1, \cdots, R^T, \Phi^T = \min_{\overline{R}^1, \overline{\Phi}^1, \cdots, \overline{R}^T, \overline{\Phi}^T} \sum_{i=1}^{N} \left\| \widehat{S}_i - \overline{R}^T\left(\overline{\Phi}^T\left(I_i, S_i^{T-1}\right)\right)\right\| \tag{5.5}$$

For the recursive form $\boldsymbol{S}$ in Eq. (5.4), it is difficult to directly optimize Eq. (5.5), similar to the square loss function in boosting. The learning algorithm is a forward distribution algorithm. Although the cascaded regression method used here cannot guarantee to achieve the optimal solution, it is sufficiently efficient to give a good enough result, which will be validated in the experiment. When learning $\boldsymbol{R}^t$, $(\boldsymbol{R}^1, \cdots, \boldsymbol{R}^{t-1})$ has already been trained. Using these trained regressors, computing the latest estimated shape $\{\boldsymbol{S}_i^{t-1}\}_{i=1}^N$ is straightforward. The input of $\boldsymbol{R}^t$ is the features generated by $\boldsymbol{\Phi}^t$, and the output needs to be fitted is the residual between $\left\{\widehat{\boldsymbol{S}}_i\right\}_{i=1}^N$ and $\{\boldsymbol{S}_i^{t-1}\}_{i=1}^N$, assuming the regression residual is $\left\{\Delta\widehat{\boldsymbol{S}}_i^t\right\}_{i=1}^N$ :

$$\Delta\widehat{\boldsymbol{S}}_i^t = \widehat{\boldsymbol{S}}_i - \boldsymbol{S}^{t-1} \tag{5.6}$$

Then we find the $\boldsymbol{R}^t$ that minimizes the total regression error:

$$\boldsymbol{R}^t = \mathbf{arg}\ \underset{\boldsymbol{R}}{\mathbf{min}} \sum_{i=1}^{N} \left\| \Delta\widehat{\boldsymbol{S}}_i^t - \boldsymbol{R}(\boldsymbol{\Phi}^t(\boldsymbol{I}_i, \boldsymbol{S}_i^{t-1})) \right\| \tag{5.7}$$

The feature generator $\Phi^t$ can be obtained without learning. For example, we can adopt SIFT [26] features, and the 128-dimensional description vector around the current estimation point can be directly calculated. However, some random binary features require a selection stage, which is also heuristic and requires finding the features with the least correlation.

## Segment endpoint locating method

We enhance the classical ESR methods with some simple and necessary preprocessing steps. These tricks improve the robustness and accuracy of the original ESR and thus achieve the practical performance in the task of localizing larval segment endpoints. We found that the regression-based method is sensitive to big rotation and transformation. To solve these problems, we designed the pipeline as shown in Fig. 5.4, including rotating the image and clustering images to train several models. For each input image, we use a threshold to detect the *Drosophila* region, and refine the region to obtain the skeleton. According to the skeleton, we align the *Drosophila* with the tail and calculate the deflection angle of the larval head. We divide the dataset based on the deflection angles, which is equivalent to making a clustering, and train several models with the cascade regression methods. In the testing period, the sample is preprocessed with the same pretreatment steps and then we determine which model it belongs to. With the corresponding model, we calculate the final results.

### *Rotate image*

We rotate the image $\boldsymbol{I}$ to preserve the tail part of the larva perpendicular to its bottom. This is helpful to overcome the sensitivity of ESR to the larva orientation in the image.

First image segmentation is used to locate the area of the object. Although the distinction between the foreground and the background of our data is clear, we use the Chan-Vese [27] algorithm to get a sharp result rather than the usual background

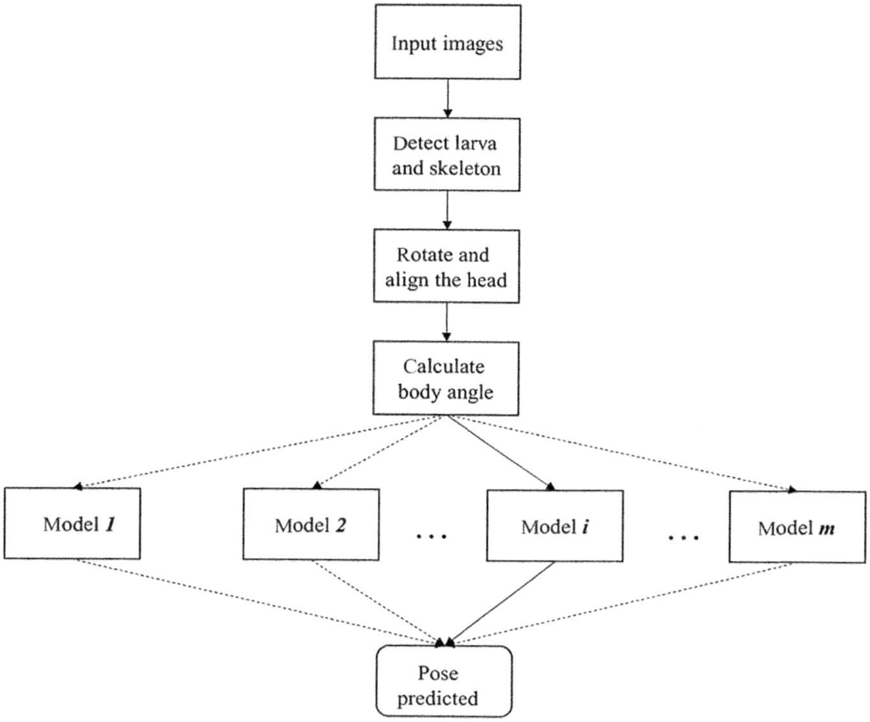


**FIGURE 5.4** Processing pipeline for larva segment endpoint locating.

subtraction. Second, based on where the area of the larva is marked, we can extract the larval skeleton directly using the conventional thinning method. Finally, we align the larva with the tail part of its skeleton. We use points around the tail to calculate the rotation matrix $\boldsymbol{M}$. By multiplying $\boldsymbol{M}$ with the original image, we get the aligned $\boldsymbol{I_r}$:

$$\boldsymbol{I_r} = \boldsymbol{M} \times \boldsymbol{I} \tag{5.8}$$

*Split dataset*

While moving in the arena, a larva might exhibit interleaved turn, cast, and crawling stages. The body shape deforms largely with these actions, resulting in considerable estimation error of the segment endpoints. To solve this problem, we split the training data into several bins by head cast $\alpha$. As shown in Fig. 5.5, we define the head deflection angle as the angle between the line connecting the head and the middle point of the skeleton and the line connecting the tail and the middle point of the skeleton. Given training images after rotation and tail alignment $\boldsymbol{I} \in \{\boldsymbol{I_1}, \cdots, \boldsymbol{I_n}\}$, we get head cast theta of $\boldsymbol{I_i}$ as $\alpha(\boldsymbol{I_i})$. The $m$-th bin of images of $\boldsymbol{I}$ (denote as $\boldsymbol{I}^m$) is determined by two scalars $c_1^m$ and $c_2^m$:

$$\boldsymbol{I}^m = \{i | i \in \boldsymbol{I}, c_1^m < \alpha(i) \le c_2^m\} \tag{5.9}$$

### *Regression-based method*

*Model.* Here we use the model described in the section "Cascaded regression model for larval segment endpoint localization" to locate larval segment landmark positions. The method combines $T$ weak regressors ($R^1,\cdots, R^t,\cdots, R^T$) as discussed in the section "Cascaded regression model for larval segment endpoint localization." Given a larva image $I$ and an initial shape $P^0$, each regressor computes a shape increment $\delta P$ from the chosen image features and then updates the shape using two-level cascaded regression with fern [8,9] as the primitive regressor. Shape-index features were calculated for each step.

*Train model.* We train the model in a cascade form. We train $t$-th regressor $R^t$ when $t-1$ regressor $R$ had been trained. At each step we learn $R^t$ by minimizing the loss function described in Eq. (5.6), and $R$ is a fern regressor which can be easily trained.

We can see that the input of the regressor consists of the aligned image and the shape estimated from the previous stage. We learn each $R$ in two steps:

$$R^t(I, S^t) = fern(\text{shape\_indexed\_feature}(I, S^{t-1})) \tag{5.10}$$

First, we extract the shape indexed features from the image and the previous shape. Then we train the fern to use the features.

*Predict.* For each testing image of larva, we can use the well-trained model to predict the location of the segment endpoints.

$$S^t = S^{t-1} + R^t(I, S^{t-1}) \tag{5.11}$$

The input is a testing image $I_{test}$, which should be classified into one of the shape bins to apply the right model. We use boosted regressors to estimate the final pose and take the median of all final shapes as the predicted pose.

# Result

## Dataset

### *Dataset property*

We have made a novel dataset for analyzing larval movements. It contains four videos with a total of 947 frames, each annotated by marking the 22 landmarks $P = [x_1, y_1,\cdots, x_N, y_N]$. The dataset can be split into training data to train the model, and use the rest to test the model. In our experiment, we used three videos (740 images) for training, and one video (203 images) for testing. Each video has 30 frames per second at a resolution of $960 \times 1280$.

Each video frame is preprocessed by the proposed framework in Fig. 5.4. We extracted the larval object from each image using the OTSU [28] algorithm and some morphological operations (eroding, hole filling, etc.), and its skeleton using the thinning algorithm. We used 20 points around the tail to estimate the orientation of the larva and rotated the image to have the tail pointing vertically downwards (Fig. 5.6).

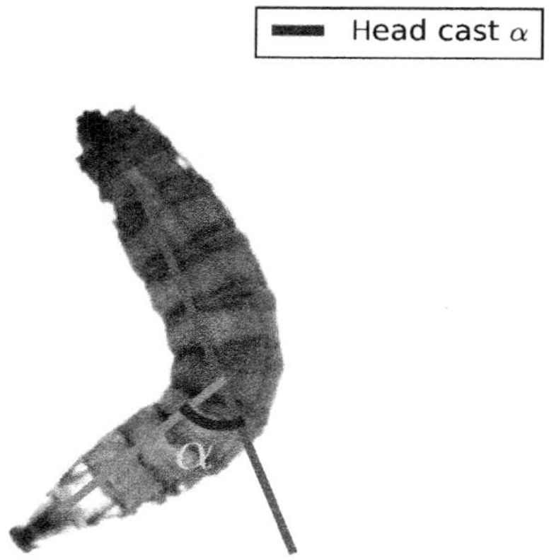


FIGURE 5.5 Calculate the head angle.

*Error metrics*

We also propose a loss function to quantify errors. To make it invariant to the larval size, the error is not in pixels but normalized by the distance between the two landmarks on the abdomen. Given that $\widehat{\boldsymbol{S}}$ is the ground truth of image $\boldsymbol{I}$, and $S$ is the predicted result, the error is defined as:

$$\frac{\left\| \boldsymbol{S} - \widehat{\boldsymbol{S}} \right\|_2}{\| p_7 - p_{17} \|_2} \tag{5.12}$$

To evaluate a model, we define the error rate as the ratio between the number of images with the error larger than 0.1 and the total image number.

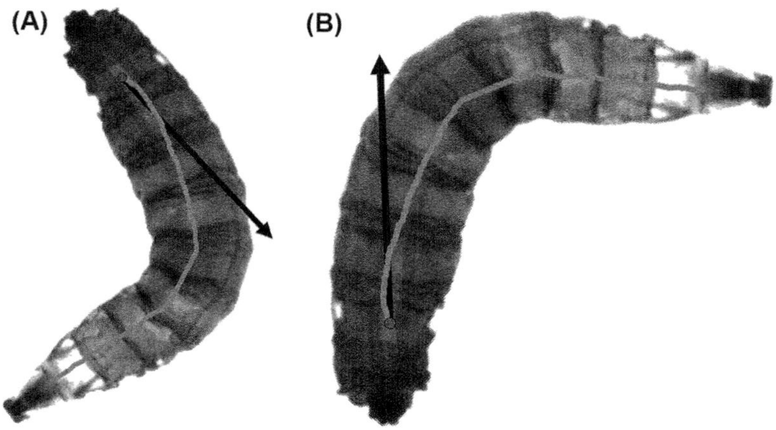


FIGURE 5.6 Rotate image: (A) calculate the angle of larva; (B) rotated larva image.

## Experiments

For the typical movement patterns of a larva, such as turning, crawling straight, and yawing, we selected three videos and drew the track diagram of the larva every 50 frames. Fig. 5.7 shows the three sections of the motion process of a larva. In each frame, we applied the alignment algorithm to the larva to locate landmarks. First, we classified the larva by the angle of its body and then initialized the regression shape with the average shape of the category. After that we used the well-trained regression model to further regress the shape step by step. Each column on the right picture shows the regression process of a frame. In the beginning, the predicted pose is far from the ground truth, but it approaches the body segments gradually in *T* steps. Finally, we achieve the estimated shape of landmarks. Fig. 5.8 shows an example of larval body segment regression. The green points represent ground-truth body segments and the red points shows the result of our method, showing that the prediction is at least qualitatively accurate.

We compared the performance of ESR, RCPR, and ERT with three configurations of dataset (original dataset D1, rotated dataset D2, and split dataset D3). As listed in Table 5.1, we found that all error rates were 100% for the original dataset D1. The reason is that these methods are sensitive to big deformations. After we aligned the larva in D1 to form the rotated dataset D2, the error rate dropped considerably. However, there still exists a technical challenge that the larva itself is very deformable. According to the

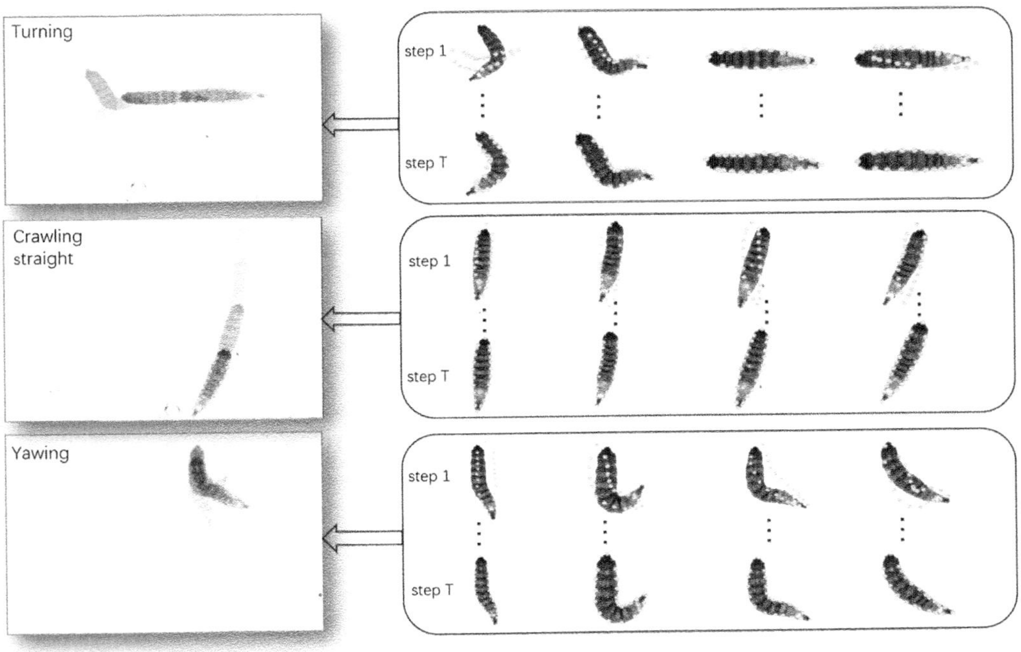


**FIGURE 5.7** Process of body segments regression.

FIGURE 5.8 An example of ground truth and prediction.

**Table 5.1** Result.

| Methods | Mean error ($\times 10^{-2}$) |
|---|---|
| ESR + alignment | 6.2 |
| ESR + alignment + clustering | 4.5 |
| RCPR + alignment | 5.9 |
| RCPR + alignment + clustering | 5.2 |
| ERT + alignment | 5.5 |
| ERT + alignment + clustering | 4.0 |

proposed pipeline, we split the video frames in the rotated dataset D2 further into four bins according to the larval head cast angle. The normalized error rate was reduced to a satisfactory level. Fig. 5.9 shows the results of different methods on D3 visually. For ESR, the final performance is 4.0%, which was the best.

We calculated cumulative error curves on the rotated dataset D2 and split dataset D3 according to the poor result of D1. We found that all D3 were better than D2, which means D3's landmark error is smaller than D2's. Fig. 5.10 illustrates the cumulative error distributions of different methods. The solid lines represent the results of the methods using clustering after alignment, while the dashed lines represent those methods merely using alignment. Among these methods, the ERT-split curve shows the least error of 0.04, resulting from the use of correlation-based feature selection method. ESR-split and RCPR-split have similar errors, which are a little larger than 0.04. In addition, all the split

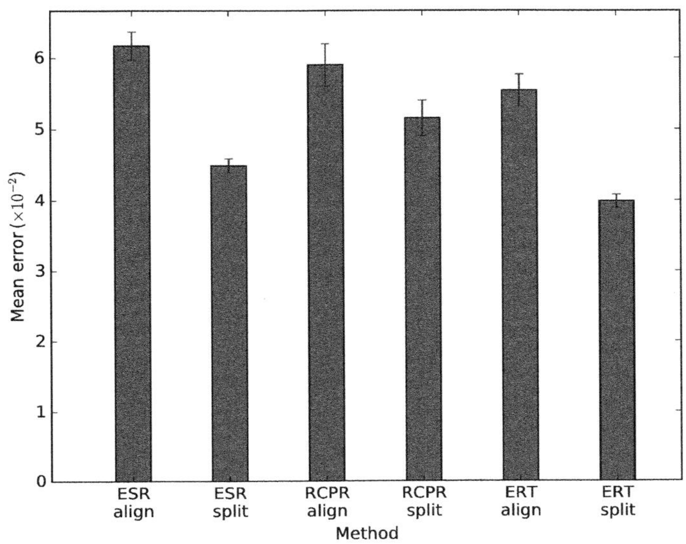


**FIGURE 5.9** Error of each algorithm.

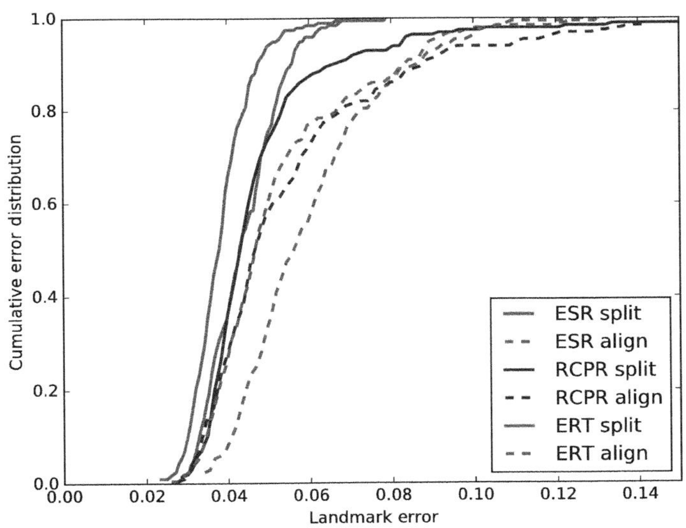


**FIGURE 5.10** Cumulative error.

methods with alignment are better than those without alignment, which proves the effectiveness of alignment with clustering.

Fig. 5.11 shows the mean error and variance of each landmark, in which the x-axis represents the index of the landmark and the y-axis is the normalized error. The first and second landmarks, corresponding to the head and tail, respectively, have the

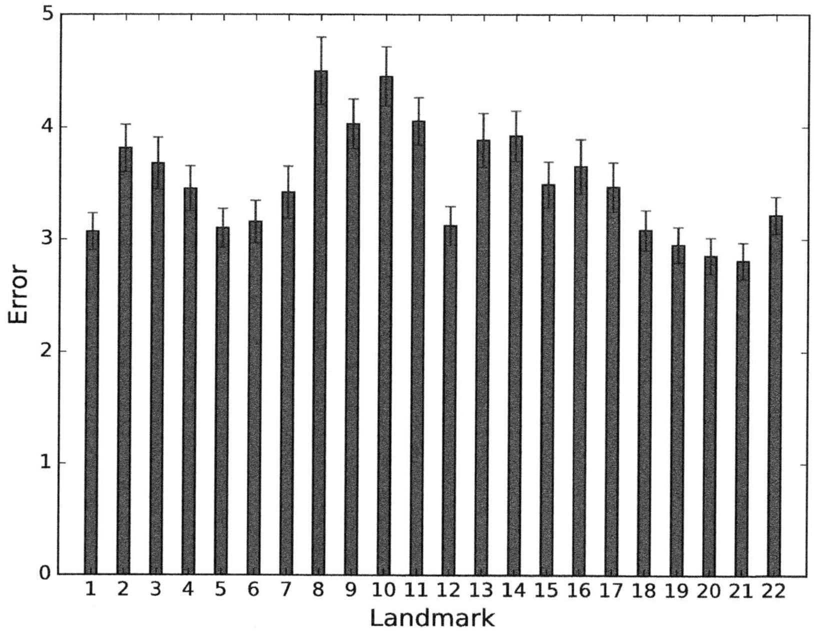


**FIGURE 5.11** Error of each landmark.

smallest errors and the most accurate alignment results. This is because these two landmarks have obvious features, making them easy to locate. The localization accuracy of the middle part of the body is also relatively high, while the errors of landmarks close to the head and tail are relatively large. This is because the parts close to the head or tail have a large deformation, frequently due to the contraction and relaxation of the body segments during the movement, making them difficult to locate. There is a linear correlation between the variance and the mean error, and a larger variance tends to lead to a larger error, because it is difficult to distinguish the landmarks.

# Conclusion

In this chapter, we propose a framework for automatically detecting body segments of larvae at an error rate of 0.49%. The problem is formulated as landmark point regression. The proposed framework improved the traditional methods on locating facial landmark points to handle large body deformations in larval movements. With the ESR algorithm, we demonstrate that the framework can accurately and efficiently calculate the detailed pose in the larva behavior videos. In the future, we expect this method to be extended to automatic behavior analysis of other animal species.

## Acknowledgment

This work is supported by the Zhejiang Provincial Natural Science Foundation (LR19F020005), National Natural Science Foundation of China (61572433, 31471063, 31671074), and thanks for a gift grant from Baidu inc. This work is also partially supported by the Fundamental Research Funds for the Central Universities.

## References

[1] Zhao W, Zhou P, Gong C, Ouyang Z, Wang J, Zheng N, Gong Z. A disinhibitory mechanism biases Drosophila innate light preference. Nature Communications 2019;10(124):1–11.

[2] Zhao W, Gong C, Ouyang Z, Wang P, Wang J, Zhou P, Zheng N, Gong Z. Turns with multiple and single head cast mediate Drosophila larval light avoidance. PLoS ONE 2017;12(7):e0181193.

[3] Burgos-Artizzu XP, Dollár P, Lin D, Anderson DJ, Perona P. Social behavior recognition in continuous video. In: Computer vision and pattern recognition (CVPR), 2012 IEEE conference. IEEE; 2012. p. 1322–9.

[4] Eyjolfsdottir E, Burgos-Artizzu XP, Branson S, Branson K, Anderson DJ, Perona P, HHMI Janelia Farm. Learning animal social behavior from trajectory features.

[5] Gomez-Marin A, Partoune N, Stephens GJ, Louis M. Correction: automated tracking of animal posture and movement during exploration and sensory orientation behaviors. PLoS One 2012;7(10).

[6] Itakura Y, Kohsaka H, Ohyama T, Zlatic M, Pulver SR, Nose A. Identification of inhibitory premotor interneurons activated at a late phase in a motor cycle during *Drosophila* larval locomotion. PLoS One 2015;10(9):e0136660.

[7] Samuel ADT, Shen K, Klein M, Tang A, Kane EA, Gershow M, Garrity P, Lahiri S. Two alternating motor programs drive navigation in *Drosophila* larva. PLoS One 2011;6.

[8] Davies A, Louis M, Webb B. A model of *Drosophila* larva chemotaxis. PLoS Computational Biology 2015;11(11):e1004606.

[9] Wystrach A, Lagogiannis K, Webb B. Continuous lateral oscillations as a core mechanism for taxis in *Drosophila* larvae. Elife 2016;5:e15504.

[10] Dollár P, Welinder P, Perona P. Cascaded pose regression. In: Computer Vision and Pattern Recognition (CVPR), 2010 IEEE conference on. IEEE; 2010. p. 1078–85.

[11] Cao X, Wei Y, Wen F, Sun J. Face alignment by explicit shape regression. International Journal of Computer Vision 2014;107(2):177–90.

[12] Gomez-Marin A, Paton JJ, Kampff AR, Costa RM, Mainen ZF. Big behavioral data: psychology, ethology and the foundations of neuroscience. Nature Neuroscience 2014;17(11):1455.

[13] Dankert H, Wang L, Hoopfer ED, Anderson DJ, Perona P. Automated monitoring and analysis of social behavior in *Drosophila*. Nature Methods 2009;6(4):297.

[14] Branson K, Robie AA, Bender J, Perona P, Dickinson MH. High-throughput ethomics in large groups of *Drosophila*. Nature Methods 2009;6(6):451.

[15] Ramanan D, Forsyth DA, Barnard K. Building models of animals from video. IEEE Transactions on Pattern Analysis and Machine Intelligence 2006;28(8):1319–34.

[16] Swierczek NA, Giles AC, Rankin CH, Kerr RA. High- throughput behavioral analysis in *C. elegans*. Nature Methods 2011;8(7):592.

[17] Veeraraghavan A, Chellappa R, Srinivasan M. Shape-and-behavior encoded tracking of bee dances. IEEE Transactions on Pattern Analysis and Machine Intelligence 2008;30(3):463–76.

[18] Fontaine EI, Zabala F, Dickinson MH, Burdick JW. Wing and body motion during flight initiation in *Drosophila* revealed by automated visual tracking. Journal of Experimental Biology 2009;212(9): 1307–23.

[19] Hong W, Kennedy A, Burgos-Artizzu XP, Moriel Z, Navonne SG, Pietro P, Anderson DJ. Automated measurement of mouse social behaviors using depth sensing, video tracking, and machine learning. Proceedings of the National Academy of Sciences of the United States of America 2015;112(38): 5351–60.

[20] Fischler MA, Elschlager RA. The representation and matching of pictorial structures. IEEE Transactions on Computers 1973;100(1):67–92.

[21] Felzenszwalb PF, Girshick RB, McAllester D, Ramanan D. Object detection with discriminatively trained part-based models. IEEE Transactions on Pattern Analysis and Machine Intelligence 2010; 32(9):1627–45.

[22] Cootes TF, Edwards GJ, Taylor CJ. Active appearance models. IEEE Transactions on Pattern Analysis and Machine Intelligence 2001;(6):681–5.

[23] Matthews I, Baker S. Active appearance models revisited. International Journal of Computer Vision 2004;60(2):135–64.

[24] Cootes TF, Taylor CJ, Cooper DH, Graham J. Active shape models—their training and application. Computer Vision and Image Understanding 1995;61(1):38–59.

[25] Cristinacce D, Cootes TF. Boosted regression active shape models. BMVC 2007;2:880–9. Citeseer.

[26] Nasrabadi NM. Pattern recognition and machine learning. Journal of Electronic Imaging 2007;16(4): 049901.

[27] Burgos-Artizzu XP, Perona P, Dollár P. Robust face landmark estimation under occlusion. In: Proceedings of the IEEE international conference on computer vision; 2013. p. 1513–20.

[28] Lowe DG. Distinctive image features from scale-invariant keypoints. International Journal of Computer Vision 2004;60(2):91–110.

[29] Chan T, Vese L. An active contour model without edges. In: International conference on scale-space theories in computer vision. Springer; 1999. p. 141–51.

[30] Otsu N. A threshold selection method from gray-level histograms. IEEE Transactions on Systems, Man, and Cybernetics 1979;9(1):62–6.

PART II

# Biomechatronic Systems Integrated with the Human Body

6

# Bionic robotics for post polio walking

King-Pong Yu[1,2], Ling-Fung Yeung[1], Sin-Wa Ng[2], Kai-Yu Tong[1]

[1]DEPARTMENT OF BIOMEDICAL ENGINEERING, THE CHINESE UNIVERSITY OF HONG KONG, HONG KONG; [2]COMMUNITY REHABILITATION SERVICE SUPPORT CENTER, HOSPITAL AUTHORITY, HONG KONG

## Chapter outline



Intelligent Biomechatronics in Neurorehabilitation. https://doi.org/10.1016/B978-0-12-814942-3.00006-4

## Background

Poliomyelitis is a deadly infectious disease caused by poliovirus. This virus spreads from one individual to another and can attack the nervous system, including the brain and spinal cord, resulting in several different symptoms. Poliomyelitis has been a statutory notifiable infectious disease in Hong Kong since 1948. According to a report from The National Committee for the Certification of Wild Poliovirus in Hong Kong (2014), the estimated number of poliomyelitis victims from 1948 to 1963 had a peak incident rate at 11.0 with 0.1 million in 1962. The majority of the individuals with a history of acute paralytic poliomyelitis infection in their youth perceive new or increased impairments after a stable period of at least 15 years, referred to as the "late effects of polio" or "post-polio syndrome" (PPS) [1].

The most common impairments reported by individuals with late effects of polio are muscle weakness, muscle fatigue, general fatigue, and musculoskeletal pain [2]. These impairments can affect an individual's balance and walking ability [3] and lead to an increased risk of falling. A previously published chapter has shown that 50%–84% of individuals with late effects of polio reported at least one fall during the previous year [4]. This percentage is considerably higher than in the general elderly people, whose fall frequency is about 20%–40%, depending on their age [5]. Moreover, many individuals with late effects of polio have osteopenia or osteoporosis, and it is reported that between 35% and 40% sustained a fracture as a result of their fall(s) [6]. Decreased mineral bone density and osteopenia are described in individuals with late effects of polio, both in men [7] and postmenopausal women, especially in smokers and those with a reduced level of mobility [6].

Reducing the falling risk and fear of falling for people with poliomyelitis is an important target for rehabilitation, as these individuals are considered to be a high-risk group for fractures. Many individuals with late-onset poliomyelitis sequelae report a decline in walking ability [8]. Limitation in walking activity is one of the most prominent problems of clients with post-poliomyelitis syndrome (PPS) [9]. Limited walking capacity has been reported as an indicator of poor performance and low activities of daily living (ADLs) in clients with PPS [10]. Horemon's article found a strong relationship between walking test performance and walking in daily life for individuals with PPS having lower walking ability [11]. Another article showed that using a stance-control knee joint in a knee-ankle-foot orthosis (KAFO) appears to improve gait biomechanics and energy efficiency compared with a locked knee [12]. From the developed electromechanical KAFO, the three-dimensional gait analysis on poliomyelitis clients revealed a considerable amount of knee flexion during swing phase in controlled-knee gait, which results in approximately 33% less energy consumption than in a locked-knee gait [13]. Thus, the development of assistive devices for enhancing controlled-knee gait leading to an improvement in the individual's walking abilities is essential and contributive in rehabilitation science.

## Current status of individuals with poliomyelitis

Currently, there are no specific exoskeleton robots for rehabilitation of poliomyelitis [14]; the common aids used are KAFO calipers or powered wheelchairs. In Hong Kong, most individuals with poliomyelitis were infected in childhood, resulting in chronic muscle atrophy and weakness in the lower limbs. At the start of infection, they are independent in daily activities and are still able to walk with unsupported legs using a limping gait due to their light body weight and relatively less atrophy. With growing body size and increased demands of daily living, their atrophic leg muscles and particularly their joints are usually overused because of abnormal limping gait, and so they need to adopt a conventional KAFO in early middle age [15]. A slide lock KAFO is a caliper that fully extends and locks the knee at any time to provide stability during ambulation with an abnormal gait pattern [13]. With a KAFO caliper, they are able to walk with a "pole jump" gait, to compensate for the knee hyperextension by setting the pelvis as a pivot and kicking the leg by swaying their trunk backward, which overuses their pelvis and hip joint. With aging, their muscles and joints begin to degenerate and consequences of joint overuse begin to emerge [16]. Usually no regular medical treatment or physical therapy would be prescribed to persons with poliomyelitis because of their irreversible deterioration in lower limb performance, the only medical support they would normally receive is the maintenance of KAFO by lower limb orthotic professionals.

## Robotic knee orthosis design

The orthotic design of a robotic knee orthosis is modified from an exoskeleton ankle robot designed by Yeung's team in Hong Kong (Fig. 6.1) [17]. It consists of five components: thermal plastic mold; motor integrated hinge joint on the lateral knee; electromechanical lock knee joint on the medial knee; sensor embedded insole; and controller box.

### Thermal plastic mold KAFO

The orthosis is built based on a thermal plastic mold, which is tailormade to perfectly fit with the client's lower limb by the prosthetic and orthotic therapist in a rehabilitation hospital. The knee cap is fabricated to guide the knee in order to prevent the knee from slipping out during ambulation and to provide a secure feeling to the client.

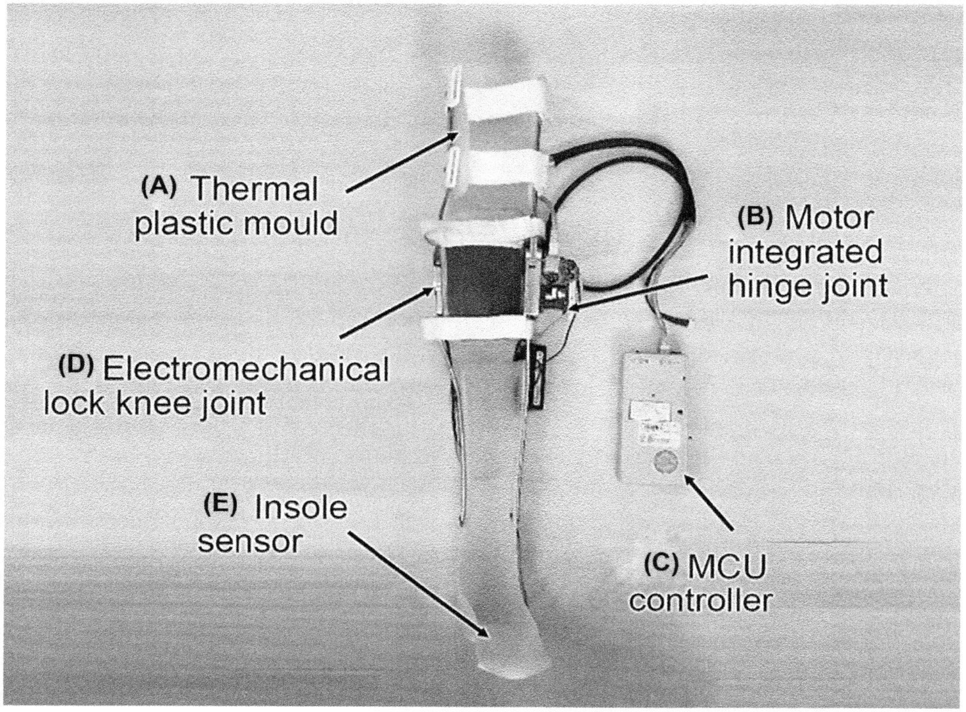


**FIGURE 6.1** Structure of the robotic knee orthosis: (A) thermal plastic mold; (B) motor integrated hinge joint; (C) MCU controller; (D) electromechanical lock knee joint; and (E) insole sensor.

## Sensory system

The robotic knee orthosis is driven by two embedded force sensitive resistors (FSRs) in the 3D printed insole. The specific FSR embedded location is determined by the sensor location identification module (Fig. 6.2A). The resistance of the FSR is decreased when pressure is sensed, which can be measured and digitalized by a microcomputer controller through loading on the client's foot. Hysteresis was added to the threshold crossing detection algorithm to avoid unstable polarity switching in foot loading [17]. The force reaction level is detected using FSRs which can measure 10 N forces. The force is then normalized into 40 units and is shown in different colors for better illustration in Table 6.1.

## Electromechanical lock knee joint

The electromechanical lock knee joint is installed on the medial side of the knee joint (Fig. 6.3). This controllable electromechanical lock knee joint is required to allow the knee movement during the swing phase and to lock the knee during the stance phase. A system knee joint (Neurotronic, Germany) is chosen with maximum loading of 200 kg

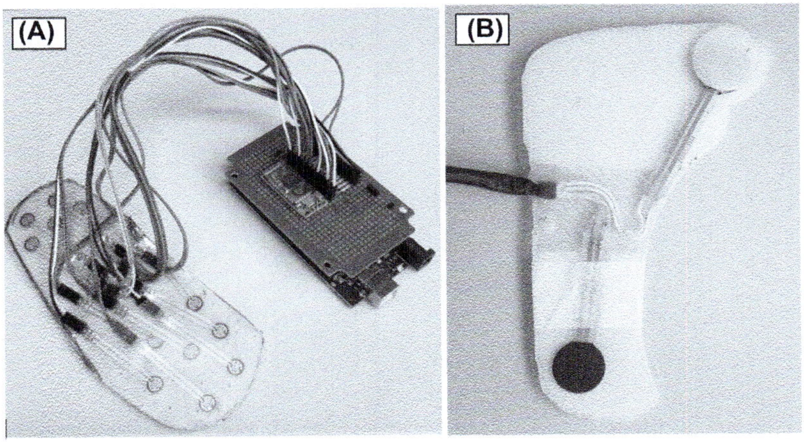


**FIGURE 6.2** (A) Sensor location identification module to identify the optimum sensing location of FSRs during ambulation. (B) FSR embedded three-dimensional printed insole according to the identification module's output.

**Table 6.1** Sensor location identification module output with thirty-step trial of client A and client B.

| Force reaction level | | Heel strike (initial contact) | Toes off (preswing) |
|---|---|---|---|
| 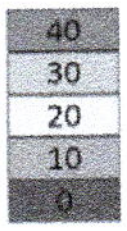  | Client A | 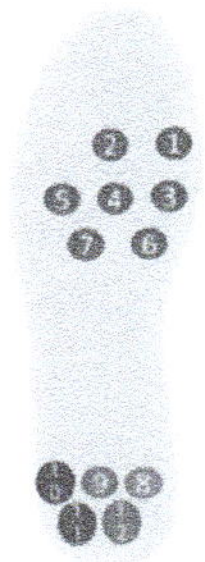 | 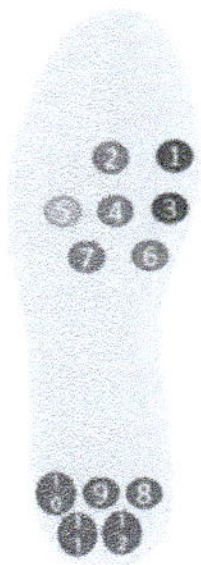 |
| | Client B | 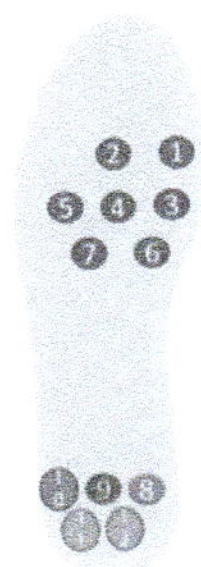 | 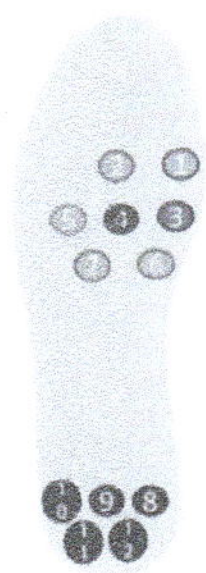 |

The force reaction level divides FSR loading into 40 units for better illustration , ranging from 0 to 40, 0 represents the unloaded FSR; 40 means the FSR is fully loaded. Seven sensor are arranged at the forefoot area and five sensor are located at the heel area to determine the loading pattern of the client while adopting a robotic knee orthosis. It shows a slight difference in pressure distribution between subjects in terms of optimum sensing location.

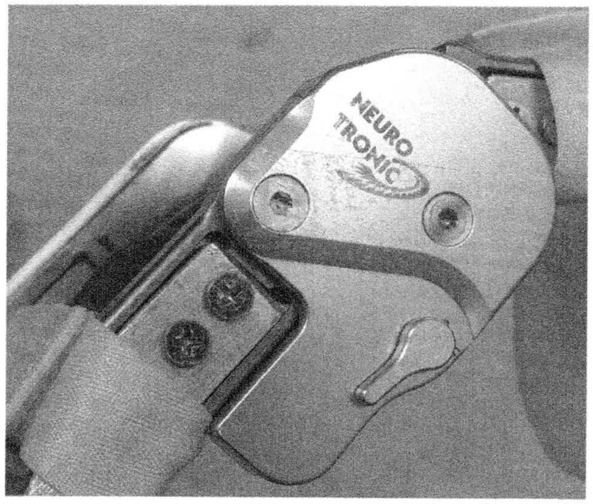


**FIGURE 6.3** Electromechanical lock knee joint on the medial side of the knee.

and with built-in solenoid to control gear lock and unlock. The control is driven to match with the locking logic, which was designed by Yeung [17].

## Actuation system

A motor integrated hinge joint is installed on the lateral side of the knee joint, which is opposite to the electromechanical lock knee joint (Fig. 6.4). The purpose of the motor is to provide power assistance to the knee joint during walking. An off-the-shelf brushless DC motor Dynamixel MX-106R servomotor (ROBOTIS, South Korea) was chosen, with built-in PID control, maximum torque output 10 Nm, and maximum angular speed 55 rpm with gear reduction ratio 1:225, operating at 12V, 5.2A [17].

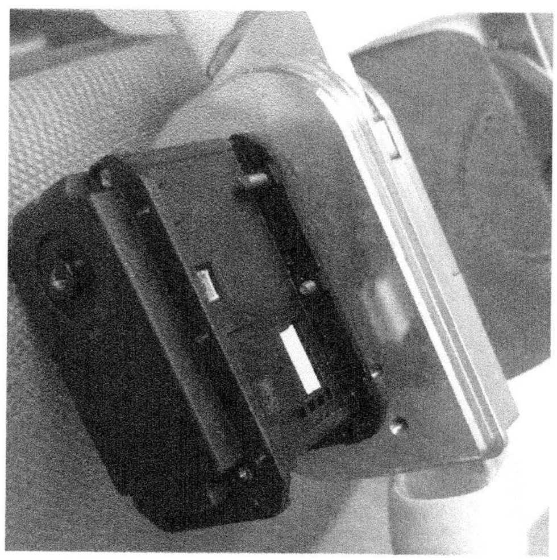

**FIGURE 6.4** Motor integrated hinge joint knee on the lateral side of the knee.

### Control algorithm

The user is recommended to initiate walking with the robotic knee orthosis instead of the good leg. The robotic knee orthosis would recognize the gait phase with the FSR reading (heel and forefoot). Since the gait pattern of people with poliomyelitis has a distinct initial contact and preswing pattern at 0-degree knee flexion of thermal plastic [18], the classification of initial contact and preswing phase can be detected by heel FSR activated and forefoot FSR deactivated, respectively, as shown in Table 6.2. The motor integrated hinge joint triggered immediately after unloading of the forefoot FSR to allow leg clearance during the swing phase. The motor would first flex to attain foot clearance and then extend after the midstance to bring the calf forward for another step, instead of kicking the calf forward by trunk movement and circumduction.

Computations are performed using an Arduino Pro Mini with ATmega328-5V-16 MHz microprocessor (Atmel, USA), which is placed inside the control box with the lithium polymer battery (12 V, 1800 mAh). The control box is connected with sensors via a cable that hangs at the back of the user to prevent disturbance. The robotic knee orthosis can communicate with the computer wirelessly to alter parameters, including swing time, FSR threshold, and power assistance level (Fig. 6.5) [17].

## Training program

The eight-session robotic knee orthosis training intervention was carried out by a physiotherapist and a healthcare assistant (Fig. 6.6). The program is designed with a preparation phase and a training phase for a total of eight sessions. During the preparation phase, there were two sessions done in a Community Rehabilitation Service Support Center and the training included aerobic training, affected lower limb stretching and strengthening, weight shifting training, and trunk and pelvic control training. After the preparation phase, clients carried out four sessions of indoor ambulation training (gait re-education and indoor walking) and two sessions of outdoor ambulation training (outdoor walking and across kerb training) using the robotic knee orthosis with and without power assistance.

The training program obtained ethics approval from the Research Ethics Committee, Hospital Authority, with the project title: Application of knee robotic orthosis to improve the walking performance of people with poliomyelitis (Ref: KC/KE-17-0256/FR-1). Written consent was collected from all clients to participate in the study.

### Case description

Two clients were recruited to the study to examine the effectiveness of robotic knee orthoses compared with conventional KAFO. Before adopting a robotic knee orthosis, they were outdoor walkers with modified functional ambulation category (MFAC) scale

**Table 6.2** The relationship between FSR, motor, and knee joint in the gait cycle, which is the basic logic of the control algorithm.

| | Stance phase | | | Swing phase | |
|---|---|---|---|---|---|
| |  |  |  |  |  |
| | Heel strike | Midstance | Preswing | Toe-off | Terminal swing |
| Heel sensor | **Triggered** | | | | |
| Forefoot sensor | | | **Triggered** | | |
| Joint | **Lock** | Lock | **Unlock** | Free | Free |
| Motor | Free | Free | **Flexion** | **Flexion** | **Extension** |

In the stance phase, after the heel strike which triggered the heel sensor, the electrotechnical joint is locked throughout the whole stance phase to provide support for weight shifting, until the forefoot sensor is triggered at preswing, which represents the end of stance, the electromechanical lock is unlocked, and the motor integrated hinge joint is then triggered by MPU to perform knee flexion to obtain foot clearance during midstance and knee extension at terminal swing in order to prepare for another heel strike, which represents the beginning of the next gait cycle. The bold letters mean action activated.

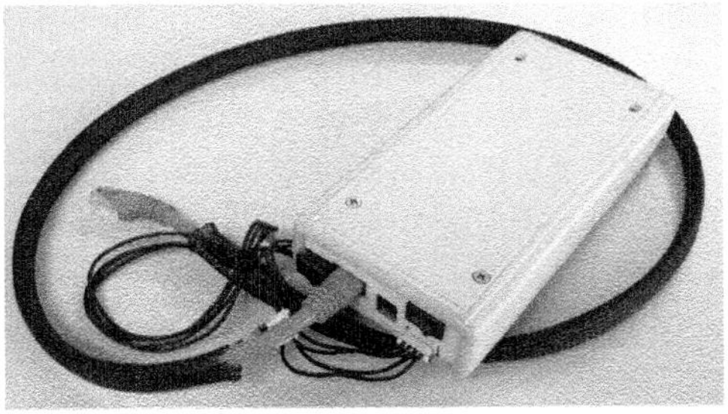

**FIGURE 6.5** Control box with RJ45 cable containing battery and MPU for communication.

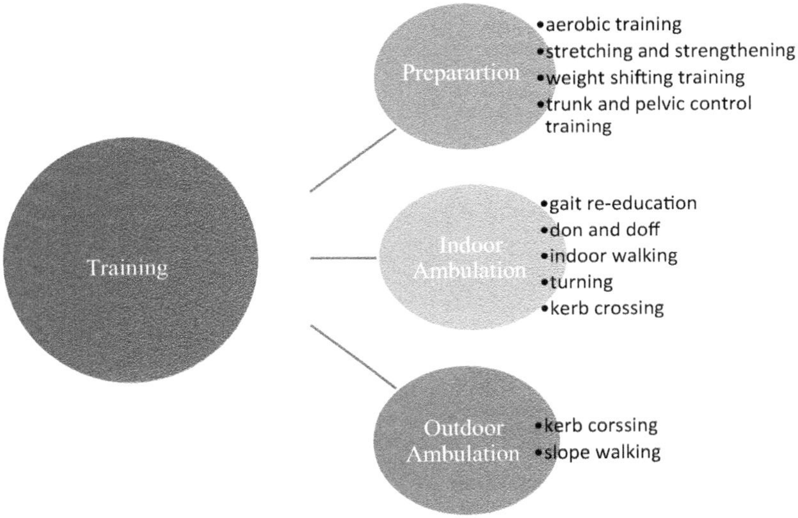


**FIGURE 6.6** Training protocol of the study; the training is divided into three phases, and each phase has unique milestones with respective to an environmental scenario.

level 5. Excessive trunk movement, left hip hiking, and left circumduction were observed during ambulation with conventional KAFO.

Client A (male, 65 years of age) was infected in childhood, with muscle atrophy gradually in the left lower limb. His body weight was 55 kg, and his height was 153 cm. He is totally independent in all activities of daily living (ADLs). After infection in childhood, the virus attacked his nervous system and caused muscle atrophy and partial paralysis, and he adopted a conventional KAFO in his early middle age. Due to his "pole jump" gait, he could walk fast and finished a 10-meter walk test in 12 s with an affected leg step length of 39.49 cm. He is totally independent for KAFO don on and off (timed up and go test [TUGT] = 19 s).

Client B (female, 57 years of age) was also infected in childhood with muscle atrophy gradually in the left lower limb. Her body weight was 57.5 kg, and her height was 157 cm. She was totally independent in all ADLs. Client B was different from client A, and had adopted a conventional slide lock KAFO in her late middle age. The negative effects of polio virus on client B were not as significant as for client A. The degree of muscle atrophy was not as severe as in client A, and client B was able to move her lower limb using the hip joint and hip muscle. With KAFO, client B walked in the same way as client B but with more controllable body movement, such as an adjustable gait speed. She completed a 10-meter walk test in 11 s with an affected leg step length 49.0 cm. She could don and doff independently (TUGT = 13 s).

## Don and doff

The clients were instructed to sit in a "90 degrees position", with 90 degrees flexion in the knee to facilitate robotic knee orthosis adoption, with the lower strap as tight as possible to provide a feeling of security, and the upper strap should be as loose as possible in order to promote knee flexion during ambulation (Fig. 6.7). The sensor embedded in the insole (Fig. 6.2B) should be sealed to the bottom of the orthosis to prevent insole displacement in order to provide optimum sensing efficacy after using the identification module. Before training, a safety belt had to be adopted on the waist to secure the client against losing their balance or collapsing during training.

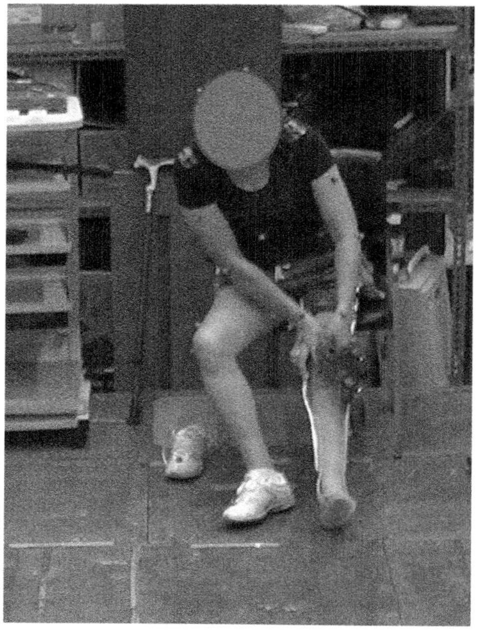

**FIGURE 6.7** Donning of the robotic knee orthosis. The client is instructed to don the orthosis with an "up to bottom" approach, first with the strap on the upper thigh and then the lower thigh, and finally strapping the knee cap.

## Sit-to-stand

Before standing up, the microcomputer controller was powered on and the knee was locked manually to trigger the sensor-locked mode of the electromechanical knee lock (Fig. 6.8). The clients were instructed to stand up with the aid of a stick, weight bearing on the unaffected side, and then distribute the weight to both sides after fully standing.

## Walking preparation

After standing up, the clients were instructed to perform weight shift training (with the locked knee) by weight bearing on both legs alternately to experience the balance of the newly adopted orthosis. The weight shifting training was followed by step training with the sensor-driven lock knee. The clients were instructed to weight shift on the unaffected side and to step forward and backward alternately using the leg with the orthosis, allowing the clients to familiarize themselves with the whole locking and unlocking mechanism (heel locking, forefoot unlocking).

In the meantime, the sensor thresholds of both FSRs were adjusted—usually the threshold does not vary much with the same shoe. If the FSR level without loading is about half the ceiling, the tightness of shoelaces was checked as this may triggered the FSR. In addition, sitting too long may be the cause, as the weight of the orthosis may temporarily change the shape of the sensor embedded insole as well as the long term stationary of orthosis which makes two soles touching each other due to the robotic knee

**FIGURE 6.8** Sit to stand. The client is instructed to first switch the electromechanical lock into the locked mode manually and then to stand up with the weight distributed evenly, then pick up the walking aids such as walking stick after fully extending the knee.

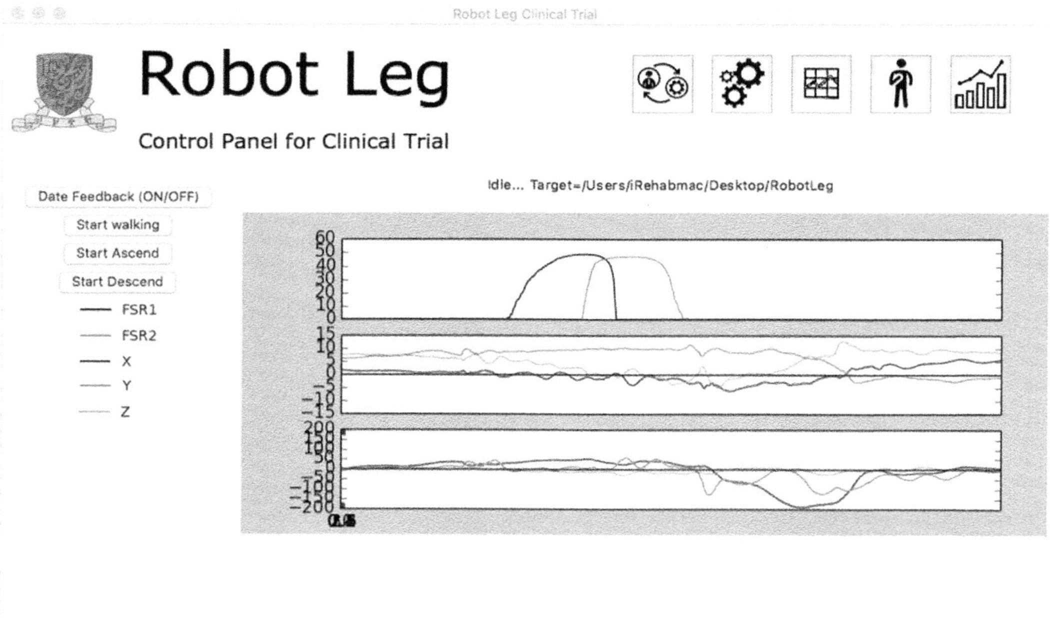


**FIGURE 6.9** Wireless Bluetooth control interface of the robotic knee orthosis software. The upper graph represents the FSR reading of the sensor embedded in the 3D printed insole ranged from 0 to 40; the middle and lower graphs represent the readings of the accelerometer and gyroscope installed at the lateral side of the motor integrated hinge joint, respectively, in order to monitor the operation and status of the robotic knee orthosis.

orthosis weight, e.g., after resting during training. If this happens, the user can slightly twist the shoe to allow the two soles to separate.

For the threshold level of FSR, the locking threshold (heel FSR) has to be low enough to guarantee rapid locking of the knee joint when there is heel strike (initial contact in the gait cycle) to provide stability during the stance phase (weight bearing). The unlock threshold has to be high enough to ensure sensitive unlocking when toeing off (preswing in the gait cycle) to make sure the knee joint is able to be flexed (floor clearance) by the motor and walk through a step during the swing phase (knee extension) (Fig. 6.9).

## Level ground walking

During level ground walking practice, the client walked on a clear walkway without obstacles (Fig. 6.10). The clients were instructed to walk with a stick held by the unaffected hand in order to perform a three-point gait which is more stable [19]. The swing time was adjusted according to the clients' training performance. In the first few sessions, the swing time was longer (0.7 s) to enable the client to familiarize with the robotic knee orthosis (with slower walking speed and more weight bearing on the unaffected side). After the client adapted to the new gait pattern, the swing time could be

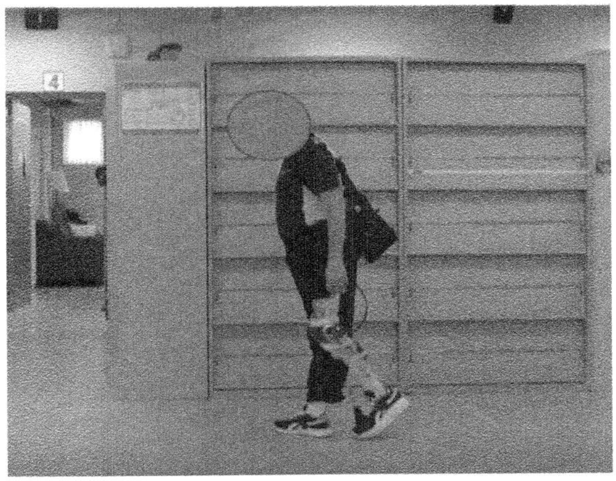

**FIGURE 6.10** Level ground walking.

shortening gradually to match the clients' gait speed so as to obtain a more natural and symmetric gait pattern.

## Turning

After level ground walking, turning is also a significant scenario in ADL. When the client turned around, he/she was instructed to bend toward the unaffected side in order to allow more clearance for the robotic knee orthosis to swing without using the hip joint to compensate for the step length, which is abnormal compared to a natural gait.

## Slope walking

While walking up a slope, the gait pattern is the same as level ground walking with a distinct heel strike and toe-off pattern (Fig. 6.11). As the affected ankle joint is fixed at a right angle by the thermal plastic of the KAFO, the client cannot perform dorsiflexion to compensate for the inadequate knee flexion. Therefore, the assistance level has to be increased to obtain better foot clearance during midstance in order to avoid the foot tip tripping on the ground and causing a fall. The gait pattern for walking down a slope is different from normal walking.

In a slope-descending movement, the clients were instructed to put their body weight on the affected leg during the terminal stance to ensure locking of the electromechanical lock and to perform a quick weight shift to the good side in order to release the lock by unloading the forefoot FSR and allowing the motor to flex the robotic knee orthosis. Since at the terminal stance of the affected side, the heel FSR may still be triggered because of the slope, leaning forward could lead to better unlocking.

**FIGURE 6.11** (A) Walking up a slope. (B) Walking down a slope.

## Kerb crossing

Kerb crossing is another common ADL scenario (Fig. 6.12). In contrast to stair climbing, a kerb is a single obstacle with a flat platform. While going up a kerb, the unaffected leg should walk first so that the robotic knee orthosis can be brought upward with knee flexion by the motor at the foot clearance phase of stair climbing, instead of compensating the fully extended knee of a conventional slide lock KAFO by hip hiking or circumduction, and it is easier for the client to perform toe-off at the end of the forward continuance phase.

While going down a kerb, the affected leg should initiate and perform heel strike with a stick as a three-point gait [19]. If the unaffected leg goes first, the situation is similar to going down a slope but with a greater gradient, also it is difficult for the client to perform toe-off at the end of the forward continuance phase due to a lack of plantarflexion. The client also finds it difficult to perform a heel strike at the foot placement phase through controlled lowering due to low or even no muscle power.

## Outdoor walking

While outdoor walking, the client was instructed to walk along a relatively smooth path without any large obstacles or soft ground, as the hardness of the floor could change the

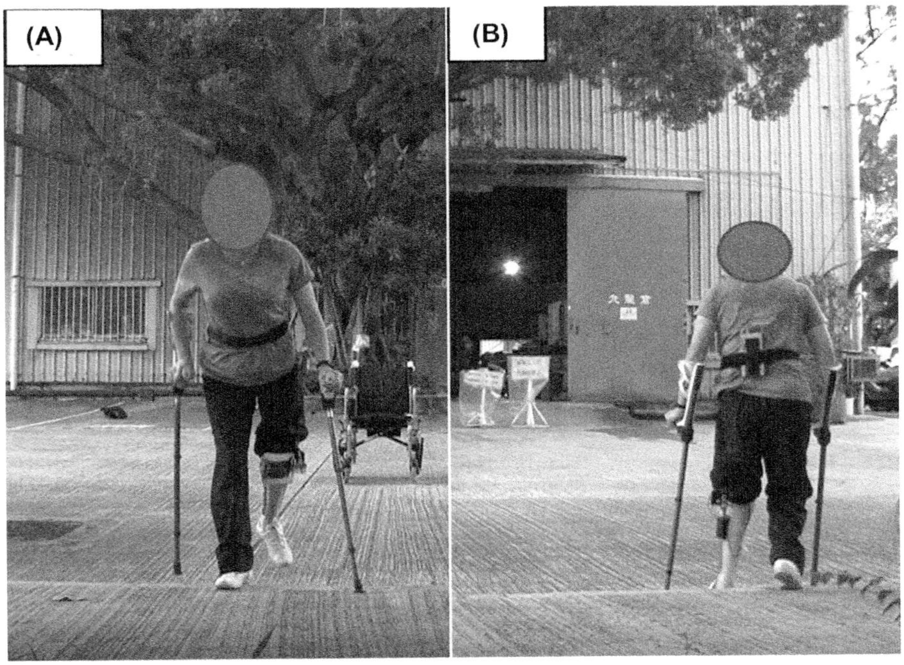


**FIGURE 6.12** (A) Going up a kerb. (B) Going down a kerb.

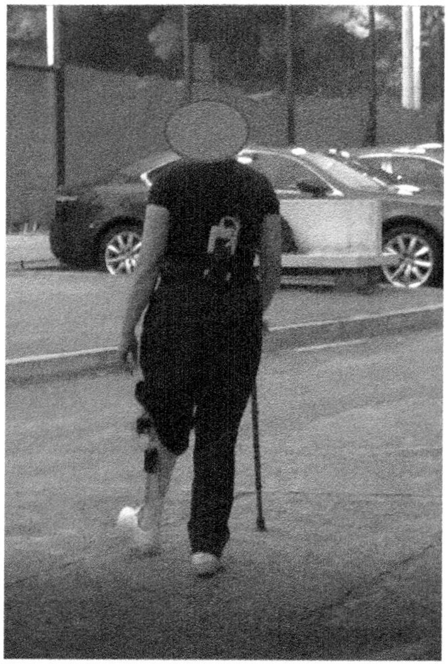

**FIGURE 6.13** Outdoor walking.

FSR threshold, especially affecting the unlock threshold (Fig. 6.13). The client should be walking under supervision of a therapist or carer, and the threshold of the FSR may need to be adjusted according to the environment.

# Method

## Clinical performance

Client A received robotic knee orthosis training for 6 months, with 13 sessions. For the first five sessions (preparation phase), Client A was instructed to perform aerobic training, affected leg stretching and strengthening, sit-to-stand training, weight shifting training, and trunk and pelvic control training; the main reason for the extended preparation phase was to ameliorate the locking logic as well as the 3D printed sensor embedded insole. As Table 6.1 shows, the optimum sensing location for client A was the lateral side of the heel and the medial side of the front of the forefoot. The three-dimensional printed insole is made according to the optimum sensing location using TPU with 25% density to ensure good durability and flexibility, which embedded the FSR inside [20].

For the indoor walking ambulation training (eighth to 11th sessions), training was carried out alternately with and without power assistance to investigate the effectiveness of the robotic knee orthosis under the two modes of operation. Client A started with a swing time of 0.8 s, heel FSR threshold 15/50, forefoot threshold 35/50, and power assistance 35%. He could finish 30 minutes ambulation training without requesting a break and had no complaints of back or hip pain. At the end of each session, stretching and strengthening were carried out by a physiotherapist and doffing of the robotic knee orthosis was done with the aid of a client care assistant. At the second indoor session, client A could perform don and doff independently. The robotic knee orthosis parameters were been adjusted according to his gait performance, the swing time was reduced from 0.8 to 0.7 s, the heel FSR threshold was reduced from 15/50 to 5/50, forefoot threshold increased from 35/50 to 45/50, and power assistance reduced from 35% to 25%. These parameters remained unchanged from the second indoor session until the end of the robotic knee orthosis training except in slope walking.

Client A started turning at the second indoor ambulation training sessions. He was instructed to bend toward the unaffected side with a large turning radius, using a traffic cone as the center of the turning orbit which was later removed when not needed. The client should not perform an about turn, to prevent undesirable unlocking of the knee joint due to unstable twisting of the shoe sole.

After the first two indoor sessions, client A was instructed to walk up and down a kerb with the procedure mentioned above to perform a steady kerb crossing with one physiotherapist standing next to him as a precaution. In the outdoor ambulation training, he was instructed to walk in an outdoor environment with kerb crossing and slope walking. While he was walking up the slope, the power assistance level was increased from 25% to 40% so as to ensure foot clearance and to prevent the robotic knee orthosis from hitting the ground.

Other observed progress during the training sessions included less hip hiking and circumduction while ambulating with without power assistance; trunk sway on both sagittal and coronal plane was also reduced. On the last training session, client A was able to ambulate handfree under supervision with the aid of a stick to perform all the tasks (level ground walking, turning, slope walking, kerb crossing, and outdoor walking).

Client B received a similar training protocol to client A. She received robotic training for 6 months, with eight sessions. Without locking logic adjustment, the training was able to maintain the scheduled protocol and the 3D printed sensor embedded insole (TPU in nature, 25% density) [20].

The progress of client B was faster than in client A. She performed kerb crossing and even stair climbing in the second indoor session. She also had better endurance with respect to conventional slide lock KAFO, not requesting a break within the 30 minutes of training (Table 6.3).

## Outcome measures

### *Clinical assessments*

The clinical assessments including subjective and objective measures, namely Visual Analogue Scale of Pain (VAS), Rated Perceived Exertion Scale (RPE), Goal Attainment Scale (GAS), 10-Meter Walk Test (10 MWT), and Timed Up and Go Test (TUGT) [21–26] were carried out in the Community Rehabilitation Service Support Center by a physiotherapist (Table 6.4).

**Table 6.3** Progress summary of clients A and B.

| | Client A | Client B |
|---|---|---|
| 1st session | Preparation | Preparation |
| 2nd session | Preparation | Preparation |
| 3rd session | Preparation | Indoor ambulation (LGW, TN) |
| 4th session | Preparation | Indoor ambulation (LGW, TN) |
| 5th session | Preparation | Indoor ambulation (LGW, TN, KC) |
| 6th session | Preparation | Indoor ambulation (LGW, TN, KC) |
| 7th session | Preparation | Outdoor ambulation (LGW, TN, KC, SW) |
| 8th session | Indoor ambulation (LGW, TN) | Outdoor ambulation (LGW, TN, KC, SW) |
| 9th session | Indoor ambulation (LGW, TN) | Nil |
| 10th session | Indoor ambulation (LGW, TN, KC) | Nil |
| 11th session | Indoor ambulation (LGW, TN, KC) | Nil |
| 12th session | Outdoor ambulation (LGW, TN, KC, SW) | Nil |
| 13th session | Outdoor ambulation (LGW, TN, KC, SW) | Nil |

LGW: level ground walking; TN: turning; KC: kerb crossing; SW: slope walking.

**Table 6.4** Summary of assessments that were taken into account in the study to access client performance.

| **Clinical assessment** | |
|---|---|
| Subjective: | 1. Visual Analogue Scale of Pain (VAS) |
| | 2. Rated Perceived Exertion Scale (RPE) |
| | 3. Goal Attainment Scale (GAS) |
| Objective: | 1. 10-Meter Walk Test (10 MWT) |
| | 2. Timed Up and Go Test (TUGT) |
| **Gait parameter** | |
| | 1. Left swing percentage |
| | 2. Left stance percentage |
| | 3. Left step length |
| | 4. Left stride velocity |
| | 5. Left single support time |
| | 6. Double support time |
| **Gait kinematics** | |
| | 1. Inclined sagittal angle, trunk movement in AP |
| | 2. Inclined coronal angle, trunk movement in ML |
| | 3. Left hip coronal, left hip circumduction |
| | 4. Left vertical movement, left hip hiking |
| | 5. Trunk transverse angle, trunk rotation |
| | 6. Shoulder coronal angle, shoulder tilting |

*Gait analysis*

After robotic knee orthosis training sessions, clients A and B attended a gait analysis at the Sports Biomechanics Laboratory in the Chinese University of Hong Kong. Both performed a trial sequence of (1) slide lock KAFO, (2) robotic knee orthosis without power assistance, and (3) robotic knee orthosis with power assistance, in both the Vicon system [27] and GAITRite [28] map to investigate the effectiveness of different calipers.

The GAITRite map is 7 meters long and allows the clients to walk around five strides, the middle five steps are taken as a sample with five trials, in total 25 steps are taken to calculate the average gait parameter [28,29]. The clients were asked to walk a few trials on the GAITRite map to familiarize themselves with the walkway, then they were instructed to complete five trials at a self-selected comfortable speed. They were positioned 1 meter behind the map for acceleration and stopped 1 meter after the edge of the

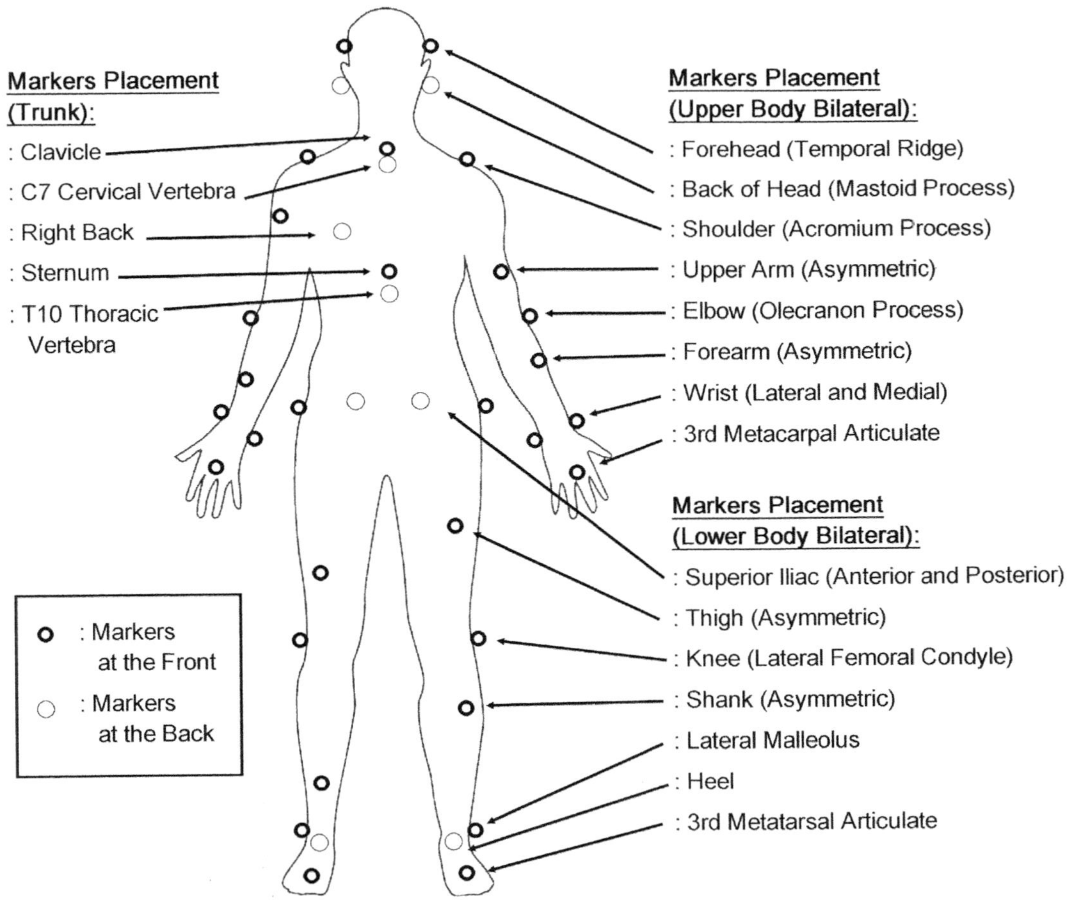


**FIGURE 6.14** Marker placement modified based on the Plug-in Gait Model. [30].

map for deceleration and they were instructed to walk with a stick. The gait parameter collection was triggered by the first pressure contact on the map by the client, with the carer and physiotherapist walking next to the clients as a precaution. The gait parameters were stored for further analysis.

A 12-camera motion capture system (Vicon Nexus, Oxford Metrics) was used to capture the kinematic gait pattern of the clients, the output parameters were computed using the Plug-in Gait Model of the Vicon system (Figs. 6.14 and 6.15) [27]. Prior to trial, the Plug-in Gait requires client body measurements to enable direct computation of kinematics and kinetics from the measured XYZ marker positions, including body height, body weight, leg length, knee width, ankle width, shoulder offset, elbow offset, wrist width, and hand thickness. After the body measurements, 40 marker labels were attached to the clients' bodies for the Plug-in Gait marker set) [30]. The laboratory was equipped with about six strides walkway, two embedded force plates, and 12 cameras, the middle five steps were recorded for five trials. A total of 25 steps were taken to calculate the gait pattern. The data were normalized to the gait cycle and averaged among the clients for five steps [17]. There was a break of 15 minutes between each caliper trial to ensure the trials were independent of each other.

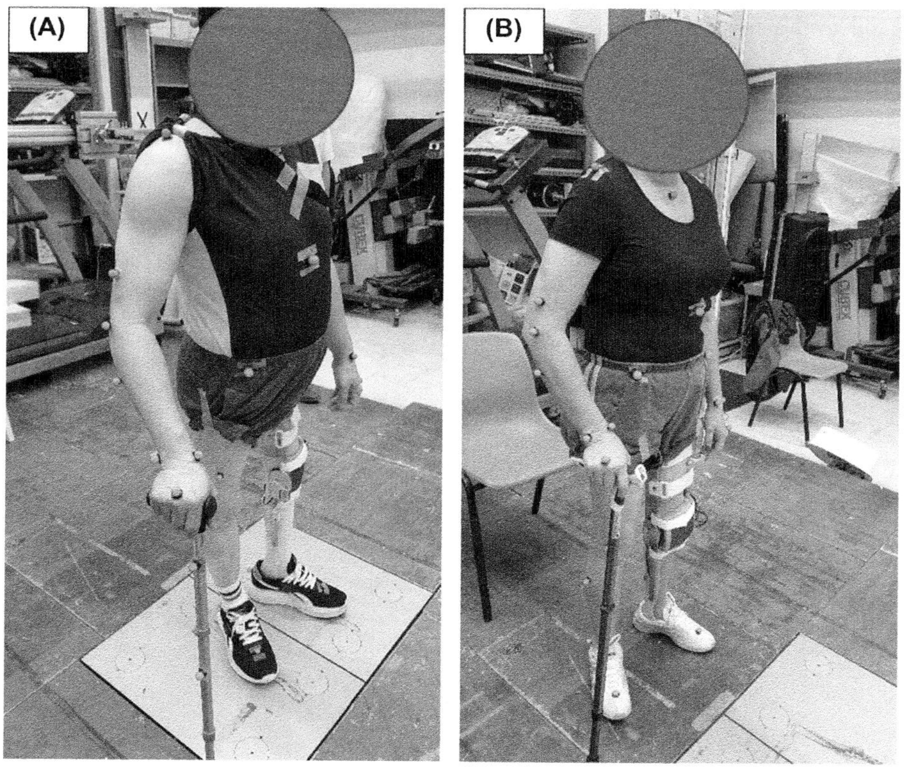


**FIGURE 6.15** (A) Client A with Plug-in Gait marker. (B) Client B with Plug-in Gait marker.

## Results

For clinical assessment, subjective and objective clinical scores were taken into account as mentioned above; for gait analysis, gait parameters and peak-to-peak kinematics were considered in the chapter.

After 6 months training, client A stated that he was capable of walking with higher endurance without complaint of back or hip pain, VAS of pain and RPE were also reduced from 7/10 to 1/10 and 8/10 to 3/10, respectively, and GAS increased from 37.6 to 68.6. In addition, there was no fall accident throughout the whole training schedule. However, the subjects walked with a slower walking speed in power-assisted robotic knee orthoses comparing to walking in a conventional slide lock KAFO: stride velocity decreased from 85.7 cm/s to 18.0 cm/s, step length was also reduced from 39.5 to 23.4 cm; for clinical test, the 10 m walk test time increased from 12 to 40 s, and the timed up and go test increased from 19 to 33 s. Additionally, the stance phase percentage of client A's affected leg increased from 61.4% to 63.6%.

Client B stated that she was not confident to use the robotic knee orthosis independently due to lack of training. She was able to finish each trial without complaints of back or hip pain, VAS of pain and RPE were also reduced from 7/10 to 4/10 and 6/10 to 5/10, respectively, and GAS increased from 36.4 to 71.3. Similar to client A, she had no falls throughout the entire training and walked with a slower walking speed: stride velocity decreased from 71.3 cm/s to 31.7 cm/s, step length also reduced from 49 to 27.5 cm; for clinical test, the 10 m walk test time increased from 11 to 26 s; timed up and go test increased from 13 to 24 s.

In addition to the clinical score and gait parameters, the kinematics was also recorded and analyzed within this chapter. The excessive trunk movement in both sagittal and coronal planes and shoulder tilting were improved while adopting the robotic knee orthosis compared with walking in a conventional slide lock KAFO for both clients A and B. However, client A's affected leg circumduction while adopting the robotic knee orthosis without power assistance was increased, and client B's adopting robotic knee orthosis with or without power assistance, left hip hiking and trunk rotation increased.

## Discussion

The results showed that the robotic knee can reduce excessive trunk movement, pain, and metabolic consumption in terms of endurance by evidence of gait analysis and clinical assessment. Most importantly, the robotic knee orthosis aids the client in resuming a normal gait pattern and is evidenced using clinical assessment scores and gait analysis. Moreover, client B mentioned that she was more confident in using the robotic knee orthosis with power assistance than without, due to the power assistance boost her ambulation performance, and she did not need to compensate for the extended knee during swing phase.

For clinical assessment, as shown in Table 6.5, both clients A and B needed more time to perform a timed task after adopting a robotic knee orthosis as compared to a conventional KAFO—this was because they were accustomed to conventional slide lock

KAFO for more than 20 years and it had become ingrained into their gait meaning they needed extra time to relearn the new gait pattern and posture, and we can speculate that intensive training is required in order to achieve greater confidence in gait independency while walking in the robot. However, the stance phase percentage of affected leg for both clients increased, which implies they were more willing to place body weight on the affected leg.

For kinematics, as shown in Table 6.6, client A's affected leg circumduction while adopting robotic knee orthosis without power assistance was increased, because client A

**Table 6.5** Comparison of clinical assessment of clients A and B using the investigated orthoses.

| | | Visual Analogue Scale of Pain | Rated Perceived Exertion Scale | Goal Attainment Scale | 10-Meter Walk Test | Timed Up and Go Test |
|---|---|---|---|---|---|---|
| Client A | SL | 7/10 | 8/10 | 37.6 | 12 s | 19 s |
| | NoM | 1/10 | 3/10 | 68.6 | 40 s | 33 s |
| | RKO | 1/10 | 2/10 | 74.8 | 40 s | 33 s |
| Client B | SL | 7/10 | 6/10 | 36.4 | 11 s | 13 s |
| | NoM | 3/10 | 5/10 | 71.3 | 26 s | 24 s |
| | RKO | 3/10 | 5/10 | 77.1 | 26 s | 24 s |

For both clients the subjective clinical assessment improvement indicates that both were positive with the robotic knee orthosis and the increased time for 10 MWT and TUGT indicates that training is essential in using robotic knee orthoses in order to correct their accustomed gait pattern. SL: slide lock caliper; NoM: robotic knee orthosis without power assistance; RKO: robotic knee orthosis with power assistance.

**Table 6.6** Comparison of kinematics of clients A and B using the investigated orthoses.

| | | Trunk sway in AP | Trunk sway in ML | Left circumduction | Left hip hiking | Trunk rotation | Shoulder tilting |
|---|---|---|---|---|---|---|---|
| Client A | SL | 16.5 degrees | 11.5 degrees | 15.9 degrees | 60.4 | 9.2 degrees | 16.5 degrees |
| | NoM | 15.8 degrees | 11.2 degrees | 20.1 degrees | 53.4 | 7.3 degrees | 8.3 degrees |
| | NoM wrt SL | −4.42% | −3.73% | +26.80% | −11.50% | −20.60% | −50.00% |
| | RKO | 11.1 degrees | 9.1 degrees | 13.2 degrees | 35.2 | 5.2 degrees | 7.2 degrees |
| | RKO wrt SL | −32.80% | −21.00% | −16.90% | −41.70% | −43.60% | −56.70% |
| Client B | SL | 10.5 degrees | 11.8 degrees | 16 degrees | 70.3 | 5.97 degrees | 8.6 degrees |
| | NoM | 9.44 degrees | 6.85 degrees | 11.8 degrees | 118 | 11.0° | 3.6 degrees |
| | NoM wrt SL | −9.83% | −40.40% | −26.30% | +68.30% | +86.90% | −58.10% |
| | RKO | 5.26 degrees | 8.87 degrees | 11.2 degrees | 87.4 | 18.8 degrees | 3.8 degrees |
| | RKO wrt SL | −49.80% | −24.80% | −30% | +24.30% | +133% | −55.80% |

For both clients, excessive trunk sway and shoulder tilting significantly decreased, implying that the robotic knee orthosis can return client gait to a more natural pattern with less excessive body movement. SL: slide lock caliper; NoM: robotic knee orthosis without power assist; RKO: robotic knee orthosis with power assist.

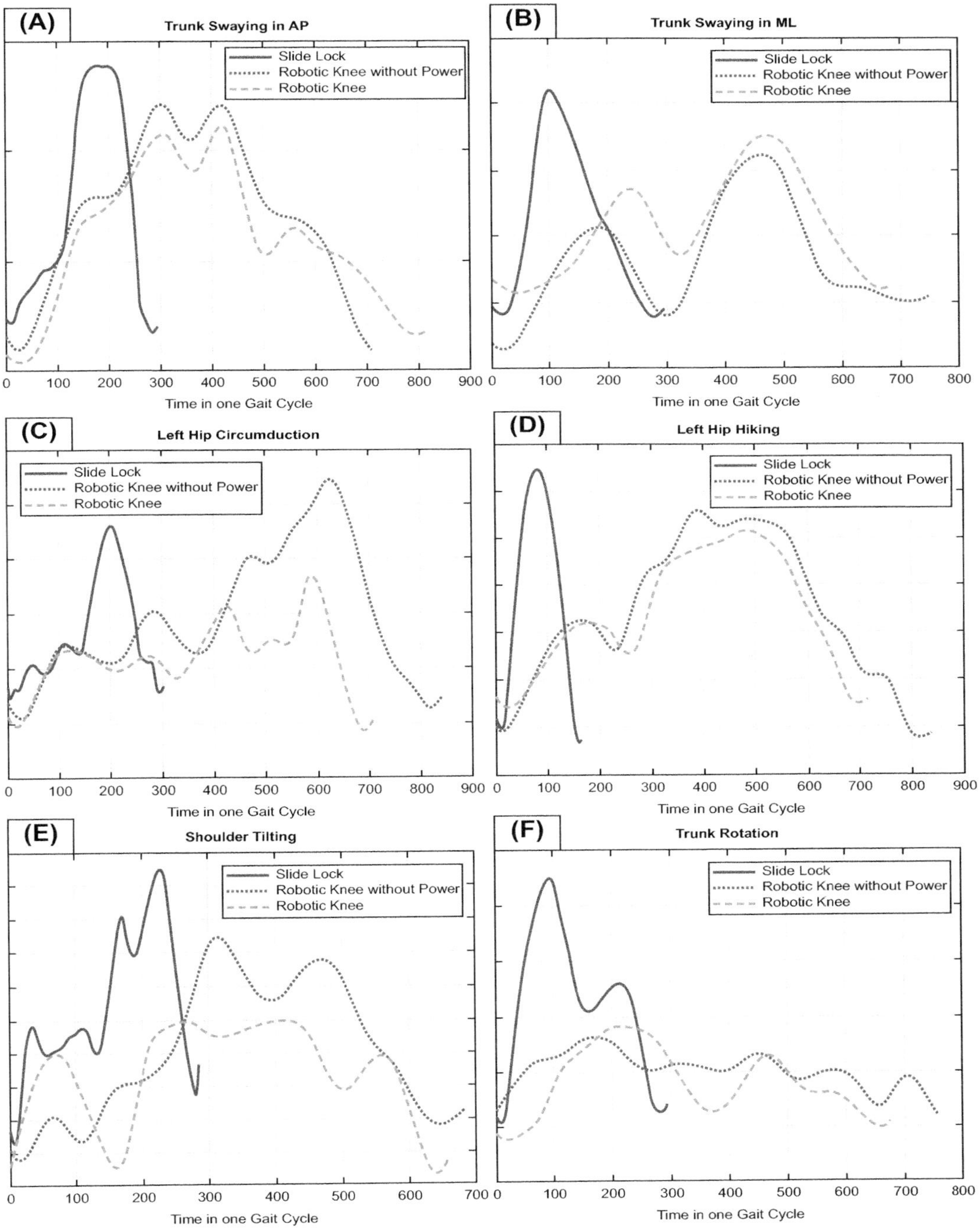


**FIGURE 6.16** Comparison of kinematics in one step of client A, the kinematics of client A show differences between three orthoses: slide lock KAFO (*blue solid line* [dark gray in print version]); robotic knee orthosis without power assistance (*red dotted line* [gray in print version]); robotic knee orthosis with power assistance (*yellow dash-dot line* [light gray in print version]). (A) Trunk sway in a sagittal plane, (B) trunk sway in a coronal plane, (C) left leg circumduction, (D) left hip hiking, (E) shoulder rotation, (F) shoulder tilt.

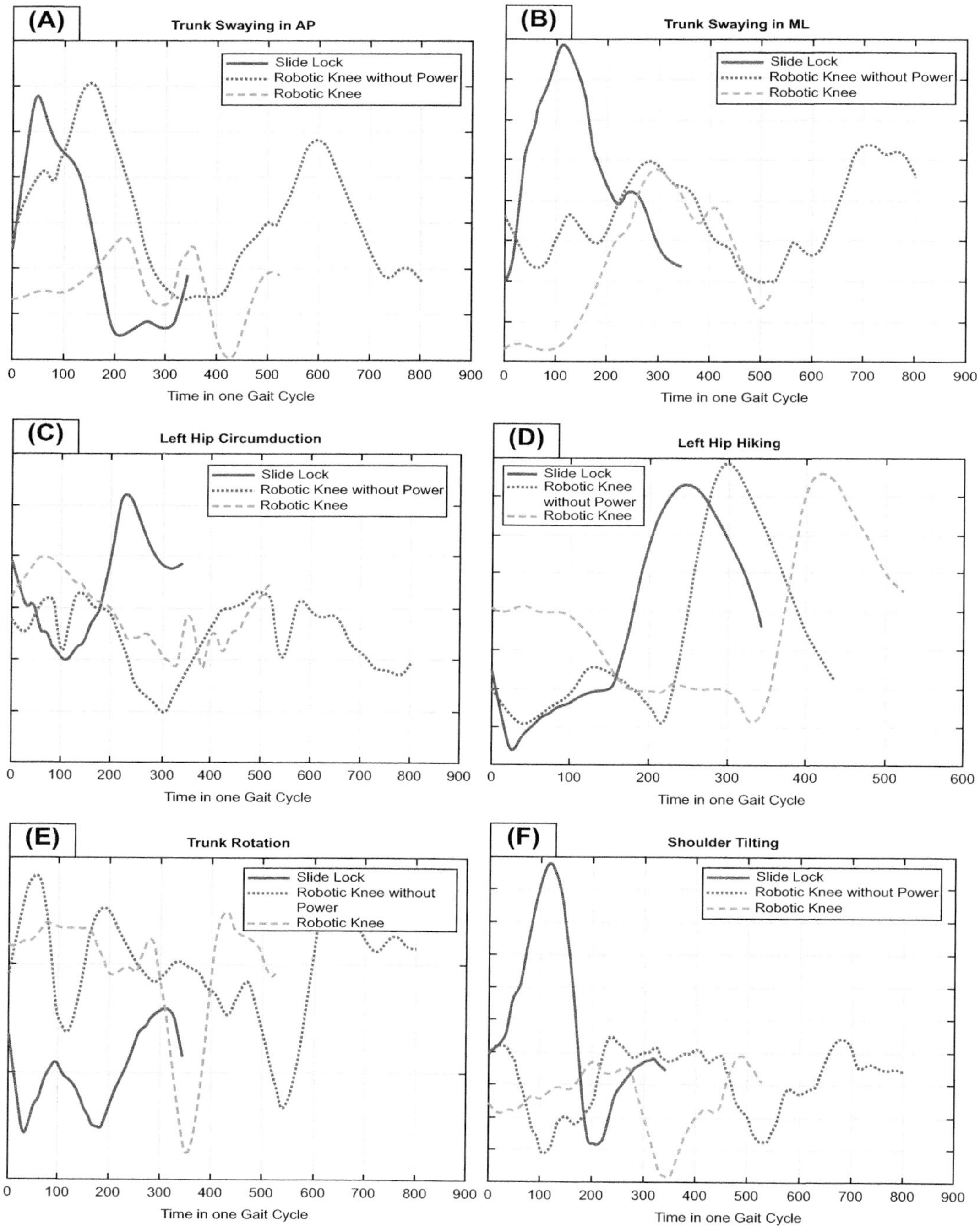


**FIGURE 6.17** Comparison of kinematics in one step of client B, the kinematics of client B show differences between three orthoses: slide lock KAFO (*blue solid line* [dark gray in print version]); robotic knee orthosis without power assistance (*red dotted line* [gray in print version]); robotic knee orthosis with power assistance (*yellow dash-dot line* [light gray in print version]). (A) Trunk sway in a sagittal plane, (B) trunk sway in a coronal plane, (C) left leg circumduction, (D) left hip hiking, (E) shoulder rotation, (F) shoulder tilt.

had a weak hip muscle, and so he had to perform circumduction in order to kick the lower limb forward for another step (Fig. 6.16). In addition, client A tended to use circumduction gait and trunk sway while facing obstacles, which might be the main reason for client A's back and hip pain. Client B's left hip hiking and trunk rotation increased when adopting the robotic knee orthosis with or without power assistance, for reasons similar to client A, that she was accustomed to a conventional slide lock KAFO, and due to the insufficient training period, she intended to compensate the extended knee by hip hiking, which is unnecessary (Fig. 6.17). Client B also mentioned that she was confident to use the orthosis if an extension of her training had been possible.

## Conclusion

In this chapter, the exoskeleton robotic knee system was introduced for people with poliomyelitis. The robotic knee orthosis reduced excessive trunk movement, pain, and metabolic consumption in terms of endurance with evidence from gait analysis and clinical assessment. Most importantly, the robotic knee orthosis aided clients to resume a normal gait pattern and has potential to be used in aiding in the activities of daily living. In this chapter, both clients were satisfied with the effect of walking using the robotic knee orthosis. By performing more training in common activities of daily living scenarios while using the robotic knee orthosis, the control algorithm can be ameliorated to be closer to the clients' daily needs. Apart from individuals with poliomyelitis, other individuals with lower limb weakness or stroke may benefit from this orthosis for training or even daily use with more research and trials. Further studies need to be carried out to determine the long-term effects of walking with the robotic knee orthosis.

## Acknowledgments

We would like to express our gratitude to Mr Richy Ng and Mr Eric Ng, the physiotherapists; Mr Marko Chan, the senior occupational therapist of Community Rehabilitation Service Support Center, Hospital Authority, in being the clinical assessors and in designing and carrying out the training program for the clients.

## References

[1] Winberg C, Flansbjer UB, Carlsson G, Rimmer J, Lexell J. Physical activity in persons with late effects of polio: a descriptive study. Disability and Health Journal 2014;7(3):302–8.

[2] Farbu E, Rekand T, Gilhus NE. Post-polio syndrome and total health status in a prospective hospital study. European Journal of Neurology 2003;10(4):407–13.

[3] Lena LW, Johan KS, Bjørn L, Steinar Ø. The Norwegian Polio Study 1994: a nation-wide survey of problems in long-standing poliomyelitis. Spinal Cord 1998;36(4):280–4.

[4] Silver JK, Aiello DD. Polio survivors: falls and subsequent injuries. American Journal of Physical Medicine & Rehabilitation 2002;81(8):567–70.

[5] Franco MR, Pereira LSM, Ferreira PH. Exercise interventions for preventing falls in older people living in the community. British Journal of Sports Medicine 2014;48(10):867–8.

[6] Mohammad AF, Khan KA, Galvin L, Hardiman O, O'connell PG. High incidence of osteoporosis and fractures in an aging post-polio population. European Neurology 2009;62(6):369–74.

[7] Haziza M, Kremer R, Benedetti A, Trojan DA. Osteoporosis in a postpolio clinic population. Archives of Physical Medicine and Rehabilitation 2007;88(8):1030–5.

[8] Ivanyi B, Nollet F, Redekop WK, de HR, Wohlgemuht M, van WJK, et al. Late onset polio sequelae: disabilities and handicaps in a population-based cohort of the 1956 poliomyelitis outbreak in The Netherlands. Archives of Physical Medicine and Rehabilitation 1999;80(6):687–90.

[9] Marincek HBC. The influence of post-polio syndrome on independence and life satisfaction. Disability & Rehabilitation 2000;22(7):318–22.

[10] Thorén JAL, Grimby G. Ability and perceived difficulty in daily activities in people with poliomyelitis sequelae. Journal of Rehabilitation Medicine 2001;33(1):4–11.

[11] Horemans HLD, Bussmann JBJ, Beelen A, Stam HJ, Nollet F. Walking in postpoliomyelitis syndrome: the relationships between time-scored tests, walking in daily life and perceived mobility problems. Journal of Rehabilitation Medicine 2005;37(3):142–6.

[12] Hebert JS, Liggins AB. Gait evaluation of an automatic stance-control knee orthosis in a patient with postpoliomyelitis. Archives of Physical Medicine and Rehabilitation 2005;86(8):1676–80.

[13] Hwang S, Kang S, Cho K, Kim Y. Biomechanical effect of electromechanical knee–ankle–foot-orthosis on knee joint control in patients with poliomyelitis. Medical, & Biological Engineering & Computing 2008;46(6):541–9.

[14] Genêt F, Schnitzler A, Mathieu S, Autret K, Théfenne L, Dizien O, et al. Orthotic devices and gait in polio patients. Annals of Physical and Rehabilitation Medicine 2010;53(1):51–9.

[15] Mehndiratta MM, Mehndiratta P, Pande R. Poliomyelitis: historical facts, epidemiology, and current challenges in eradication. The Neurohospitalist 2014;4(4):223–9.

[16] Jensen MP, Alschuler KN, Smith AE, Verrall AM, Goetz MC, Molton IR. Pain and fatigue in persons with postpolio syndrome: independent effects on functioning. Archives of Physical Medicine and Rehabilitation 2011;92(11):1796–801.

[17] Yeung LF, Ockenfeld C, Pang MK, Wai HW, Soo OY, Li SW, et al. Design of an exoskeleton ankle robot for robot-assisted gait training of stroke patients. In: Rehabilitation Robotics (ICORR), 2017 international conference on, 2017; 2017. p. 211–5.

[18] Kim JH, Ji SG, Jung KJ, Kim JH. Therapeutic experience on stance control knee-ankle-foot orthosis with electromagnetically controlled knee joint system in poliomyelitis. Annals of Rehabilitation Medicine 2016;40(2):356–61.

[19] Li S, Armstrong CW, Cipriani D. Three-point gait crutch walking: variability in ground reaction force during weight bearing. Archives of Physical Medicine and Rehabilitation 2001;82(1):86–92.

[20] Andrysek J, Redekop S, Matsui NC, Kooy J, Hubbard S. A method to measure the accuracy of loads in knee-ankle-foot orthoses using conventional gait analysis, applied to persons with poliomyelitis. Archives of Physical Medicine and Rehabilitation 2008;89(7):1372–9.

[21] Tiggemann C, Dias C, Radaelli R, Massa J, Bortoluzzi R, Schoenell M, et al. Effect of traditional resistance and power training using rated perceived exertion for enhancement of muscle strength, power, and functional performance. Age 2016;38(2):1–12.

[22] Collen FM, Wade DT, Bradshaw CM. Mobility after stroke: reliability of measures of impairment and disability. International Disability Studies 1990;12(1):6–9.

[23] Shumway CA, Brauer S, Woollacott M. Predicting the probability for falls in community-dwelling older adults using the timed up & go test. Physical Therapy 2000;80(9):896–903.

[24] Ohnhaus EE, Adler R. Methodological problems in the measurement of pain: a comparison between the verbal rating scale and the visual analogue scale. Pain 1975;1(4):379–84.

[25] Zacharias B, Kannenberg A. Clinical benefits of stance control orthosis systems: an analysis of the scientific literature. JPO: Journal of Prosthetics and Orthotics 2012;24(1):2–7.

[26] Turner SL. Goal attainment scaling (GAS) in rehabilitation: a practical guide. Clinical Rehabilitation 2009;23(4):362–70.

[27] Dongo R, Moscoso M, Callupe R, Pajaya J, Elías D. Normal human gait patterns in Peruvian individuals: an exploratory assessment using VICON motion capture system. In: 13th international conference on medical information processing and analysis, vol. 10572. International Society for Optics and Photonics; November 2017. p. 1057209.

[28] Mcdonough AL, Batavia M, Chen FC, Kwon S, Ziai J. The validity and reliability of the GAITRite system's measurements: a preliminary evaluation. Archives of Physical Medicine and Rehabilitation 2001;82(3):419–25.

[29] Bilney B, Morris M, Webster K. Concurrent related validity of the GAITRite® walkway system for quantification of the spatial and temporal parameters of gait. Gait & Posture 2003;17(1):68–74.

[30] Clark CCT, Barnes CM, Holton M, Summers HD, Stratton G. A kinematic analysis of fundamental movement skills. Sport Science Review 2016;25(3–4):261–75.

7

# Voluntary intention-driven rehabilitation robots for the upper limb

Yao Huang[2], Steven W. Su[2], Rong Song[1]
[1]SCHOOL OF BIOMEDICAL ENGINEERING, SUN YAT-SEN UNIVERSITY, GUANGZHOU, PR CHINA; [2]BIOMEDICAL ENGINEERING SCHOOL, FACULTY OF ENGINEERING AND INFORMATION TECHNOLOGY, UNIVERSITY OF TECHNOLOGY SYDNEY, SYDNEY, AUSTRALIA

## Chapter outline



## Abbreviations

**MFE** Root means squared model fitting error between the needed cable forces and actual forces
**MMA** Mean values of muscle activation
**MNMA** Mean values of the normalized muscle activation
**MVR** Mean velocity ratio in each main motion direction
**NJS** Normalized jerk score of the actual trajectory
**PCC** Pearson correlation coefficients
**RMSE** Root mean square error between the target and actual tracking trajectories
**sEMG** surface electromyography

**Intelligent Biomechatronics in Neurorehabilitation. https://doi.org/10.1016/B978-0-12-814942-3.00007-6**

## Introduction

As one of the most severe human diseases, stroke always causes permanent disability, which impeded the ability of survivors to conduct activities during daily living independently. Generally, the affected motor function of patients can be regained by rehabilitation training with the brain plastic reorganization theory [1]. Voluntary involvement during rehabilitation training is vital for patients who still have the residual motor ability [2].

During the arm movements, the activities of arm muscles are indicated both against gravity and for executing actual motion intention [3]. Due to the weakness of muscles gravity is not naturally resisted, which further impedes the performance of the patients' voluntary involvement for active upper limb movement [4]. With their superiority in efficiency, accuracy, and controllability, many robot systems have been used for post-stroke patients' motor rehabilitation training [2,5]. However, in most rehabilitation robot studies, upper limb gravity is not properly compensated for at real-time, and the patients' efforts with the affected limbs for voluntary training have not been fully revealed or supported by a robot.

Although position control during robot-aided rehabilitation can offset the influence of upper limb gravity, it mainly pays attention to the control accuracy but neglects the patients' voluntary involvement, and its effectiveness needs to be further improved. For patients with muscle weakness, gravity compensation shows meaningful results for minimizing the impedance of gravity on task performance and for further exploring the patient's residual motor ability [6–8]. The strategies for compensating arm gravity are generally offered to participants by specialized devices for arm support [9,10] or robotic systems [11–13], and can be roughly classified into three kinds.

First, fixed and manual adjustable gravity compensation is usually provided to offset the effects of upper limb gravity [14–16]. Similarly, a vertical force provided by a motorized perpendicular cabling system is applied for compensating the gravity of the upper arm by the robot [12]. Second, flexible forces as passive compensation can be provided by elastic materials [17,18]. T-WREX, which is an arm exoskeleton, can compensate for the gravity of the arm with different numbers of elastic bands to adjust the level of support [19]. Freebal is a spring-based device which can provide adjustable passive gravity compensation during arm movement tracking tasks both in the horizontal and perpendicular planes [10,20,21]. Some arm orthoses with a spring can also provide passive compensation and aim to help patients with residual motor ability to move and reach [9,22]. Third, the gravity may be compensated according to the physical characteristics of the human arm, because of the highly coupled relationship between the arm dynamics and gravity torque during changes to arm positions and postures. A control strategy which includes a compensation part for gravity which can be adjusted by arm dynamics has been proposed by Hsu et al. [23]. A compliant beam developed by Cheng et al. for simulating arm movements can also compensate the upper limb gravity according to the arm dynamic [7]. Lin et al. developed a torque-angle model for compensating upper limb gravity during the arm motion quality evaluation [24].

With the gravity torque closely correlated with the upper limb positions during arm movement, the fixed and elastic compensation, which do not consider the position-coupling effect, need to be improved. Although Hsu and Li et al. proposed robot control strategies that included compensation based on human characteristics, gravity compensation is only a part of their strategies [23,25]. Meanwhile, robot-aided strategies with compensating gravity should not affect the voluntary participation during arm rehabilitation, even with consideration of the arm dynamics. Furthermore, it is rarely reported how movement outcomes and muscle activation are affected by arm gravity during arm movements.

Nowadays, the control strategies of rehabilitation robots for balancing motor ability and assistance from robots to participants are mostly based on kinematic or kinetic signals [26,27], e.g., impedance and admittance control. These strategies mostly focus on accurate control of the robot for finishing tasks and partly involving the voluntary movements of participants, but are less concerned with the residual motor ability of patients' affected arms. Surface electromyography (sEMG) signals, as conveniently captured physiological, embedded human motion intentions, are also highly correlated with muscle forces and joint torques during limb movements [27]. Robot control strategies based on sEMG signals can reflect the motion intention of the human body and are more anthropomorphic and more suitable for patients who retain motor ability and can carry out voluntary training tasks. Previous EMG-based control strategies have tried to map sEMG signals to the signals for binary control of the robot [27]. Advanced EMG-based control strategies have been further developed for estimating a single joint torque during arm movements [28,29]. Artur et al. proposed an EMG analytical model that maps muscle activations through autoregressive functions and self-organizing mapping algorithms to the position of joint motion [30].

One of the goals is to propose a novel gravity compensation method strategy which varies with changes in arm position based on arm dynamics. The strategy is realized by establishing a gravity torque estimation model for estimating the arm gravity torque in real-time [31]. Four different directions of human–machine cooperative motion-tracking tasks are designed for evaluating the movement performance of participants under assistance from a robot with the proposed compensation strategy. As for comparisons, participants are required to finish the same tasks under no compensation and with fixed compensation from the robot. The root means square error (RMSE) of actual trajectory away from the target trajectory, the normalized jerk score (NJS) of actual trajectory, the mean velocity ratio (MVR) along the target movement direction, and mean activations of six muscles (MMA) are used to evaluate the performance of different compensation strategies.

Since few control strategies based on sEMG signals can continuously estimate the positions or torques for multijoint upper-limb movements, the other goal of this study is to propose a novel sEMG-based motion intention estimation model during human–machine cooperation movements. The estimation model is built up by a state-space model whose inputs are muscle activations and outputs are movement-needed forces [32]. Therefore, the participants' voluntary motor intention can be estimated and used for controlling a robot for assisting in what participants need. A vertical human–machine cooperative motion tracking task is designed for evaluating the accuracy of the estimation model and movement performance of participants under assistance from a robot with the proposed sEMG-based control strategy. The model accuracy is evaluated by model fitting error (MFE) and Pearson correlation coefficients (PCC) between targeted forces and actual force provided by the robot, while movement performance is evaluated by RMSE, and mean values of the normalized muscle activations (MNMA) of the six main contributing muscles are also used.

All of the researches in this chapter are supported by a cable-based upper limb rehabilitation robot [33].

# Methodology

## Participants

For research into gravity compensation, a total of seven healthy participants (males, aged 23.7 ± 1.1 years) were invited into this study. For research into the EMG-based control strategy, a total of 10 healthy participants (males, aged 22.0 ± 1.3 years) were invited.

All participants had no experience with this robot and were able to move their right arms without any musculoskeletal or neurological system problems. All participants were required to sign individual informed consent prior to taking part in any experiments. The human ethics committee approved all experimental procedures at the Sun Yat-sen University.

## Experimental platform

To provide assistance for compensating the gravity of multiple joint arm movements and to provide the cable forces estimated by sEMG signals during arm movements, both strategies were applied on a cable-based robot for arm rehabilitation training (Fig. 7.1). The robot platform consists of four parts: a frame component, a motion capture system (Flex 3, OptiTrack, NaturalPoint, USA), an sEMG capture system, and a computer. The frame component consists of an aluminum link cube frame, a splint as the end effector, three cables, and a motor group [including three motor pairwise drivers (DM1B-045G&UB1DG3, Yokogawa, Japan)] [33]. One end of each cable was connected to the splint, thereby generating three degrees of freedom (DOF) for controlling. The cable forces produced by the motor set were applied on the splint to assist the user in performing arm motions in three-dimensional (3D) space.

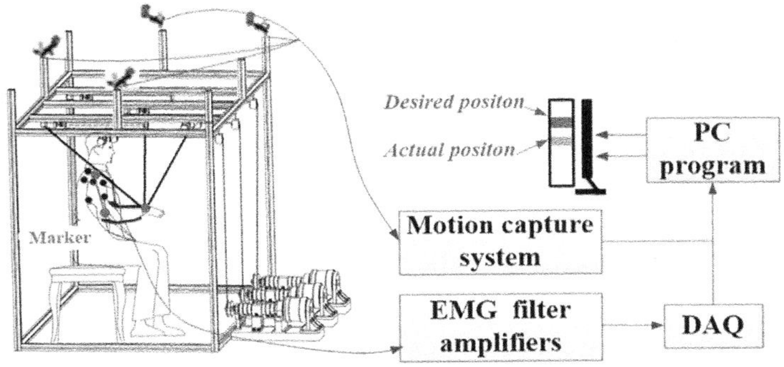


**FIGURE 7.1** Architecture of the cable-based rehabilitation robot. *Reproduced with permission from Huang Y, Yang Q, Chen Y, Song R. Assessment of motor control during three-dimensional movements tracking with position-varying gravity compensation. Frontiers in Neuroscience 2017;11:253.*

There were three infrared reflective markers stuck to the skin of the arm joint centers (i.e., wrist, elbow, and shoulder), for recording joint positions in 3D space by a motion capture system. The actual position and target position during a tracking task of the end-effector were both programmed to be displayed on the computer screen. The motion capture system has a sampling rate of 100 Hz, and the raw captured data were filtered using a second-order Butterworth filter and a cut-off filter at 6 Hz.

The sEMG capture system includes an analog-to-digital (AD) data acquisition device (PXI-6229, National Instruments, USA) and six EMG collection boards with amplifiers. For research into gravity compensation, sEMG signals of six upper limb muscles, including biceps brachii (BIC), triceps (TRI), anterior part (DA), middle part (DM), posterior part (DP) of deltoid, and upper trapezius (TRA) were captured by the sEMG capture system. For the research of EMG-based control strategy, sEMG signals of six upper limb muscles, including brachioradialis (BR), BIC, TRI, DA, DM, and DP were captured. The sampling rate of the sEMG capture system was 1000 Hz. The amplifier had a gain of 5000, and raw signals were filtered through a 10–400Hz band-filter and a fourth-order Butterworth. In order to obtain the envelope of the sEMG signals, a low-pass filter which cut off at 4 Hz with a fourth-order Butterworth filter was applied.

## Experimental procedure

In both the research into gravity compensation and the EMG-based control strategy, similar preparations and arm/trunk postures were required.

Prior to each trial, each participant was seated in a chair and strapped in to reduce additional compensation from the torso. The initial position of the arm was as follows: the upper arm was vertically suspended, and close to the torso, the shoulder joint was relaxed, the elbow joint was flexed 90 degrees, the forearm pronation was 90 degrees, the wrist joint was relaxed, and all five fingers were required to be close to each other. The computer screen displayed the actual position and the target position was put on a desk

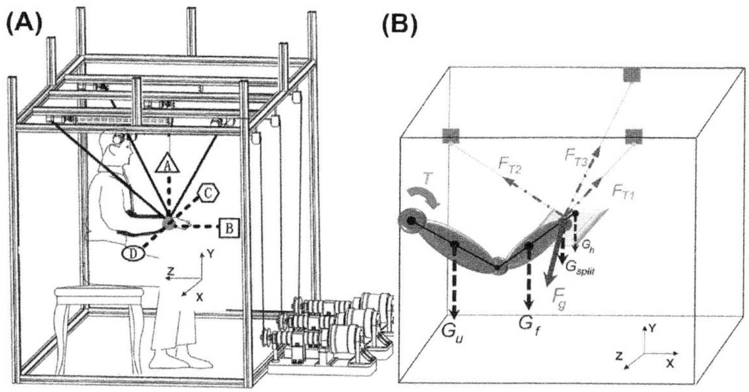


**FIGURE 7.2** The task directions (A) and static force model (B). *Reproduced with permission from Huang Y, Yang Q, Chen Y, Song R. Assessment of motor control during three-dimensional movements tracking with position-varying gravity compensation. Frontiers in Neuroscience 2017;11:253.*

in front of the participants to provide visual reality feedback in real-time. A yellow cursor indicated the actual position of the participant's wrist and a red cursor indicated the target position.

In the study of gravity compensation, each subject was required to finish the four tracking tasks shown in Fig 7.2A with the right arm. The length of all four tasks [including upward (A)/forward (B)/left (C)/right (D)] was 0.2 m from the initial point. For each task, the participants were asked to complete the tracking movements with assistance from the robot with three kinds of gravity compensation strategy. Each strategy group needed to be executed six times.

In the EMG-based control strategy study, each subject was required to finish a tracking task along the vertical direction shown in Fig 7.2A with the right arm, while the length was 0.25 m. To build up and verify the estimation model based on sEMG signals, the whole experimental procedure was separated into two rounds. In the first round, the model between muscle activations and motion-needed cable forces was built up by participants executing the tracking task independently. The modeling data were collected after executing the task three times. In the next round, the participants were required to complete the vertical tracking task without/with assistance provided by the robot controlled by the EMG-based strategy. The without/with assistance group executed the task six times.

## The dynamics of the robot

The direction of the three cables from the connecting point on the splint to the linked points on the frame was considered and the unit vector was further calculated and used for dynamic analysis of the robot. The cable-based construction required that all cables retained tension during the entire task. The tensile forces for supporting arm movements was calculated from the output forces in the X, Y, and Z axes ($\boldsymbol{F_{OUT}} = [\boldsymbol{F_X}, \boldsymbol{F_Y}, \boldsymbol{F_Z}]$) from the gravity compensation or the EMG-based control strategy as follows:

$$\boldsymbol{F_T} = J^{-1}\boldsymbol{F_{OUT}} \tag{7.1}$$

where $\boldsymbol{F_T} = [\boldsymbol{F_{T1}}, \boldsymbol{F_{T2}}, \boldsymbol{F_{T3}}]^T$ was the tensile force matrix for the cables, and $J = [\boldsymbol{u_1}, \boldsymbol{u_2}, \boldsymbol{u_3}]$ was the unit vector matrix along the three cables. The tensile force matrix $\boldsymbol{F_T}$ was further transmitted to the motor with the dynamic analysis between the motor group and cables, which had been reported in previous research [33].

In the study of gravity compensation, the tensile force matrix was calculated directly based on the static force model.

Unlike the gravity compensation, which did not consider the low-velocity motion of the splint and upper limb, the EMG-based control strategy needed to consider all forces required to complete the movement throughout the task based on the muscle activities. Therefore, the force required for the task at real-time was calculated by dynamic analysis of the moving splint and the human arm.

## Gravity compensation strategies

A real-time estimate of gravity torque during the upper limb movements was the basis of the proposed position-varying gravity compensation. The torque estimation changing with the arm movement was used to calculate the sum of the limb gravity moments to the shoulder as follows:

$$\boldsymbol{T} = G_u L_{ush} + G_f L_{fsh} + G_h L_{hsh} \tag{7.2}$$

where $\boldsymbol{T}$ was the estimated gravity torque of the hand, forearm, and upper arm to the shoulder, $G_h$ was the gravity of the hand, $G_f$ was the gravity of the forearm, $G_u$ was the gravity of the upper arm, and $L_{hsh}$, $L_{fsh}$, $L_{ush}$ were the moment arms from the mass centers of the participants' hand, forearm, and upper arm to the mass centers of the shoulder.

Then, the force needed against gravity and supported by the splint was calculated as follows:

$$\boldsymbol{F_g} = \boldsymbol{T}/L_{shw} \tag{7.3}$$

where $\boldsymbol{F_g}$ was the needed force matrix against the gravity torque, and $L_{shw}$ was the moment arm from the mass centers of the shoulder to the mass centers of the wrist where there was contact with the splint.

The tensile force matrix was calculated directly using the analyzed static force model by considering that $\boldsymbol{F_{OUT}}$ equals the sum of the equivalent needed force and gravity of the splint because both were applied on the wrist as shown in Fig. 7.2B. The resultant force ($\boldsymbol{F_{OUT}}$) for compensating the gravity was calculated as follows:

$$\boldsymbol{F_{OUT}} = -(\boldsymbol{F_g} + \boldsymbol{G_{split}}) \tag{7.4}$$

where $\boldsymbol{G_{split}}$ was the gravity of the splint applied on the wrist.

The arm movements without compensation and with fixed compensation from the robot were also executed by participants as for comparisons. The gravity torque calculated by the Equation (7.2) at the initial position was applied during the entire movement as the fixed compensation. Without compensation indicated no assistance from the robot. When participants were finishing movements without compensation conditions, to guaranteeing extra weight conditions were the same, they were required to take on an extra weight equal to the mass of the splint.

## An EMG-based control strategy

To assist participants based on their own motion intentions, an EMG-based control strategy was used to control the robot. The strategy was based on models for revealing the relationship between real-time sEMG signals and the forces needed to execute the movements. The voluntary intention of the participants implicit in the activation of muscles was used as the input for estimating the volunteered forces.

To obtain muscle activations, an sEMG to activation model was used based on previous studies [34]. A second-order discrete linear model was used for the neural activation from the envelopes of raw sEMG signals of each muscle.

$$u(t) = \alpha e(t-d) - \beta_1 u(t-1) - \beta_2 u(t-2) \tag{7.5}$$

where $e(t)$ was the envelope of raw EMG signals at time $t$, $u(t)$ was the neural activation at time $t$, $d$ was the electromechanical delay (80 ms), $\alpha$ was the gain coefficient, and $\beta_1$ and $\beta_2$ were the recursive coefficients.

The muscle activation $\alpha(t)$ was then calculated as follows [35]:

$$\alpha(t) = \begin{cases} d\ln(cu(t)+1), & 0 \le u(t) < 0.3 \\ mu(t) + b, & 0.3 \le u(t) < 1 \end{cases} \tag{7.6}$$

where $c$, $m$, and $b$ were all constants, at 0.3512, 0.8854, and 0.1155, and $d = 1 + \beta_1 + \beta_2$.

A state-space model was trained to continuously estimate the force that participants voluntarily used in moving the arm in 3D space.

$$\begin{aligned} x_{k+1} &= \boldsymbol{A}x_k + \boldsymbol{B}a_k + w_k \\ F_k &= \boldsymbol{C}x_k + v_k \end{aligned} \tag{7.7}$$

where $k$ was the sampling point, $a_k$ was the normalized muscle activation matrix, $F_k$ was the estimated voluntary force matrix, and $w_k$ and $v_k$ were zero-mean-Gaussian noises.

The characteristics $\boldsymbol{A}$, $\boldsymbol{B}$, and $\boldsymbol{C}$ were gained during the modeling phase using the prediction-error minimization algorithm. That is, to map the relationship of the normalized muscle activations and voluntary force calculated based on the dynamic analysis of the human–machine cooperation motion. The resultant forces for driving the arm movement of the splint and arm were calculated as follow:

$$\boldsymbol{F_N} + \boldsymbol{F_g} + \boldsymbol{G_{split}} = Ma \tag{7.8}$$

where $\boldsymbol{F_N}$ was the participants' voluntary forces for driving the arm to complete the task, $M$ was the sum mass of the split and arm, and $a = [\boldsymbol{a_x}, \boldsymbol{a_y}, \boldsymbol{a_z}]$ was the acceleration calculated by the changes in movement trajectory.

## Data analysis

Root mean square error (RMSE) of trajectories was calculated for evaluating tracking accuracy.

$$\mathrm{RMSE} = \sqrt{\sum_{i=1}^{N}\left((X_{a_i} - X_{t_i})^2 + (Y_{a_i} - Y_{t_i})^2 + (Z_{a_i} - Z_{t_i})^2\right)/N} \tag{7.9}$$

where $i$ was the sampling point, $X_{a}$, $Y_{a_i}$, $Z_{a_i}$ were the actual XYZ values of coordinates, $X_{t_i}$, $Y_{t_i}$, $Z_{t_i}$ were the target XYZ values of coordinates, and $N$ was the number of sampling points.

The normalized jerk score (NJS) of trajectories was used to represent the motion smoothness for evaluating the arm control abilities [36].

$$NJS = \sqrt{\frac{1}{2} \times \frac{T^5}{D^2} \times \int \left( \dddot{s}(t) \right)^2 dt} \tag{7.10}$$

where $t$ was the time series, $s(t)$ was the wrist position in 3D space at the time $t$, $T$ was the whole sampling time, and $D$ was the length of the actual trajectory.

The mean velocity ratio (MVR) along each straight direction proposed for evaluating the efficiency of the tracking movement was calculated as:

$$MVR = \sum_{i}^{N}\left(V_{m_i}\Big/\sqrt{V_{x_i}^2 + V_{y_i}^2 + V_{z_i}^2}\right)\Big/N \tag{7.11}$$

where $V_{m_i}$ was the resultant velocity along the main motion direction at the $i$ th point, $V_{x_i}$, $V_{y_i}$, $V_{z_i}$ were the velocity along the $X$, $Y$, and $Z$ axes at the $i$ th point.

The mean muscle activation (MMA) of each muscle can be directly calculated based on the sEMG envelope during every movement.

The mean value of the normalized muscle activation (MNMA) of each muscle which was normalized by the maximum voluntary contractions (MVCs) of each muscle was further investigated.

The root means squared MFE is calculated between the needed cable forces from the arm/robot dynamic model and actual forces provided by the motor.

$$\mathrm{MFE}_k = \sqrt{\sum_{i=1}^{N}\left((F_{ck_i} - f_{ck_i})^2\right/N} \tag{7.12}$$

where $F_{ck_i}$ was the actual force of the $k$ th cable, and $f_{ck_i}$ was the needed cable forces from the arm/robot dynamic model of the $k$ th cable.

Furthermore, the Pearson correlation coefficients (PCC) were used to measure how strong the relationships between the needed cable forces and actual forces were.

In the study of gravity compensation, RMSE, NJS, MVR, and MMA were used to assess the effects of the three gravity compensation methods. Two-way analysis of variance (ANOVA) was applied for assessing the main effects of the two factors (i.e., different compensation methods and different tasks in four directions) and their interaction to these four selected characteristics. A multiple comparison test, post hoc Tukey test, was applied to examine the differences between the RMSE, NJS, and MVR values. Paired t-test was subsequently applied to examine the difference in the values of RMSE, NJS, MVR, and MMA during movements with the three compensation strategies per task.

In the study of the EMG-based control strategy, RMSE, MNMA, MFE, and PCC were used to evaluate its performance. Paired t-test was also applied to evaluate the differences in the values of RMSE and MNMA, and one-way ANOVA was applied for assessing the differences in MFE and PCC for each cable during task execution without and with robot assistance.

All statistical tests were analyzed using SPSS (SPSS, Inc., Chicago, IL, version 22.0), and the significance level was set as 0.05.

## Results

Fig. 7.3A–C shows the performance of RMSE, NJS, and MVR during arm tracking movements with the three different gravity compensations. The actual trajectories of one participant completing the four movements are displayed in Fig. 7.3D.

The results of a two-way ANOVA test are summarized in Table 7.1. As the results in Table 7.1 illustrate, the effects on RMSE, NJS, and MVR caused by the compensation methods and the task directions are both significant ($P < .05$). The effect caused by the interaction of the two main factors was not found. The effect of the compensation method resulted in a higher rank of the values of RMSE and NJS in this order: without (highest), fixed, and position-varying compensation; and the rank of the MVR values were in the order: position-varying (highest), fixed, and without compensation. According to the results of the Tukey test, the RMSE values of upward and forward movements were significantly higher than those in leftward and rightward directions. In contrast to RMSE, the MVR values of movements in upward and forward directions were significantly less than those of movements in leftward and rightward directions. The NJS values of upward movements were found to be significantly lower than those of rightward movements. According to the results of the paired t-test, the RMSE values of leftward and rightward movements with position-varying compensation strategy were significantly lower than the other two compensations, and the values of all four movements with a position-varying compensation strategy were significantly lower than those with fixed compensation. For NJS, the values of upward and forward movements with the position-varying compensation were significantly lower than those without compensation, and the values of leftward movements with the position-varying compensation were significantly lower than those with fixed compensation. The MVR values of upward, leftward, and rightward movements with position-varying compensation were significantly higher than the other two compensations.

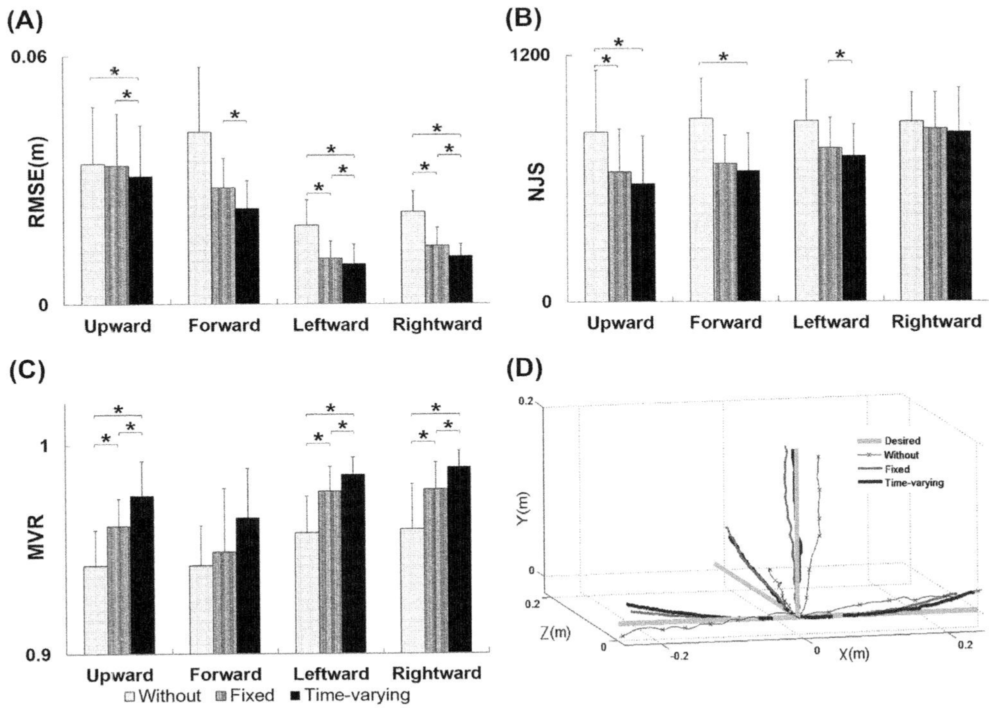


**FIGURE 7.3** The performance of (A) RMSE, (B) NJS, and (C) MVR during movements with different gravity compensation strategies, the target and actual trajectories during movement tracking with different gravity compensation strategies (D). *Significant difference was found between the two kinds of gravity compensation strategies ($P < .05$). *Reproduced with permission from Huang Y, Yang Q, Chen Y, Song R. Assessment of motor control during three-dimensional movements tracking with position-varying gravity compensation. Frontiers in Neuroscience 2017;11:253.*

**Table 7.1** The results for all factors involved in ANOVA tests.

| **F-value** | | **Main effects** | | **Interaction effect** |
|---|---|---|---|---|
| Outcome measures | | Compensation method (DOF = 2) | Target direction (DOF = 3) | Compensation method × target direction (DOF = 6) |
| RMSE | | 9.823 ($P$ = .000)* | 25.986 ($P$ = .000)* | 1.032 ($P$ > 05) |
| NJS | | 6.625 ($P$ = .002)* | 2.872 ($P$ = .042)* | 0.495 ($P$ > 05) |
| MVR | | 18.483 ($P$ = .000)* | 8.793 ($P$ = .000)* | 0.276 ($P$ > 05) |
| Muscle activation | BIC | 54.228 ($P$ = .000)* | 0.091 ($P$ > 05) | 0.842 ($P$ > 05) |
| | TRI | 4.415 ($P$ = .020)* | 0.928 ($P$ > 05) | 0.362 ($P$ > 05) |
| | DA | 49.943 ($P$ = .000)* | 7.628 ($P$ = .000)* | 1.622 ($P$ > 05) |
| | DM | 19.882 ($P$ = .000)* | 4.197 ($P$ = .009)* | 1.201 ($P$ > 05) |
| | DP | 1.031 ($P$ > 05) | 2.594 ($P$ > 05) | 0.664 ($P$ > 05) |
| | TRA | 18.705 ($P$ = .000)* | 3.963 ($P$ = .011)* | 0.429 ($P$ > 05) |

*Indicates significant difference ($P < .05$).

Reproduced with permission from Huang Y, Yang Q, Chen Y, Song R. Assessment of motor control during three-dimensional movements tracking with position-varying gravity compensation. Frontiers in Neuroscience 2017;11:253.

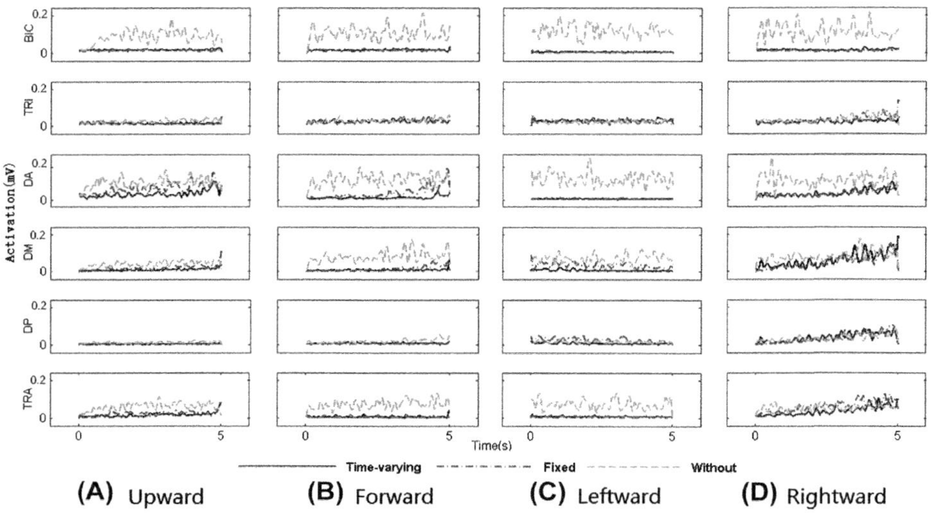


**FIGURE 7.4** EMG envelope time series of one participant for all muscles monitored during the study. The data are shown for three gravity compensation strategies (without, fixed, and position-varying) and for the following six muscles: BRI, TRI, DA, DM, DP, and TRA. *Reproduced with permission from Huang Y, Yang Q, Chen Y, Song R. Assessment of motor control during three-dimensional movements tracking with position-varying gravity compensation. Frontiers in Neuroscience 2017;11:253.*

As shown in Fig. 7.4, the sEMG envelopes of six muscles of one participant were displayed during the four tracking movements with the three compensations. The results of the envelope showed when the participant finished upward, leftward, and rightward movements with gravity compensation, the activations of four muscles (BIC, DA, DM, and TRA) were lower, and when the participant finished rightward movements with gravity compensation, only the activations of BIC and DA were lower. The MMA values of each muscle during movement in four directions with different gravity compensation methods are shown in Fig. 7.5. The results of the two-way ANOVA indicated that the effect on MMA of BIC, TRI, DA, DM, and TRA caused by the compensation methods were significant, and the effect on MMA of DA, DM, and TRA caused by tracking directions were significant ($P < .05$). The effect caused by the interaction of the two main factors was not found. The MMA values of BIC and DA during the four direction movements, the values of both DM and TRA during forward, leftward, and rightward movements, and the values of TRI during upward and rightward movements with gravity compensations were found to be significantly lower than during movements without compensation. The results showed that the MMA values of BIC, TRI, DA, and DM during upward movements with position-varying compensation were significantly reduced compared to those with fixed compensation.

The modeling performance of the proposed state-space model for real-time estimating the voluntary motion forces during the tracking movements from six muscle

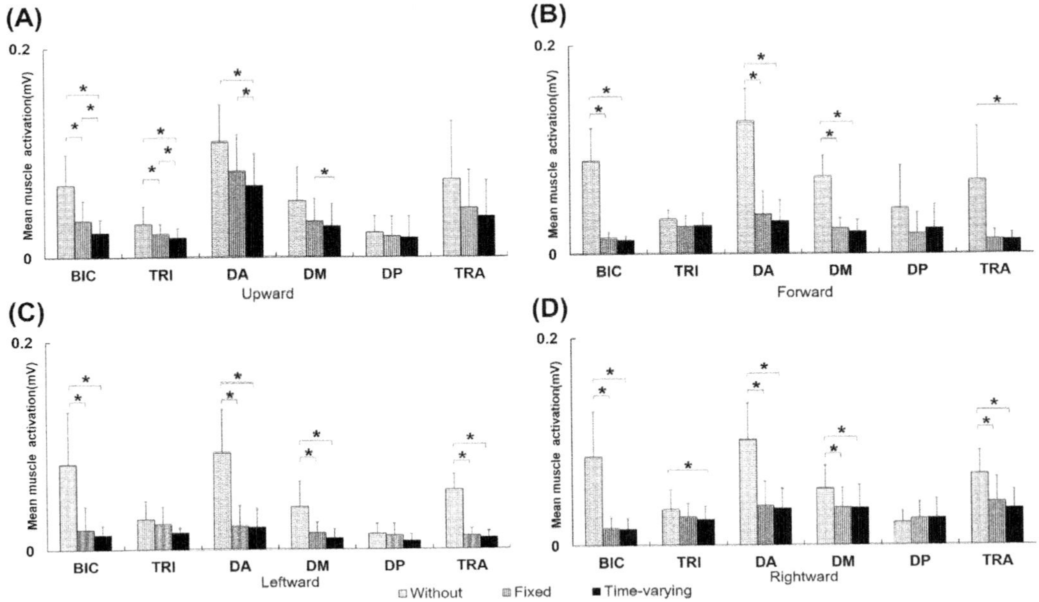


**FIGURE 7.5** The mean activation of six muscles during four direction movements with different gravity compensation strategies. *Significant difference was found between the two kinds of gravity compensation strategies ($P < .05$). *Reproduced with permission from Huang Y, Yang Q, Chen Y, Song R. Assessment of motor control during three-dimensional movements tracking with position-varying gravity compensation. Frontiers in Neuroscience 2017;11:253.*

activations is shown in Fig. 7.6. The gray dashed line is the target forces based on the dynamic analysis, and the solid dark line is the actual force based on the state-space model. The target and actual forces of the three cables are similar, and their values are very close. The further results of the MFE and PCC values tested by one-way ANOVA test are shown in Table 7.2. Only the difference between the values of COC in cable 2 during both without/with robot assistance movements showed significance.

The RMSE results in three-dimensional space of the movements without and with robot assistance are presented in Fig. 7.7. The results of the paired t-test on RMSE in each dimension between no robot assistance and with assistance from the robot controlled by an EMG-based strategy showed that no significant difference was found. The MNMA values of the six muscles (BR, BIC, TRI, DA, DM, and DP) when participants finished the tracking task without and with assistance from the robot are shown in Fig. 7.8. The comparison between the MNMA of the six muscles in movements without robot assistance and with robot assistance indicated that DA and DM primarily contributed to the task. In addition to TRI, MNMA values of all other muscles were reduced with robot assistance, but no significant difference was found.

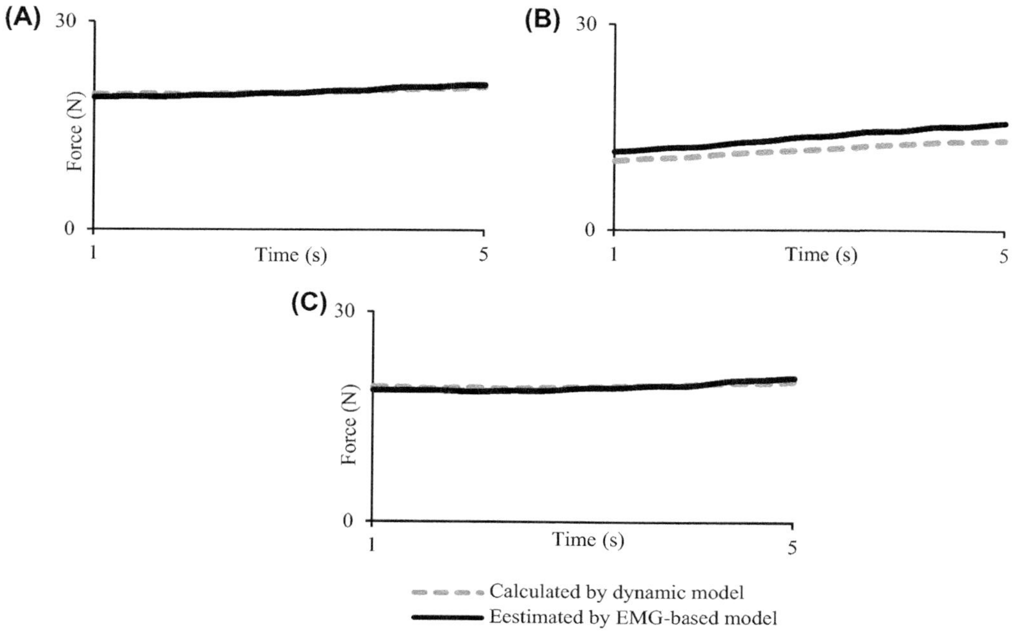


**FIGURE 7.6** The movement-needed forces calculated by the dynamic model (the *dashed gray line*) and forces estimated by the EMG-based model (the *solid dark line*): (A) cable 1, (B) cable 2, (C) cable 3. *Reproduced with permission from EMG-based control for three-dimensional upper limb movement assistance using a cable-based upper limb rehabilitation robot. In: Huang Y, Chen Y, Niu J, Song R, editors. International conference on intelligent robotics and applications. Springer; 2017.*

**Table 7.2** The results for MFE and COC involved in one-way ANOVA tests.

| Phase | Model accuracy | Cable 1 | Cable 2 | Cable 3 |
|---|---|---|---|---|
| Modelling | PCC (%) | 0.86(0.15) | 0.95(0.07)* | 0.90(0.14) |
| | MFE (N) | 0.69(0.54) | 0.67(0.43) | 1.16(1.74) |
| Verifying | PCC (%) | 0.84(0.17) | 0.86(0.15)* | 0.79(0.26) |
| | MFE (N) | 1.10(0.69) | 1.41(0.51) | 1.88(1.75) |

*Indicates significant difference is found between these two groups of values ($P < .05$).

# Discussion

One of the purposes of this study was first to propose a novel position-varying gravity compensation and compare it with another two different gravity compensations. Three kinematic parameters, RMSE, NJS, and MVR, and physiologic parameters, MMA, were applied to evaluate the participants' performance during movements with different gravity compensations. Generally, RMSE can be used to reflect accuracy during tracking

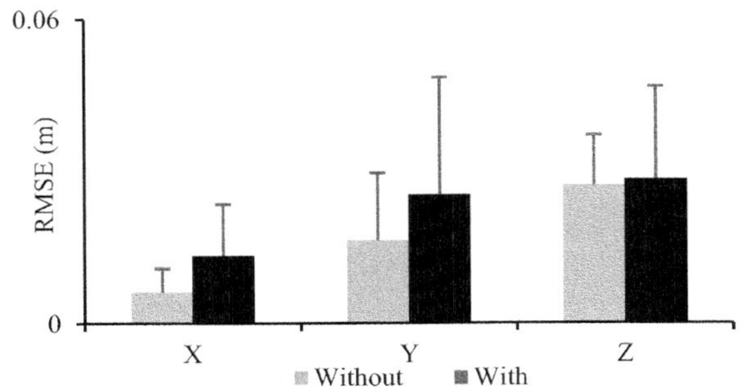


**FIGURE 7.7** The group mean of RMSE in three-dimensional space, the first task without assistance (the *gray bar*); the second task with assistance (the *black bar*). *Reproduced with permission from EMG-based control for three-dimensional upper limb movement assistance using a cable-based upper limb rehabilitation robot. In: Huang Y, Chen Y, Niu J, Song R, editors. International conference on intelligent robotics and applications. Springer; 2017.*

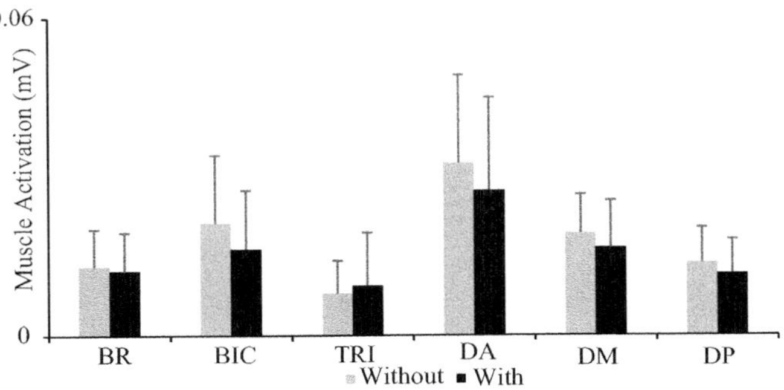


**FIGURE 7.8** The group mean muscle activations [brachioradialis (BR), biceps (BIC), triceps (TRI), and anterior (DA), middle (DM), posterior (DP) parts of the deltoid] of subjects while performing two tasks; the first task without assistance (the *gray bar*); the second task with assistance (the *black bar*). *Reproduced with permission from EMG-based control for three-dimensional upper limb movement assistance using a cable-based upper limb rehabilitation robot. In: Huang Y, Chen Y, Niu J, Song R, editors. International conference on intelligent robotics and applications. Springer; 2017.*

movements and can be identified by sensory perception, planning ability during movements, and task performing [37]. The lower values of RMSE during movements with position-varying gravity compensation, which is consistent with previous researches [38], indicate the improvement of tracking accuracy. NJS, which can be used for reflecting the smoothness of movements, is often used to assess the motion control ability [39]. Coscia et al. also reported a significant reduction in the values of NJS during reaching tasks without and with gravity compensation [21]. MVR is used for evaluating the relative speed deviation along the desired direction. Kim et al. applied this parameter along the angular moving direction for evaluating motion precision [40]. A higher value of MVR during those movements with position-varying gravity compensation indicates that participants can pay more efforts along the desired movement direction. The effect caused by task direction on kinematics characteristics is significant and is consistent with previous studies in sagittal [41], positive [42], or horizontal [43] panels. RMSE and MVR can both be used for assessing motion accuracy from different points, and a higher RMSE value corresponds to a lower MVR value during movements in the upward and forward directions.

Compared to movements with no compensation from the robot, the main contributed muscle activations were significantly reduced during movements with both the fixed and position-varying gravity compensation. Some researches proposed BIC, DA, and TRA are the main muscles for use against the gravity effect on the arm. The activations of these three muscles were also reduced when finishing the tasks in horizontal [20,44] and frontal panels [10,21]. Moreover, McCrea et al. proposed a recruit theory that when the main antigravity muscles cannot be used during upper limb movements, DM will be recruited to supplement them [45]. The results of this study are consistent with previous studies. Although the activations of the main contributed muscles of the upper limb are affected by direction [46], the activations of DP and TRI are direction-independent. This may be because they are not main muscles during all four tracking tasks. It has been reported that by properly activating the arm muscles, participants can maintain dynamic criteria during movements, such as optimizing motion commands [47] or energy consumption [48]. With assistance for compensating gravity, participants have a wider range of motion during movements in the horizontal panel [19,49]. The reduction in MMA caused by gravity compensation indicates that participants can focus more along the target direction during the movements, have a wider motion range, and more completed movements. In addition, muscle activation is associated with forces for executing movements [34], the lower values of MMA may reflect the ability of the robot to share the load of gravity on the participant's arm, and further, uncover the muscle forces for primarily concentration on the required exercise execution.

In all, the study is a first exploratory study of position-varying gravity compensation during arm-tracking tasks. This compensation can improve human–machine coordinated motion in 3D space by compensating gravity to uncover the active motion intention and focus voluntary residual motor efforts on task completion. Moreover, the proposed gravity compensation can reduce the activation of muscles for antigravity, which can help patients who suffer from muscle weakness.

The other purpose of this study was to propose and verify a model for continuously estimating voluntary motion intention based on sEMG signals. Regarding the results of model accuracy and kinematics characteristics, the proposed model in estimating the necessary force for motion is possible and accurate based on a state-space model regarding the MFE and PCC results. Although a slight decrease in the PCC was found when comparing the modeling and verifying phases, it is acceptable with the only limited effect on cable 2 but not all cables. The ability of the model to continuously estimate voluntary motion based on real-time EMG signals in 3D space is novel, while most other human–machine interactions by EMG-based control can only give continuous control in one-dimensional space or control in a discrete mode [28,29]. Although

Artemiadis et al. [50] also used state-space models for estimating movement-needed force, their models were not used for rehabilitation robots, nor did they consider cooperation between people and machines.

Statistical results of RMSE and MNMA demonstrate that participants can perform the tracking task with similar positional accuracy and similar muscle recruitments, no matter whether there was assistance from the robot controlled by the EMG-based strategy or not. The results of MNMA indicate that assistance from the estimation model only slightly reduced the activation of the six primarily movement-responsible muscles, but did not affect their contributions. The slight reduction of MNMA, which is consistent with previous studies [51], demonstrated that healthy participants showed better control ability on activating muscles and strength with robot assistance [52]. These results also demonstrate that the robot control strategy with a state-space model for estimating motion-needed forces from sEMG signals, can provide appropriate assistance to participants based on their voluntary motion intention. The strategy for robots can realize robots working for rehabilitation in conjunction with the revealed voluntary residual motion ability of patients. Moreover, the brain reorganization after stroke is encouraged by active training with the revealed voluntary motion ability [53].

In the future, more parameters and diversity tasks for rehabilitation should be adopted for clarifying the overall performance of both the position-varying gravity compensation and the EMG-based control strategy. Clinical trials will be conducted to verify the feasibility of robot-assisted rehabilitation with the position-varying gravity compensation and the EMG-based control strategy.

## Conclusion

This study first discussed the effects of position-varying gravity compensation on kinematics and muscle activation and compared them with no compensation movements and with fixed compensation movements. The improvement in kinematics and fewer activations of those antigravity muscles when participants executed the movements with this compensation indicated that this novel compensation method has potential in robot-assisted rehabilitation.

The second part of the study proposed an active control algorithm based on sEMG signals. A linear vertical motion task is designed to verify the strategy. The results show the feasibility to estimate the motion-needed forces from muscle activations during voluntary movements of participants based on the state-space model. The activation-forces model is supposed to help participants to reduce the excitability of multiple muscles in the upper limb without affecting the position control performance and does not change the contributions of the main muscles.

More studies on pathologies and more gender-matched participants are needed for evaluating whether rehabilitation training with the position-varying gravity compensation and the EMG-based control strategy is clinically feasible and practical.

# References


[1] Rossini PM, Calautti C, Pauri F, Baron J-C. Post-stroke plastic reorganisation in the adult brain. The Lancet Neurology 2003;2(8):493–502.

[2] Krebs HI, Volpe BT, Aisen ML, Hogan N. Increasing productivity and quality of care: robot-aided neuro-rehabilitation. Journal of Rehabilitation Research and Development 2000;37(6):639–52.

[3] Uno Y, Kawato M, Suzuki R. Formation and control of optimal trajectory in human multijoint arm movement. Biological Cybernetics 1989;61(2):89–101.

[4] Beer RF, Ellis MD, Holubar BG, Dewald JPA. Impact of gravity loading on post-stroke reaching and its relationship to weakness. Muscle & Nerve: Official Journal of the American Association of Electrodiagnostic Medicine 2007;36(2):242–50.

[5] Riener R, Nef T, Colombo G. Robot-aided neurorehabilitation of the upper extremities. Medical, & Biological Engineering & Computing 2005;43(1):2–10.

[6] A motorized gravity compensation mechanism used for active rehabilitation of upper limbs. In: van Elk MG, Driessen BJ, Dorrepaal M, van der Werff JJ, van der Meché EG, Aulbers AP, editors. 9th international conference on rehabilitation robotics, 2005 ICORR 2005. IEEE; 2005.

[7] Towards a multi-DOF passive balancing mechanism for upper limbs. In: Cheng Z, Foong S, Sun D, Tan U-X, editors. 2015 IEEE international conference on rehabilitation robotics (ICORR). IEEE; 2015.

[8] Runnalls KD, Anson G, Byblow WD. Partial weight support of the arm affects corticomotor selectivity of biceps brachii. Journal of Neuroengineering and Rehabilitation 2015;12(1):94.

[9] Herder JL, Vrijlandt N, Antonides T, Cloosterman M, Mastenbroek PL. Principle and design of a mobile arm support for people with muscular weakness. Journal of Rehabilitation Research and Development 2006;43(5):591.

[10] Kloosterman MG, Snoek GJ, Kouwenhoven M, Nene AV, Jannink MJ. Influence of gravity compensation on kinematics and muscle activation patterns during reach and retrieval in subjects with cervical spinal cord injury: an explorative study. Journal of Rehabilitation Research and Development 2010;47(7):617–28.

[11] Kahn LE, Zygman ML, Rymer WZ, Reinkensmeyer DJ. Robot-assisted reaching exercise promotes arm movement recovery in chronic hemiparetic stroke: a randomized controlled pilot study. Journal of Neuroengineering and Rehabilitation 2006;3(1):12.

[12] MEDARM: a rehabilitation robot with 5DOF at the shoulder complex. In: Ball SJ, Brown IE, Scott SH, editors. 2007 IEEE/ASME International Conference on Advanced Intelligent Mechatronics. IEEE; 2007.

[13] Ladenheim B, Altenburger P, Cardinal R, Monterroso L, Dierks T, Mast J, et al. The effect of random or sequential presentation of targets during robot-assisted therapy on children. NeuroRehabilitation 2013;33(1):25–31.

[14] Nef T, Mihelj M, Riener R. ARMin: a robot for patient-cooperative arm therapy. Medical, & Biological Engineering & Computing 2007;45(9):887–900.

[15] Stopforth R. Control investigation of a customizable/adjustable exoskeleton upper-limb. Industrial Robot: International Journal 2013;40(2):132–42.

[16] Lenzo B, Fontana M, Marcheschi S, Salsedo F, Frisoli A, Bergamasco M. Trackhold: a novel passive arm-support device. Journal of Mechanisms and Robotics 2016;8(2):021007.

[17] De Luca A, Siciliano B, Zollo L. PD control with on-line gravity compensation for robots with elastic joints: theory and experiments. Automatica 2005;41(10):1809–19.

[18] Stienen AHA, Hekman EEG, Prange GB, Jannink MJA, Aalsma AMM, van der Helm FCT, et al. Dampace: design of an exoskeleton for force-coordination training in upper-extremity rehabilitation. Journal of Medical Devices 2009;3(3):031003.

[19] Housman SJ, Scott KM, Reinkensmeyer DJ. A randomized controlled trial of gravity-supported, computer-enhanced arm exercise for individuals with severe hemiparesis. Neurorehabilitation and Neural Repair 2009;23(5):505–14.

[20] Prange GB, Kallenberg LAC, Jannink MJA, Stienen AHA, van der Kooij H, Ijzerman MJ, et al. Influence of gravity compensation on muscle activity during reach and retrieval in healthy elderly. Journal of Electromyography and Kinesiology 2009;19(2):e40–9.

[21] Coscia M, Cheung VCK, Tropea P, Koenig A, Monaco V, Bennis C, et al. The effect of arm weight support on upper limb muscle synergies during reaching movements. Journal of Neuroengineering and Rehabilitation 2014;11(1):22.

[22] Design of a dynamic arm support (DAS) for gravity compensation. In: Kramer G, Romer GR, Stuyt HJ, editors. 2007 IEEE 10th international conference on rehabilitation robotics. IEEE; 2007.

[23] A gravity compensation-based upper limb rehabilitation robot. In: Hsu L-C, Wang W-W, Lee G-D, Liao Y-W, Fu L-C, Lai J-S, editors. 2012 American Control Conference (ACC). IEEE; 2012.

[24] NTUH-II robot arm with dynamic torque gain adjustment method for frozen shoulder rehabilitation. In: Lin C-H, Lien W-M, Wang W-W, Chen S-H, Lo C-H, Lin S-Y, et al., editors. 2014 IEEE/RSJ international conference on intelligent robots and systems. IEEE; 2014.

[25] Li Z, Milutinovic D, Rosen J. Spatial map of synthesized criteria for the redundancy resolution of human arm movements. IEEE Transactions on Neural Systems and Rehabilitation Engineering 2015;23(6):1020–30.

[26] Kong K, Tomizuka M. Proxy-based impedance control of a cable-driven assistive system for upper extremity rehabilitation. IFAC Proceedings Volumes 2011;44(1):2871–6.

[27] Dipietro L, Ferraro M, Palazzolo JJ, Krebs HI, Volpe BT, Hogan N. Customized interactive robotic treatment for stroke: EMG-triggered therapy. IEEE Transactions on Neural Systems and Rehabilitation Engineering 2005;13(3):325–34.

[28] Song R, Tong K-y, Hu X, Li L. Assistive control system using continuous myoelectric signal in robot-aided arm training for patients after stroke. IEEE Transactions on Neural Systems and Rehabilitation Engineering 2008;16(4):371–9.

[29] Potvin JR, Norman RW, McGill SM. Mechanically corrected EMG for the continuous estimation of erector spinae muscle loading during repetitive lifting. European Journal of Applied Physiology and Occupational Physiology 1996;74(1–2):119–32.

[30] High-level controller for an arm rehabilitation robot-positioning algorithms with respect to EMG data. In: Gmerek AJ, editor. 2011 16th international conference on methods & models in automation & robotics. IEEE; 2011.

[31] Huang Y, Yang Q, Chen Y, Song R. Assessment of motor control during three-dimensional movements tracking with position-varying gravity compensation. Frontiers in Neuroscience 2017;11:253.

[32] EMG-based control for three-dimensional upper limb movement assistance using a cable-based upper limb rehabilitation robot. In: Huang Y, Chen Y, Niu J, Song R, editors. International conference on intelligent robotics and applications. Springer; 2017.

[33] Yang J, Su H, Li Z, Ao D, Song R. Adaptive control with a fuzzy tuner for cable-based rehabilitation robot. International Journal of Control, Automation and Systems 2016;14(3):865–75.

[34] Lloyd DG, Besier TF. An EMG-driven musculoskeletal model to estimate muscle forces and knee joint moments in vivo. Journal of Biomechanics 2003;36(6):765–76.

[35] Manal K, Buchanan TS. A one-parameter neural activation to muscle activation model: estimating isometric joint moments from electromyograms. Journal of Biomechanics 2003;36(8):1197–202.

[36] Hogan N, Sternad D. Sensitivity of smoothness measures to movement duration, amplitude, and arrests. Journal of Motor Behavior 2009;41(6):529–34.

[37] Seppänen TM, Alho O-P, Seppänen T. Reducing the airflow waveform distortions from breathing style and body position with improved calibration of respiratory effort belts. BioMedical Engineering Online 2013;12(1):97.

[38] Grimm F, Naros G, Gharabaghi A. Compensation or restoration: closed-loop feedback of movement quality for assisted reach-to-grasp exercises with a multi-joint arm exoskeleton. Frontiers in Neuroscience 2016;10:280.

[39] Adamovich SV, Fluet GG, Merians AS, Mathai A, Qiu Q. Incorporating haptic effects into three-dimensional virtual environments to train the hemiparetic upper extremity. IEEE Transactions on Neural Systems and Rehabilitation Engineering 2009;17(5):512–20.

[40] Kim SC, Kim JY, Lee HN, Lee HH, Kwon JH, beom Kim N, et al. A quantitative analysis of gait patterns in vestibular neuritis patients using gyroscope sensor and a continuous walking protocol. Journal of Neuroengineering and Rehabilitation 2014;11(1):58.

[41] Papaxanthis C, Pozzo T, Stapley P. Effects of movement direction upon kinematic characteristics of vertical arm pointing movements in man. Neuroscience Letters 1998;253(2):103–6.

[42] d'Avella A, Fernandez L, Portone A, Lacquaniti F. Modulation of phasic and tonic muscle synergies with reaching direction and speed. Journal of Neurophysiology 2008;100(3):1433–54.

[43] Beer RF, Dewald JPA, Dawson ML, Rymer WZ. Target-dependent differences between free and constrained arm movements in chronic hemiparesis. Experimental Brain Research 2004;156(4): 458–70.

[44] Sabatini AM. Identification of neuromuscular synergies in natural upper-arm movements. Biological Cybernetics 2002;86(4):253–62.

[45] McCrea PH, Eng JJ, Hodgson AJ. Saturated muscle activation contributes to compensatory reaching strategies after stroke. Journal of Neurophysiology 2005;94(5):2999–3008.

[46] Hughes A-M, Freeman CT, Burridge JH, Chappell PH, Lewin PL, Pickering RM, et al. Shoulder and elbow muscle activity during fully supported trajectory tracking in neurologically intact older people. Journal of Electromyography and Kinesiology 2009;19(6):1025–34.

[47] Nakano E, Imamizu H, Osu R, Uno Y, Gomi H, Yoshioka T, et al. Quantitative examinations of internal representations for arm trajectory planning: minimum commanded torque change model. Journal of Neurophysiology 1999;81(5):2140–55.

[48] Soechting JF, Buneo CA, Herrmann U, Flanders M. Moving effortlessly in three dimensions: does Donders' law apply to arm movement? Journal of Neuroscience 1995;15(9):6271–80.

[49] Wang W, Dounskaia N. Load emphasizes muscle effort minimization during selection of arm movement direction. Journal of Neuroengineering and Rehabilitation 2012;9(1):70.

[50] Artemiadis PK, Kyriakopoulos KJ. EMG-based control of a robot arm using low-dimensional embeddings. IEEE Transactions on Robotics 2010;26(2):393–8.

[51] Kwon S, Kim Y, Kim J. Movement stability analysis of surface electromyography-based elbow power assistance. IEEE Transactions on Biomedical Engineering 2014;61(4):1134–42.

[52] Suzuki K, Mito G, Kawamoto H, Hasegawa Y, Sankai Y. Intention-based walking support for paraplegia patients with Robot Suit HAL. Advanced Robotics 2007;21(12):1441–69.

[53] Cauraugh J, Light K, Kim S, Thigpen M, Behrman A. Chronic motor dysfunction after stroke: recovering wrist and finger extension by electromyography-triggered neuromuscular stimulation. Stroke 2000;31(6):1360–4.

8

# Artificial sensory feedback for bionic hands

John E. Downey, Jack Brooks, Sliman J. Bensmaia
*DEPARTMENT OF ORGANISMAL BIOLOGY AND ANATOMY, UNIVERSITY OF CHICAGO, CHICAGO, IL, UNITED STATES*

**Chapter outline**



## Introduction

Manual interactions with objects give rise to a barrage of neural signals from the hand that convey tactile information about the objects—their shape, size, texture, etc.—and about contact events—contact timing, contact location, contact pressure, etc. This tactile information supports our ability to recognize objects by touch, when vision is not available, and is required for the dexterous manipulation of objects [1].

**Intelligent Biomechatronics in Neurorehabilitation. https://doi.org/10.1016/B978-0-12-814942-3.00008-8**

For example, neural responses from the hand signal the establishment of contact with an object and prompt the transition from reach to grasp. Information about object contact is not always visually available, for example when the object is obscured by the hand or vice versa, and even when vision is available, the resulting sensory signals are often ill suited to support grasp and manipulation. As a result, the abolition of tactile signals leads to severe impairments in hand use, particularly with regards to object manipulation [1].

In addition to tactile information about object interactions, limbs provide information about their posture and movements, a sense known as proprioception. Proprioceptive information is carried primarily by signals stemming from the muscles and tendons [2]. Loss of proprioception results in the inability to track the limb without attending to it visually and a loss of coordination in movements. Patients without proprioception report that the concentration required to generate purposeful movements is highly effortful [3,4].

To create a dexterous bionic hand, then, requires not only establishing a means to move it volitionally but also the restoration of sensory signals that convey information about the consequences of those movements and about object interactions. Indeed, myoelectric prostheses—robotic arms controlled by activating muscles above the level of amputation—are often rejected in favor of less-advanced cable-driven prostheses because the latter provide information about the force applied through the cable, whereas a myoelectric device conveys no such information [5].

Restoring sensory feedback through a bionic arm requires sensors in the robotic limb that encode the state of the limb—its posture and movements—typically by tracking the angle of individual joints and the torques exerted at those joints, as well as contact with objects, with force sensors on the fingertips. The output of these sensors must then be converted to signals that are interpretable by the user, and this can be achieved using several approaches that vary in invasiveness. The least invasive approach provides physical stimulation to intact portions of the body based on the sensor output. Another approach is to implant electrodes in the nerve through which electrical currents can be delivered, thereby eliciting artificial sensory percepts. For people with spinal cord injuries, however, peripheral stimulation cannot convey information beyond the site of injury, so these patients require direct electrical stimulation of the central nervous system. These different approaches involve a variety of technologies, as described in more detail below.

A central principle in the development of approaches to convey sensory feedback is that of biomimicry: if we could mimic natural patterns of neural activation perfectly by stimulating the nervous system, we could evoke completely natural and thus intuitive and useful sensory percepts. While the current state of neural interface technology limits the verisimilitude of the evoked neural activity, most efforts are driven by the belief that the artificial percepts will be more natural and useful the closer the evoked activity is to its natural counterpart.

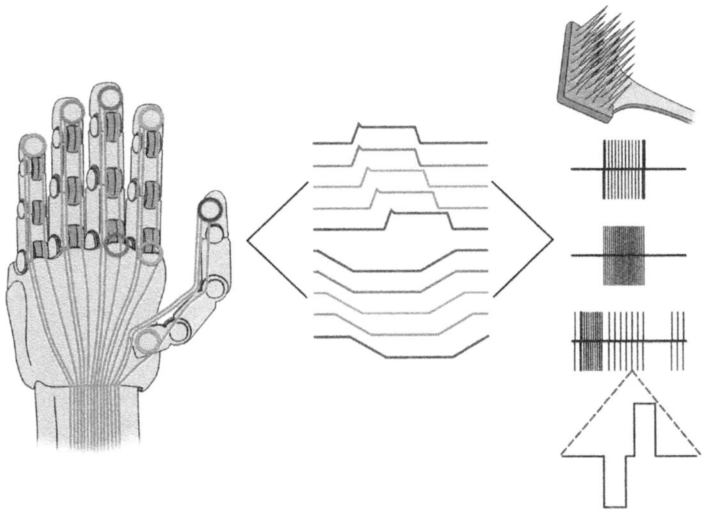

**FIGURE 8.1** (Left) Example of the transformation from sensors in the bionic hand to electrical stimulation of neurons. (Middle) Example traces from force (top) and position (bottom) sensors in each of the five fingers sequentially grasping and releasing an object. (Right) A bionic system could transform these traces into electrical stimulation of the nerve or the brain.

## Sensors

Bionic hands are instrumented with a variety of sensors that convey information about the state of the hand and about its interactions with objects (Fig. 8.1). Information about contact with objects can be collected by sensing grasp force from the torque exerted by the motors that drive the fingers [6] or by directly sensing the force imparted on the fingertips [7]. Sensors are also being developed to extract texture information, for example by attaching thin, flexible layers of sensors to the surface of the bionic fingers [8]. These sensors are compatible with any robotic hand since they can conform to any shape, but have yet to reach biological levels of sensitivity. A sensorized fingertip, the BioTac, provides a high level of sensitivity, comparable in some respects to that of the human finger [9], but it is more difficult to integrate into a bionic hand [10]. Currently, the bottleneck for sensitization of bionic hands is not the sensor but rather the neural interface. As interfaces with the nervous system get more sophisticated, greater emphasis will be placed on improvements in sensory technology.

## Interfaces with the peripheral nervous system

The individuals that stand to benefit most from bionic hands are amputees. The level of the amputation varies from individual to individual—from distal, near the wrist, to proximal, at the shoulder—and these different amputations involve different approaches to restore sensation. One of the major differences is that the nerves, which carry sensory signals from the hand and also motor signals to the muscles, are much more sensory distally (near the hand) than they are proximally (near the shoulder). Thus, stimulating a nerve in the forearm mostly activates sensory neurons from the hand but stimulating a nerve at the shoulder activates motor and sensory neurons for the entire limb. A variety of technologies are being developed to interface with the peripheral nervous system, each with its own advantages and disadvantages and better suited for some levels of amputation (distal vs. proximal) than others [11,12].

## Targeted sensory reinnervation

Some tactile sensation can be restored by rerouting the nerves that previously supplied the hand to other skin patches. This phenomenon was discovered incidentally during a procedure designed for control of myoelectric prosthetics, called targeted muscle reinnervation (TMR). In this procedure, the residual nerves are typically rerouted to a large muscle, for example the pectoral muscle in upper limb amputees. The muscle activity, ectopically induced in the pectoral muscles when the subject generates a motor command to move the arm, is sensed with electromyographic electrodes and used to control the prosthetic limb [13]. In one such procedure, the nerve from a patient's arm stump was rerouted to the pectoral muscles, and the patient reported feeling a sensation on his amputated hand when he was touched on the chest over the site of reinnervation [14]. These sensations elicited by touching small patches of the chest were referred to multiple fingers or to large swaths of the hand and arm. A second subject reported similar touch sensations, confined entirely to the hand [14].

Spurred by this incidental finding, the technique of targeted sensory reinnervation (TSR) was developed, which consists of splitting the motor and sensory fascicles so that sensory fascicles can be rerouted to the skin while the motor fascicles are routed to the muscle. With this technique, the referred sensations can be confined to small areas, for example a single finger segment [15]. Using TMR and TSR, a subject was able to detect the timing of contact with an object and judge its size and compliance [15]. While TSR is a promising approach to convey sensory feedback, the spatial layout of the referred sensations is idiosyncratically determined by the pattern of reinnervation, a problem that might be solved with a more refined surgical approach [15]. Furthermore, tactors used to stimulate the reinnervated skin are power-hungry and bulky, which limits the spatial resolution of the resulting sensory feedback, and these two factors currently limit the clinical viability of this approach [16].

## Electrical interfaces with the somatosensory nerves

Another approach to artificial somatosensation consists in directly activating the nerve through electrical stimulation delivered through electrodes positioned around or in the nerve. Three nerves innervate the palmar surface of the hand (Fig. 8.2): the median nerve innervates the first three digits (thumb, index, and middle fingers), the lateral aspect of

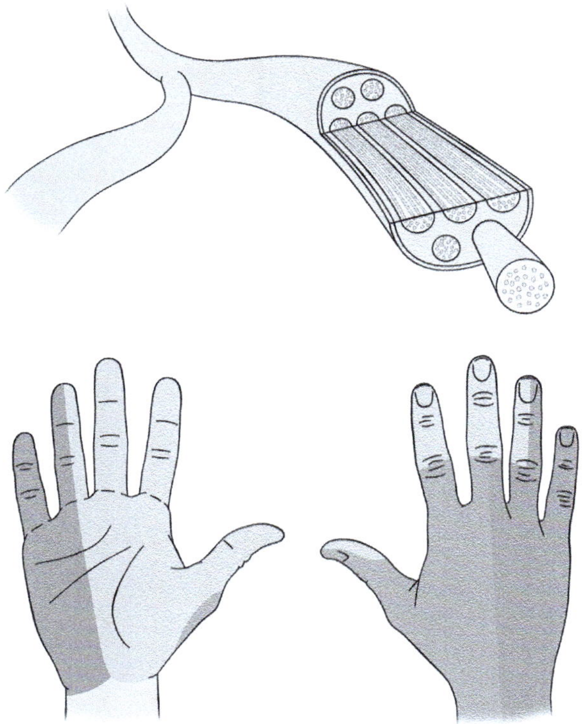

**FIGURE 8.2** (Top) Cross-section of a nerve showing multiple fascicles comprising many nerve fibers. (Bottom) Innervated by the hand: *light green* [light gray in print version]—median nerve, *dark green* [dark gray in print version]—ulnar nerve, *purple* [gray in print version]—radial nerve.

the fourth digit (ring finger), and the underlying palm, the ulnar nerve innervates the rest of the fourth digit and the fifth digit (little finger) and the underlying palm, and the radial nerve innervates a small patch of skin under the thumb (but primarily the dorsal surface of the hand). Each nerve is composed of multiple segregated bundles of axons, called fascicles (Fig. 8.2) and individual sensory fascicles tend to innervate a restricted patch of skin. Multiple electrode technologies have been developed to interface with the nerve. These technologies can be divided into two broad categories: extrafascicular electrodes, which surround the nerve and intrafascicular electrodes, which penetrate it (Fig. 8.3).

*Extrafascicular electrodes* wrap around the outside of the nerve to stimulate sensory fibers without breaching the sheath surrounding the nerve. The two most commonly used extrafascicular electrodes are the spiral cuff electrode [17], which coils around the nerve, and the flat interface nerve electrode (FINE; Fig. 8.3) [18–20], which flattens the nerve to provide access to a greater number of fascicles, including those that are near the center of the nerve in its undeformed state. Extrafascicular electrodes have been shown to provide stable percepts over the span of years [21].

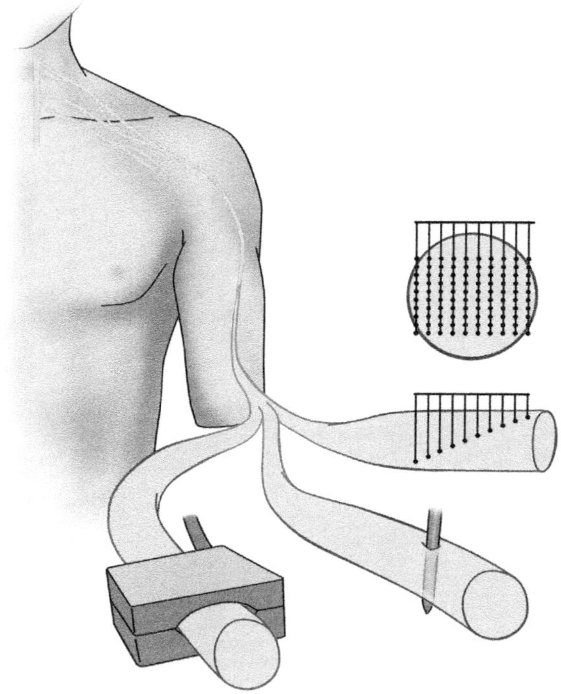

**FIGURE 8.3** Example electrodes for peripheral nerve stimulation. (From left to right) Cuff electrodes stimulate from outside the nerve, transverse and longitudinal intrafascicular electrodes pass through the nerve and stimulate from the inside (transverse is shown here), and microelectrode arrays (inset shows transverse view) put large numbers of electrodes at varying depths across the width of the nerve.

*Intrafascicular electrodes*, which penetrate the nerve to make direct contact with nerve fibers, fall into two categories. One approach consists of implanting one or a small number of leads either parallel to the nerve—the longitudinal intrafascicular electrode [22,23]—or across it—the transverse intrafascicular multichannel electrodes [24] (Fig. 8.3). Each lead has multiple contacts [24] to allow for stimulation of multiple fascicles. Another approach consists in implanting into the nerve an array of 100 electrodes arranged in a 10 × 10 grid over a 4 × 4 mm area—the Utah slant electrode array [25,26] (Fig. 8.3). The electrodes get progressively shorter (or longer) along the nerve to ensure that electrode tips—where the electrical contact is located—impinge on different fascicles.

Whether intrafascicular electrodes can provide stable and robust feedback over years has yet to be demonstrated. However, stimulation thresholds have been shown to increase rapidly, which casts doubt on their long-term viability [25,27].

## Perceptual effects of nerve stimulation

Stimulation of individual leads on an extrafascicular cuff evokes sensations spanning one or two phalanges or similar-sized patches of skin on the palm and wrist [17,28].

In principle, intrafascicular electrodes allow for more selective stimulation since electrical contact is made directly with the nerve. In fact, at low currents, one or a handful of nerve fibers can be activated [29]. As might be expected, more distinct projected fields of sensations are evoked through intrafascicular electrodes than through extrafascicular ones [12]. However, the currents that are typically delivered through intrafascicular electrodes are sufficiently high to activate a comparable number of nerve fibers as do their extrafascicular counterparts.

Increasing the frequency or amplitude of nerve stimulation increases the intensity of the sensation [18,27] and changing the patterning of the stimulation affects the quality of the sensation, transforming it from a paresthesia into a more natural sensation [17]. Sensations evoked through sustained electrical stimulation tend to become weaker over a period of minutes, a phenomenon known as sensory adaptation that is also observed with natural mechanical stimulation [21].

## Functional tests

Bionic hands that provide sensory feedback through nerve stimulation convey information about objects with which they interact. For example, subjects can discriminate the shape and stiffness of grasped objects [7,26,28] in the absence of vision.

Another way to test hand function is to have a subject move wooden blocks from one box to another; performance is assessed by counting the number of blocks the subject can move within a given period of time. Blindfolded subjects can remove twice as many blocks with feedback as without [7]. A more challenging version of this task consists in moving "fragile" objects from one location to another—too much pressure "breaks" the object—and task performance is greatly improved with somatosensory feedback, even with vision [30]. Sensory feedback also improves the ability to pluck cherries from their stems [17], a challenging task without touch, because either too much force is exerted and the cherry is crushed or too little force is exerted and it slips from the hand.

## Biomimicry

Typically, the frequency or amplitude of stimulation is modulated in proportion to the output of sensors on the bionic hand. While this approach provides useful sensory feedback, the resulting patterns of nerve activation are fundamentally different from those evoked during manipulation with a native hand. For example, the initiation and termination of contact evokes activity that dwarfs that during maintained contact, and these phasic bursts of afferent activation convey precise information about the timing and nature of contact events [31]. This aspect of natural neural activity is not incorporated into most sensory feedback algorithms that convert sensor output into electrical stimulation pulse trains. Some evidence suggests that more naturalistic—biomimetic—sensory feedback algorithms lead to more natural percepts and improved function [30,32,33]. These early successes are leading to the development of an increasingly biomimetic approach to artificial sensory feedback.

### Embodiment, phantom pain, and patient acceptance

One remarkable phenomenon engendered by sensitization of a bionic hand is that it starts to feel like a part of the body. Indeed, the visual experience of the hand being touched paired with a tactile sensation projected to the corresponding location on the phantom hand leads to embodiment of the bionic hand. This phenomenon was first demonstrated with TSR [34] but since has been reported with other peripheral nerve interfaces [17]. Another benefit of sensory restoration is that it reduces phantom limb pain, a common sequelae of amputation [26]. The embodiment and reduction of phantom pain afforded by artificial sensory feedback improve the acceptance of bionic hands above and beyond their functional benefits described above.

### Lower limb prostheses

Somatosensation plays an important role in walking, particularly on uneven surfaces or in the dark. Sensory feedback can be conveyed to lower limb amputees by implanting electrodes in the sciatic nerve [20]. While stimulation of this nerve evokes sensations in the phantom leg and foot [19], the functional benefits of sensitization of lower limb prostheses have yet to be demonstrated.

## Interfaces with the central nervous system

While interfaces with the peripheral nervous system have shown great promise to restore sensory feedback for amputees, these approaches cannot be applied to patient populations for whom the connection between the peripheral and central nervous systems is severed, for example, patients with spinal cord injuries or brainstem strokes. In these cases, the neural interface must be with the central nervous system (CNS) itself. The control signals, rather than stemming from the muscles or the nerves, are obtained by recording from brain areas involved in motor control, typically primary motor cortex, and sensory stimulation is delivered to brain areas that are involved in processing tactile and proprioceptive information, typically somatosensory cortex [35] (Fig. 8.4). Electrical stimulation is delivered through electrocorticography (ECoG) electrodes [36–39] or microelectrode arrays (MEA) [6,40] (Fig. 8.5), each with its benefits and shortcomings. A major disadvantage of cortical interfaces relative to their peripheral counterparts is that many neural mechanisms that integrate sensory information into movement planning and execution are bypassed.

### Somatosensory cortex

The main target for conveying sensory feedback through an interface with the central nervous system is somatosensory cortex, because this part of the brain is more accessible than the two upstream neural structures, the cuneate nucleus (in the brainstem) and the thalamus. The somatosensory cortex is the major cortical recipient of somatosensory

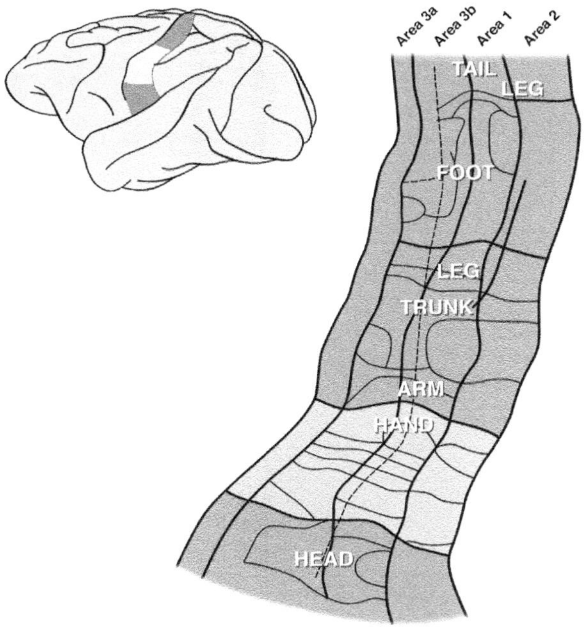


**FIGURE 8.4** Somatosensory cortex (shown here for a Rhesus macaque) is located on the posterior bank of the central sulcus and in the postcentral gyrus (top left). One of the key features of somatosensory cortex is that it is organized somatotopically. That is, it contains a complete representation of the body in each area (3a, 3b, 1, and 2), with neighboring neurons responding to neighboring and overlapping patches of skin.

signals from the body, including touch and proprioception [2]. Located in the posterior bank of the central sulcus and on the postcentral gyrus, somatosensory cortex comprises four cortical fields—Brodmann's areas 3a, 3b, 1, and 2—each of which contains a map of the body, a so-called somatosensory homunculus (shown in Fig. 8.4 for a Rhesus macaque). Area 3a processes primarily proprioceptive information, areas 3b and 1 process primarily cutaneous information, and these two streams of information converge onto area 2. One of the central organizing principles of somatosensory cortex is its somatotopic organization: neurons that are close together tend to respond to adjacent and overlapping parts of the body.

## Electrocorticography

Electrocorticography (ECoG) electrodes are small metal discs (2–3 mm in diameter) implanted inside the skull, typically in a grid pattern with electrodes spaced 3–10 mm apart (Fig. 8.5). These implants are used to localize the source of epileptic seizures but they offer the opportunity to carry out invasive studies in human subjects, including studies related to neuroprosthetics [36–39]. Stimulation of somatosensory cortex through ECoG electrodes results in unnatural sensations, typically described as buzzing,

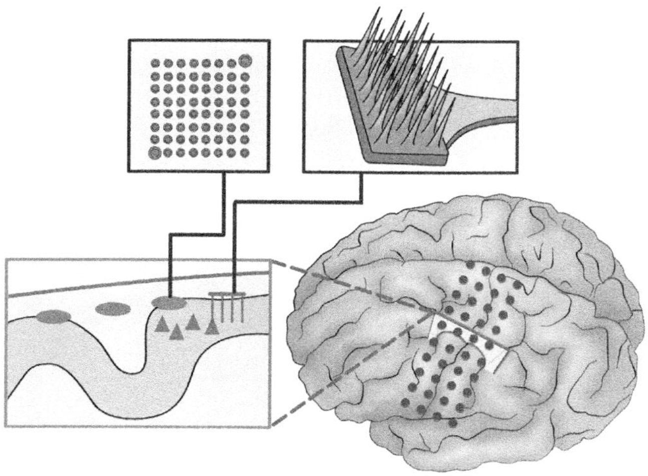

**FIGURE 8.5** Example electrodes for cortical stimulation. Electrocorticography electrodes are arranged in a grid just above the brain's surface, while microelectrode arrays penetrate the brain.

tingling, or vibration [38,39], referred to large swaths of skin, for example the entire hand [36–39]. The main challenge for ECoG is that the electrodes are placed outside of the pia mater and large currents are required to elicit detectable sensations. These large currents spread over large distances and stimulation through neighboring electrodes on the array often activates a largely overlapping population of neurons [38,39]. The combination of the current spread and the size and spacing of the electrodes [36,38,39] severely limits the spatial resolution of sensory feedback via ECoG, especially considering that finger representations are separated by only 3 mm in the human cortex [41]. The large populations of activated neurons, and the fact that ECoG preferentially activates the superficial layers of cortex, may also contribute to the paresthetic nature of the evoked sensations.

## Intracortical microstimulation

Another approach to interfacing with the CNS is with arrays of electrodes that penetrate the brain, similar to those described above for the nerve but with electrodes of equal length (rather than tapered). These multielectrode arrays can be used to record from motor structures in the brain and stimulate sensory ones (Fig. 8.5). Because these electrodes make direct contact with neurons, lower currents are required to elicit detectable sensations, so behaviorally relevant levels of stimulation spread less and therefore elicit more localized sensations than do their ECoG counterparts [6,42]. Indeed, stimulation through individual penetrating electrodes typically evokes sensations confined to small regions of skin on an individual finger, or a comparable area of the palm. Sensations evoked through intracortical microstimulation (ICMS) are reported to be "nearly natural," described as pressure, squeeze, tap, tingle, warmth, and sometimes as electrical [6,40].

The major challenge to approaches involving penetrating electrodes is that they never last beyond a few years. The problem is particularly severe for recording of neural signals to infer intended movements, as the neuronal signals degrade over time and eventually disappear, usually within a few years. Thresholds for ICMS are highly stable for years [43,44] but electrode arrays eventually fail catastrophically [45]. The longevity of penetrating electrode arrays will need to improve substantially for these to become clinically viable.

### Perceptual effects of brain stimulation

The quality of the sensation varies across electrodes, with some electrodes evoking pressure-like sensations, others more paresthetic ones. In some cases, stimulation can result in sensations of limb movement, particularly at high stimulation levels [40]. As for peripheral stimulation, the intensity of the sensations evoked through brain stimulation scale with stimulation frequency and amplitude [36,38,39]. The dependence of the detectability and discriminability of ICMS has been extensively characterized in both humans [6] and monkeys [42,46]. Changing the stimulation frequency also changes the quality of the sensation [47,48].

As with nerve stimulation, temporally congruent vision of a rubber hand being touched and somatotopically appropriate brain stimulation leads to embodiment of the hand, a phenomenon that is observed even with ECoG, which evokes paresthetic sensations [37]. Results with biomimetic nerve stimulation suggest that ICMS will lead to even greater embodiment given the greater naturalness of the evoked sensations.

### Functional tests

Importantly, brain stimulation can be used to guide behavior. For example, a subject was able to change the aperture of his hand based on the strength of ECoG stimulation [49]. Animals have been shown to be able to navigate [50,51] or find objects in virtual reality based on ICMS feedback [52]. In a more ethologically valid demonstration, a human subject performed significantly better on a clinical test of hand and arm function—the Action Research Arm Test—with sensory feedback than without [53].

## Conclusions

With the development of approaches to decode intended movements from the muscles, nerves, or brain and of anthropomorphic robotic arms that approximate the functionality of a native human arm [25,27,54,55] came the realization that somatosensation must also be restored to achieve a dexterous bionic hand. Spurred by this realization, an international effort is currently underway to develop more robust interfaces with the nerve or the brain that have higher selectivity—so that small populations of neurons can be activated—and bandwidth—so that more independent streams of sensory information can be simultaneously conveyed. For interfaces with both the peripheral and central nervous systems, choice of technology involves trade-offs between stability and selectivity. Indeed, intrafascicular and intracortical electrodes are more selective than are extrafascicular and ECoG electrodes, but the former are less stable and long lasting than the latter. Efforts are underway to correct the respective shortcomings of these technologies and to develop entirely new ones [12] in the hopes of achieving a stable and selective interface that lasts a lifetime.

The limb is a complex effector, the state of which our nervous system tracks effortlessly. After amputation or spinal cord injury, sensory feedback will be intuitive to the bionic hand user to the extent that the evoked sensations are natural. Indeed, if the sensations are completely natural, no learning would be required. In contrast, subjects will have to learn to interpret the unfamiliar artificial sensations and learn a mapping between a sensory experience and the corresponding state of the limb or contact event. Whether adults can learn to use a constant stream of completely novel sensory signals in adulthood is dubious [56]. Rather, sensory algorithms attempt to reproduce natural patterns of neuronal activation to the extent that they can. In other words, artificial somatosensation is conveyed in the brain's own language, to minimize the need to learn.

While many challenges remain, the progress in neuroprosthetics is staggering, and an increasing number of groups are working on refining sensory technologies, interface technologies, robotic arms, and sensory encoding algorithms. The first sensitized prosthetic hands have recently been deployed for clinical use [57]. While this first generation is rather simple, it constitutes a major milestone on the path to a dexterous bionic hand.

# References


[1] Johansson RS, Flanagan JR. Coding and use of tactile signals from the fingertips in object manipulation tasks. Nature Reviews Neuroscience 2009:345–59.

[2] Delhaye BP, Long KH, Bensmaia SJ. Neural basis of touch and proprioception in primate cortex. Comparative Physiology 2018;8:1575–602.

[3] Sharp FR, Rando TA, Greenberg SA, et al. Pseudochoreoathetosis: movements associated with loss of proprioception. Archives of Neurology 1994;51:1103–9.

[4] Sainburg RL, Poizner H, Ghez C. Loss of proprioception produces deficits in interjoint coordination. Journal of Neurophysiology 1993;70:2136–47.

[5] Carey SL, Lura DJ, Highsmith MJ. Differences in myoelectric and body-powered upper-limb prostheses: systematic literature review. Journal of Rehabilitation Research and Development 2015;52:247–62.

[6] Flesher SN, Collinger JL, Foldes ST, et al. Intracortical microstimulation of human somatosensory cortex. Science Translational Medicine 2016;8:1–11.

[7] Schiefer M, Tan D, Sidek SM, et al. Sensory feedback by peripheral nerve stimulation improves task performance in individuals with upper limb loss using a myoelectric prosthesis. Journal of Neural Engineering 2015;13:16001.

[8] Park M, Bok BG, Ahn JH, et al. Recent advances in tactile sensing technology. Micromachines 2018; 9. https://doi.org/10.3390/mi9070321.

[9] Delhaye BP, Schluter EW, Bensmaia SJ. Robo-psychophysics: extracting behaviorally relevant features from the output of sensors on a prosthetic finger. IEEE Transactions on Haptics 2016;9: 499–507.

[10] Fishel JA, Loeb GE. Sensing tactile microvibrations with the BioTac – comparison with human sensitivity. In: IEEE RAS/EMBS. Roma, Italy; 2012. p. 1122–7.

[11] Navarro X, Krueger TB, Lago N, et al. A critical review of interfaces with the peripheral nervous system for the control of neuroprostheses and hybrid bionic systems. Journal of the Peripheral Nervous System 2005;10:229–58.

[12] Rijnbeek EH, Eleveld N, Olthuis W. Update on peripheral nerve electrodes for closed-loop neuroprosthetics. Frontiers in Neuroscience 2018;12:1–9.

[13] Kuiken TA, Li G, Lock BA, et al. Targeted muscle reinnervation for real-time myoelectric control of multifunction artificial arms. The Journal of the American Medical Association 2009;301:619–28.

[14] Kuiken TA, Marasco PD, Lock BA, et al. Redirection of cutaneous sensation from the hand to the chest skin of human amputees with targeted reinnervation. Proceedings of the National Academy of Sciences 2007;104:20061–6.

[15] Hebert JS, Olson JL, Morhart MJ, et al. Novel targeted sensory reinnervation technique to restore functional hand sensation after transhumeral amputation. IEEE Transactions on Neural Systems and Rehabilitation Engineering 2014;22:765–73.

[16] Schultz AE, Marasco PD, Kuiken TA. Vibrotactile detection thresholds for chest skin of amputees following targeted reinnervation surgery. Brain Research 2009;1251:121–9.

[17] Tan DW, Schiefer MA, Keith MW, et al. A neural interface provides long-term stable natural touch perception. Science Translational Medicine 2014;6:257ra138.

[18] Graczyk EL, Schiefer MA, Saal HP, et al. The neural basis of perceived intensity in natural and artificial touch. Science Translational Medicine 2016;8. https://doi.org/10.1126/scitranslmed.aaf5187.

[19] Charkhkar H, Shell CE, Marasco PD, et al. High-density peripheral nerve cuffs restore natural sensation to individuals with lower-limb amputations. Journal of Neural Engineering 2018;15:56002.

[20] Freeberg MJ, Stone MA, Triolo RJ, et al. The design of and chronic tissue response to a composite nerve electrode with patterned stiffness. Journal of Neural Engineering 2017;14. https://doi.org/10.1088/1741-2552/aa6632.

[21] Graczyk EL, Delhaye BP, Schiefer MA, et al. Sensory adaptation to electrical stimulation of the somatosensory nerves. Journal of Neural Engineering 2018;15. https://doi.org/10.1088/1741-2552/aab790.

[22] Yoshida K, Stein RB. Characterization of signals and noise rejection with bipolar longitudinal intrafascicular electrodes. IEEE Transactions on Biomedical Engineering 1999;46:226–34.

[23] Lawrence SM, Dhillon GS, Jensen W, et al. Acute peripheral nerve recording characteristics of polymer-based longitudinal intrafascicular electrodes. IEEE Transactions on Neural Systems and Rehabilitation Engineering 2004;12:345–8.

[24] Boretius T, Badia J, Pascual-Font A, et al. A transverse intrafascicular multichannel electrode (TIME) to interface with the peripheral nerve. Biosensors and Bioelectronics 2010;26:62–9.

[25] Wendelken S, Page DM, Davis T, et al. Restoration of motor control and proprioceptive and cutaneous sensation in humans with prior upper-limb amputation via multiple Utah Slanted Electrode Arrays (USEAs) implanted in residual peripheral arm nerves. Journal of NeuroEngineering and Rehabilitation 2017;14:1–17.

[26] Page DM, George JA, Kluger DT, et al. Motor control and sensory feedback enhance prosthesis embodiment and reduce phantom pain after long-term hand amputation. Frontiers in Human Neuroscience 2018;12:1–16.

[27] Rossini PM, Micera S, Benvenuto A, et al. Double nerve intraneural interface implant on a human amputee for robotic hand control. Clinical Neurophysiology 2010;121:777–83.

[28] Raspopovic S, Capogrosso M, Petrini FM, et al. Restoring natural sensory feedback in real-time bidirectional hand prostheses. Science Translational Medicine 2014;6. https://doi.org/10.1126/scitranslmed.3006820.

[29] Kundu A, Harreby KR, Yoshida K, et al. Stimulation selectivity of the 'Thin-Film longitudinal intrafascicular electrode' (tfLIFE) and the 'transverse intrafascicular multi-channel electrode' (TIME) in the large nerve animal model. IEEE Transactions on Neural Systems and Rehabilitation Engineering 2014;22:400–10.

[30] Valle G, Mazzoni A, Iberite F, et al. Biomimetic intraneural sensory feedback enhances sensation naturalness, tactile sensitivity, and manual dexterity in a bidirectional prosthesis. Neuron 2018;100:37–45.e7.

[31] Callier T, Suresh AK, Bensmaia SJ. Neural coding of contact events in somatosensory cortex. Cerebral Cortex 2019. https://doi.org/10.1093/cercor/bhy337.

[32] Okorokova EV, He Q, Bensmaia SJ. Biomimetic encoding model for restoring touch in bionic hands through a nerve interface. Journal of Neural Engineering 2018;15:66033.

[33] George JA, Brinton MR, Davis TS, et al. Expanding the reach of upper-limb prosthetics: restoring sensorimotor function after complex regional pain syndrome, long-term hand disuse and an elective transradial amputation. In: Society for neuroscience; 2018.

[34] Marasco PD, Kim K, Colgate JE, et al. Robotic touch shifts perception of embodiment to a prosthesis in targeted reinnervation amputees. Brain 2011;134:747–58.

[35] Bensmaia SJ, Miller LE. Restoring sensorimotor function through intracortical interfaces: progress and looming challenges. Nature Reviews Neuroscience 2014;15:313–25.

[36] Johnson LA, Wander JD, Sarma D, et al. Direct electrical stimulation of the somatosensory cortex in humans using electrocorticography electrodes: a qualitative and quantitative report. Journal of Neural Engineering 2013;10. https://doi.org/10.1088/1741-2560/10/3/036021.

[37] Collins KL, Guterstam A, Cronin J, et al. Ownership of an artificial limb induced by electrical brain stimulation. Proceedings of the National Academy of Sciences 2017;114:166–71.

[38] Hiremath SV, Tyler-Kabara EC, Wheeler JJ, et al. Human perception of electrical stimulation on the surface of somatosensory cortex. PLoS One 2017;12:1–16.

[39] Lee B, Kramer D, Armenta Salas M, et al. Engineering artificial somatosensation through cortical stimulation in humans. Frontiers in Systems Neuroscience 2018;12:1–11.

[40] Armenta Salas M, Bashford L, Kellis S, et al. Proprioceptive and cutaneous sensations in humans elicited by intracortical microstimulation. Elife 2018;7:1–11.

[41] Roux FE, Djidjeli I, Durand JB. Functional architecture of the somatosensory homunculus detected by electrostimulation. Journal of Physiology 2018;596:941–56.

[42] Tabot GA, Kim SS, Winberry JE, et al. Restoring tactile and proprioceptive sensation through a brain interface. Neurobiology of Disease 2015;83:191–8.

[43] Callier T, Schluter EW, Tabot GA, et al. Long-term stability of sensitivity to intracortical microstimulation of somatosensory cortex. Journal of Neural Engineering 2015;12. https://doi.org/10.1088/1741-2560/12/5/056010.

[44] Flesher SN, Weiss JM, Downey JE, et al. Neurophysiological, psychophysical and electrochemical assessment of intracortical microstimulation stability in human somatosensory cortex. In: Neural interfaces conference; 2016.

[45] Barrese JC, Rao N, Paroo K, et al. Failure mode analysis of silicon-based intracortical microelectrode arrays in non-human primates. Journal of Neural Engineering 2013;10. https://doi.org/10.1088/1741-2560/10/6/066014.

[46] Kim S, Callier T, Tabot GA, et al. Behavioral assessment of sensitivity to intracortical microstimulation of primate somatosensory cortex. Proceedings of the National Academy of Sciences 2015;112:15202–7.

[47] Callier T, Kumaravelu K, Miller LE, et al. Perceptual consequences of changing the frequency of intracortical microstimulation applied to somatosensory cortex. In: Society for neuroscience; 2018.

[48] Hughes CL, Weiss JM, Bensmaia SJ, et al. Human perception of biomimetic intracortical microstimulation in somatosensory cortex. In: Society for neuroscience; 2018.

[49] Cronin JA, Wu J, Collins KL, et al. Task-specific somatosensory feedback via cortical stimulation in humans. IEEE Transactions on Haptics 2016;9:515–22.

[50] Thomson EE, Carra R, Nicolelis MAL. Perceiving invisible light through a somatosensory cortical prosthesis. Nature Communications 2013;4:1–7.

[51] Dadarlat MC, O'Doherty JE, Sabes PN. A learning-based approach to artificial sensory feedback leads to optimal integration. Nature Neuroscience 2015;18:138–44.

[52] O'Doherty JE, Lebedev M a, Ifft PJ, et al. Active tactile exploration using a brain-machine-brain interface. Nature 2011;479:228–31.

[53] Flesher SN, Downey JE, Weiss JM, et al. Effects of intracortical microstimulation feedback on functional task performance during human brain-computer interface control. In: Society for neuroscience; 2017.

[54] Wodlinger B, Downey JE, Tyler-Kabara EC, et al. Ten-dimensional anthropomorphic arm control in a human brain-machine interface: difficulties, solutions, and limitations. Journal of Neural Engineering 2015;12:16011.

[55] Ajiboye AB, Willett FR, Young DR, et al. Restoration of reaching and grasping movements through brain-controlled muscle stimulation in a person with tetraplegia: a proof-of-concept demonstration. Lancet 2017;389:1821–30.

[56] Makin TR, Bensmaia SJ. Stability of sensory topographies in adult cortex. Trends in Cognitive Sciences 2017;21:195–204.

[57] Abbas JJ, Kuntaegowdanahalli SS, Horch K, et al. Assessment of functional benefits afforded by sensory-enabled prostheses to upper-limb amputees. In: Society for neuroscience; 2018.

# 9

# Robotic and neuromuscular electrical stimulation (NMES) hybrid system

Qiuyang Qian[1], Chingyi Nam[1], Wei Rong[1], Waiming Li[1], Ziqi Guo[1], Yanhuan Huang[1], Xiaoling Hu[1], Yongping Zheng[1], Waisang Poon[2]

[1]DEPARTMENT OF BIOMEDICAL ENGINEERING, THE HONG KONG POLYTECHNIC UNIVERSITY, KOWLOON, HONG KONG; [2]DEPARTMENT OF SURGERY, PRINCE OF WALES HOSPITAL, THE CHINESE UNIVERSITY OF HONG KONG, SHATIN, HONG KONG

**Chapter outline**



## Introduction

Stroke is a leading cause of permanent upper limb deficit. Less than 25% of stroke survivors experiencing upper extremity (UE) paralysis achieve near-to-normal motor recovery [1], greatly affecting their daily living.

In recent decades, clinicians and therapists have proved that high-intensity physical practice [2] and precise repetition are effective in speeding up motor restoration after stroke [3]. However, due to the increasing incidence of stroke and constrained resources, even medical and healthcare systems in developed countries struggle to provide adequate treatments for stroke survivors [4]. Therefore, rehabilitation robotics was

Intelligent Biomechatronics in Neurorehabilitation. https://doi.org/10.1016/B978-0-12-814942-3.00009-X

proposed as an important supplement to manual therapies. In recent years, various robotic systems have been developed, leading to an associated increase in research into their clinical effects. Robots that can integrate voluntary effort from residual neuromuscular pathways have demonstrated better rehabilitation effects than those using a continuous passive motion (CPM) training mode [5]. Numerous paralysis-assisting devices incorporate electromyography (EMG) signals as markers of voluntary effort; the EMG-driven strategy is a rapidly expanding technique that maximizes the involvement of voluntary effort [6]. We previously designed a series of exoskeletal robotics that adopt an EMG-driven strategy for physical training at different UE joints [7–10]. Their training effectiveness was demonstrated in the related clinical studies, which showed significant UE motor improvements.

However, we noticed that the robot alone could not directly activate the desired muscles or muscle groups and could hardly limit the compensatory contraction in the other nontarget muscles [11]. Meanwhile, another prevalently used technique in stroke rehabilitation, the neuromuscular electrical stimulation (NMES), could apply programmed electrical currents directly to the paretic UE muscles, thereby generating limb movements. The NMES could not only effectively prevent muscle atrophy and improve muscle strength, but also precisely stimulate the target muscles, as reported in systematic reviews [12]. Based on other research, when the paralyzed or paretic muscle contracts, NEMS was also found to evoke sensorimotor feedback to the brain, and modulate the plasticity of the central nervous system [13]. Despite its advantages in stroke rehabilitation, the problem with using NMES alone is its kinematic inaccuracies, such as motion speed and trajectories. Hence, in the subsequent work, we integrated the NMES into the EMG-driven robots (i.e., the EMG-driven, NMES-robotic system) [14–16].

According to recent research, task-oriented training with multijoint coordination in the paretic upper limb has proven to be effective in converting motor improvements into meaningful limb functions for stroke rehabilitation [17]. However, it is hard for a single human therapist adopting traditional "one-to-one" manual operations, to manage simultaneously the coordinated movements of multiple joints (e.g., fingers, wrist, and elbow joints). Yet, as well as fulfilling the training requirements, the EMG-driven NMES-robotic systems developed in our previous studies could also provide different supportive schemes in multijoint coordinated UE treatments. In this chapter, we introduce two EMG-driven NMES exoskeleton robots that illustrate multijoint coordinated training effects under different supportive schemes during post-stroke UE rehabilitation.

## EMG-driven NMES-robots

The devices introduced in this chapter are wearable exoskeleton robotic devices that support distal hand/finger practice (i.e., the EMG-driven NMES-robotic hand) [18] and wrist-and-elbow practice (i.e., the EMG-driven NMES-robotic sleeve) [16].

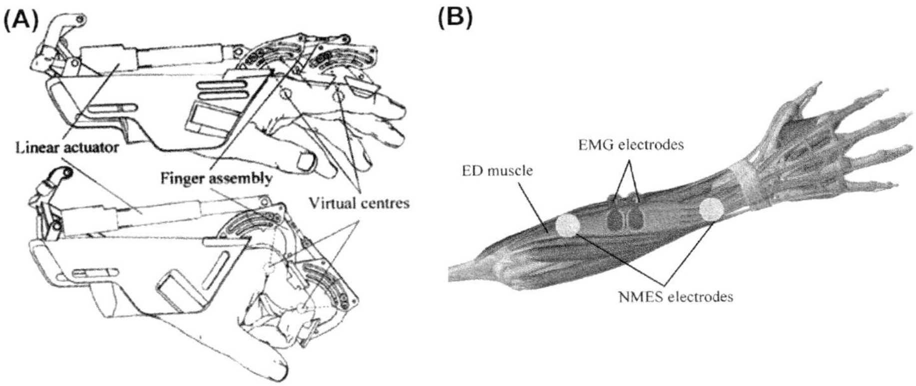


**FIGURE 9.1** The system structure of the EMG-driven NMES-robotic hand system and the configuration of the EMG and NMES electrodes [18].

## EMG-driven NMES-robotic hand

The EMG-driven NMES-robotic hand is shown in Fig. 9.1. The system harnesses voluntary EMG from paretic upper limbs to continuously and interactively assist finger extension and flexion in stroke patients.

The mechanical part of the system mainly consisted of a wearable palm–wrist module and five individual finger assemblies (Fig. 9.1A). Each assembly provides individual mechanical assistance to the finger by a linear actuator (Firgelli L12, Firgelli Technologies Inc., Ferndale, USA) [19]. The proximal and distal sections of the fingers, including the index, middle, ring and little fingers, were rotated around the virtual centers located at the metacarpophalangeal (MCP) and proximal interphalangeal (PIP) joints. The thumb was rotated around the virtual center of its MCP joint. Each finger assembly could provide a range of motion (ROM) of 55 degrees for the MCP joint and 65 degrees for the PIP joint. During the training, the angular rotation speeds of the two joints were set as 22 degrees/s and 26 degrees/s at the MCP and PIP joints, respectively. The assistance from the robotic hand was defined as [19]:

$$Assistance_{Robot,\ phase} = (R+1)\cdot V_{max}\cdot \overline{M}(t)_{i,\ phase} \tag{9.1}$$

where $Assistance_{Robot,\ phase}$ was the mechanical support to assist finger movements in each motion phase (i.e., finger extension and finger flexion in the NMES-robotic hand training). $R$ represents a constant gain factor (set from 0% to 100%) and $V_{\max}$ was the maximal moving velocity of the finger assemblies at the MCP joint. $\overline{M}(t)_{i,\ phase}$ is discussed later.

The configuration of the one-channel NMES electrode pair (Axelgaard Corp., Fallbrook, CA, USA) is demonstrated in Fig. 9.1B. Electrodes were attached to the skin surface of the extensor digitorum (ED) muscle and stimulated by square-wave electrical pulses at a constant amplitude of 70 V, 40 Hz frequency, and a manually adjustable pulse width in the range of 0–300 μs. No assistance from NMES was provided to the flexor digitorum (FD) muscle during finger flexion movements. The rationale for this design is because stroke patients commonly experience muscle weakness moving UE extensors such as opening the hand, and suffer abnormal contractions or muscle spasticity in the UE flexors (e.g., FD during hand closing). The assistance from the NMES was defined as [19]:

$$Assistance_{NMES,\ extension}\,(t) = W_{i,\ max} \cdot \overline{M}(t)_{i,\ extension} \tag{9.2}$$

where $Assistance_{NMES,\ extension}(t)$ is the real-time stimulation pulse width in the finger extension phase during the NMES-robotic hand practice. $W_{i,\ max}$ represents the maximum NMES impulse width applied to the agonist driving muscle $i$ (i.e., the ED muscle), individually set to the intensity threshold needed to achieve the maximal extended position of the fingers during NMES-robotic hand training. Stimulation also evokes sensory feedback to the brain and helps stroke subjects localize target muscles during the training [15,19]. $M(t)_{i,\ phase}$ in Eqs. (9.1) and (9.2) is defined as [19]:

$$M(t)_{i,\ phase} = \frac{EMG_i\,(t) - EMG_{i,\ Rest}}{EMG_{i,\ Max} - EMG_{i,\ Rest}} \tag{9.3}$$

where $EMG_i(t)$ represents the normalized real-time EMG level of the agonist driving muscle $i$ (i.e., the ED muscle during the NMES-robotic hand training). $EMG_{i,\ Rest}$ is the averaged EMG of the target muscle in its resting state, and $EMG_{i,\ Max}$ is the maximal EMG value of the muscle during its isometric maximum voluntary contractions (IMVCs).

To control the robot and NMES assistance in the NMES-robotic hand training, all EMG signals captured by electrode pairs (Blue Sensor N, Ambu Inc., Columbia, USA) were used as a voluntary neuromuscular driver. The EMG electrode pairs (20 × 30 mm, with a 2 cm gap between pairs) were attached as previously described to the bellies of the ED and FD muscles (Fig. 9.1B). All EMG signals obtained were first amplified 1000 times (preamplifier: INA 333; Texas Instruments Inc., Dallas, TX, USA), then band-pass filtered from 10 to 500 Hz, and subsequently sampled at 1000 Hz for digitization using a data acquisition card (DAQ, 6218 NI DAQ card; National Instruments Corp., Shatin, Hong Kong). After digitization, to obtain an envelope of EMG signals (i.e., the EMG activation level) in the real-time control, the EMG signals from the ED and FD muscles were rectified and low-pass filtered (fourth-order, zero-phase forward and reverse Butterworth filter; cutoff frequency, 10 Hz).

In each extension or flexion motion phase, once the EMG activity of a driving muscle exceeded a preset threshold [three times the standard deviation (SD) above the EMG baseline at rest], continuous assistance from the robot was provided. This was triggered by the real-time EMG of the FD muscle in the flexion phase and ED muscle in the extension phase. The NMES simultaneously delivered to the ED muscle with motor support in the finger extension phase only.

## EMG-driven NMES-robotic sleeve

The EMG-driven NMES-robotic sleeve is shown in Fig. 9.2. The robot part consists of two exoskeleton modules, the wrist module and the elbow module. To accommodate post-stroke joint stiffness and muscle spasticity, the modules were not mechanically

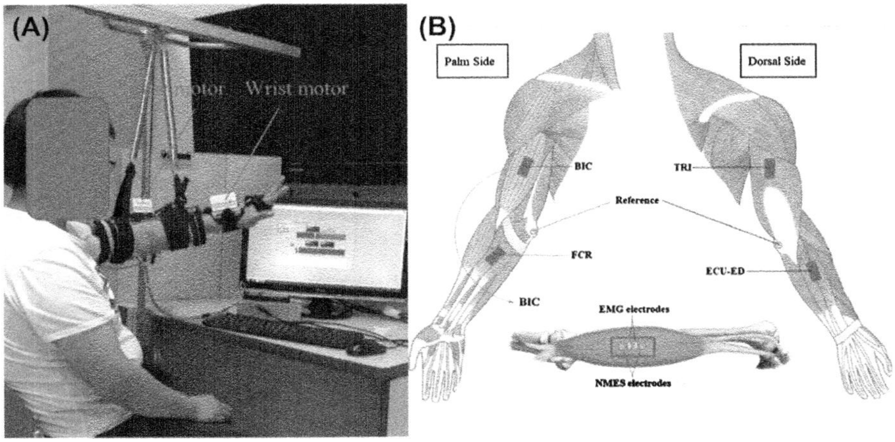


**FIGURE 9.2** The experimental setup of the EMG-driven NMES-robotic sleeve during motor training and the configuration of EMG and NMES electrodes in the sleeve-assisted training [15,16].

connected, meaning they could fit diverse ergonomic parameters (e.g., limb length and pronation angles away from the neutral position at the wrist). The movement of each robotic module was controlled by an independent servo motor (MX 106, ROBOTIS) and could provide joint flexion and extension motions [15]. The wrist module orthosis covered the hand side of the palm, leaving the fingers free for flexion and extension, while the elbow module orthosis covered the entire elbow joint, including the distal and proximal sides of the arm. The maximum motor-controlled ROM for the wrist module was from 45 degrees of wrist extension to 60 degrees of wrist flexion. For the elbow module, the maximum ROM was from 30 degrees of elbow flexion to 180 degrees of elbow extension. The ROMs for both joints were based on findings from our previous works. During the sleeve training, the servo motors would assist movement at a constant velocity of 10 degrees/s (i.e., the Vmax in Eq. 9.2).

In contrast to the NMES-robotic hand, two-channel NMES was applied in the sleeve setting and NMES electrodes were applied to the bellies of biceps brachii (BIC) for elbow flexion, triceps brachii (TRI) for elbow extension, and flexor carpi radialis (FCR) for wrist flexion; the last channel was applied to the common area of the extensor carpi ulnaris (ECU) and ED muscles for wrist extension and associated hand opening. The design for the ECU-ED muscle union was based on the proximal anatomical position of ECD and ED muscles that have narrow muscle bellies on the dorsal side of the forearm, and these muscles can be recruited synchronously by one-channel surface NMES [18]. The stimuli parameter settings (i.e., amplitude, frequency, and pulse) of the NMES applied to UE extensors were the same as for the NMES-robotic hand. Meanwhile, to evoke sensory prompting for the motion initiation, minimal dose electrical stimulation was applied to UE flexors; no NMES or robotic support was provided to the hand close motion. The design was similar for the NMES-robotic hand to accommodate the common experience of stroke survivors in having difficulties in hand opening and arm extension rather than hand closing and arm flexion. Inappropriate stimulation on the UE flexors may accelerate the frequency of muscle spasms.

The same control algorithm was used for the NMES-robotic hand and the sleeve. The mechanical assistance from both robotic modules of the robotic sleeve and assistance from the NMES part were under the control of the EMG detected from BIC, TRI, and FCR muscles and ECU-ED muscle union. Once the assistance of the system was initiated, voluntary effort from the patient was no longer required in the subsequent training. The assistance of both NMES and robotic parts was provided continuously to support sequential limb movements, i.e., (1) elbow extension; (2) wrist extension and related hand opening; (3) wrist flexion; and (4) elbow flexion, which simulates the coordination of joints in arm reaching, hand grasping, and arm withdrawing motion. The configuration of NMES and EMG electrodes is illustrated in Fig. 9.2B. The EMG electrode pair and the NMES electrode pair were placed perpendicular to each other on a target muscle belly, an empirical configuration that minimized stimulation artifacts during EMG recording.

When using the NMES-robotic sleeve, the participant's system-mounted, paretic upper limb was lifted to a horizontal position by a hanging system (Fig. 9.2A); this offset the gravity effects of the robotic system. This was essential as few participants, especially those patients with subacute stroke, could sustain the weight of both their limb and the robotic system.

## Clinical trials

Three clinical trials with ethic approval from the Human Subjects Ethics Subcommittee of Hong Kong Polytechnic University were conducted for investigation of the training effectiveness of both robots and to make the comparison between different joint-supportive strategies.

### Early stroke UE rehabilitation by the EMG-driven NMES-robotic sleeve

It is already known that after a stroke episode, spontaneous motor recovery usually occurs in the first days to weeks, reaching a plateau within 6 months (the subacute period) [20]. Physical practice during this period could optimize neural plasticity progress and speed up the recovery of UE motor functions [21]. Therefore, we applied the EMG-driven NMES-robotic sleeve to stroke in-patients with subacute staging, providing multijoint coordinated training to their paretic upper limbs. The feasibility of the system and its training effectiveness were evaluated in a pilot randomized control trial that used routine physical rehabilitation treatments as a control.

Twenty-four subacute stroke patients were recruited into the study and randomly assigned to either the experimental group ($n = 14$) or control group ($n = 10$). The training loads between the two groups were matched by the duration and frequency of upper limb training sessions. The duration of each daily session was 1 h (including rest time), totaling five sessions/week for 4 weeks (20 training sessions in total). In the experimental group, each participant received multijoint coordinated UE training assisted by the NMES-robot sleeve. The clinical assessments used in this study were the Fugl-Meyer Assessment (FMA: full score, shoulder–elbow and wrist–hand subscores), the Modified Ashworth Scale (MAS: elbow, wrist, and fingers), the Action Research Arm Test (ARAT), and the Functional Independence Measurement (FIM) to assess all participants. Clinical pretraining assessments were conducted before the training started, immediately after the 20th training session (posttraining assessment), and 3 months after the training (3-month follow-up, 3MFU). Assessments were performed by a therapist who was blinded to the training progress.

The training effects of early stroke UE rehabilitation by the EMG-driven NMES-robotic sleeve were reflected by the clinical evaluations (Fig. 9.3). Significant

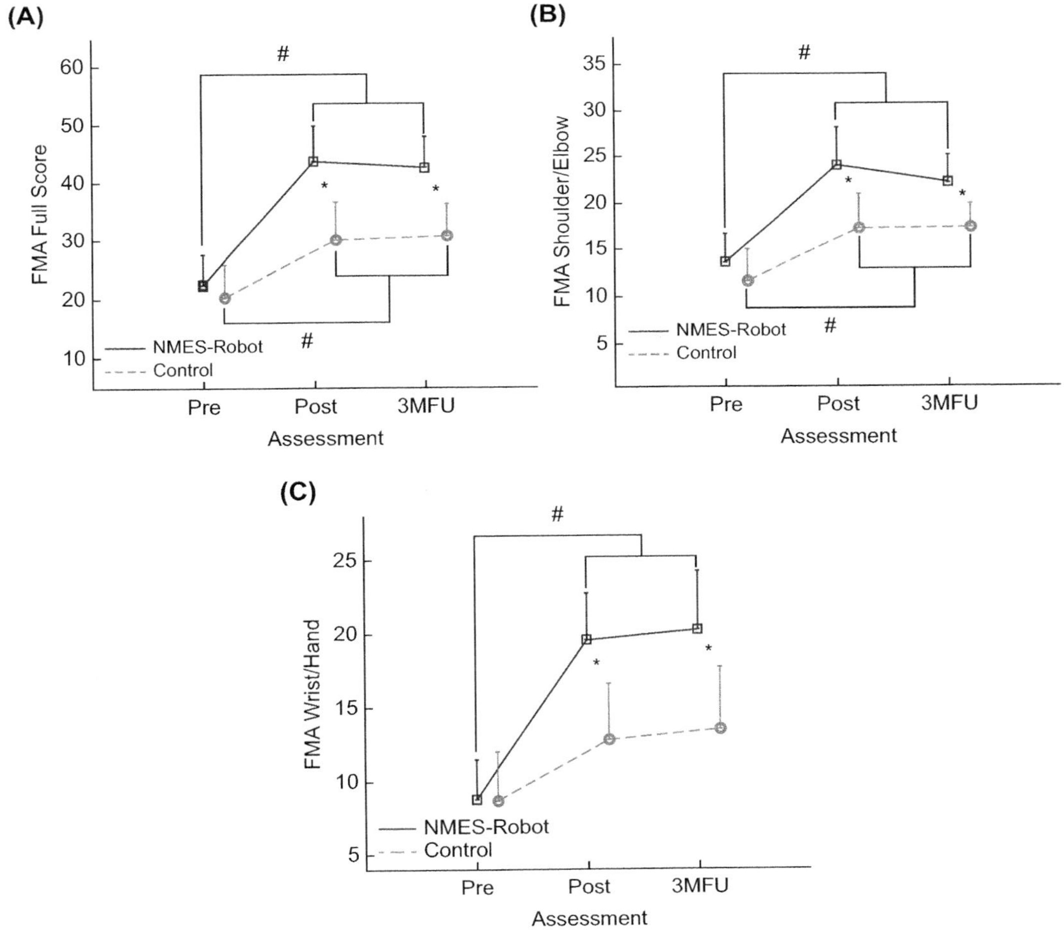


**FIGURE 9.3** Overview of the variations in clinical assessments in early stroke training by NMES-robotic sleeve [16]. The '#'s indicate significant intra-group differences and the '*' show the significant inter-group differences.

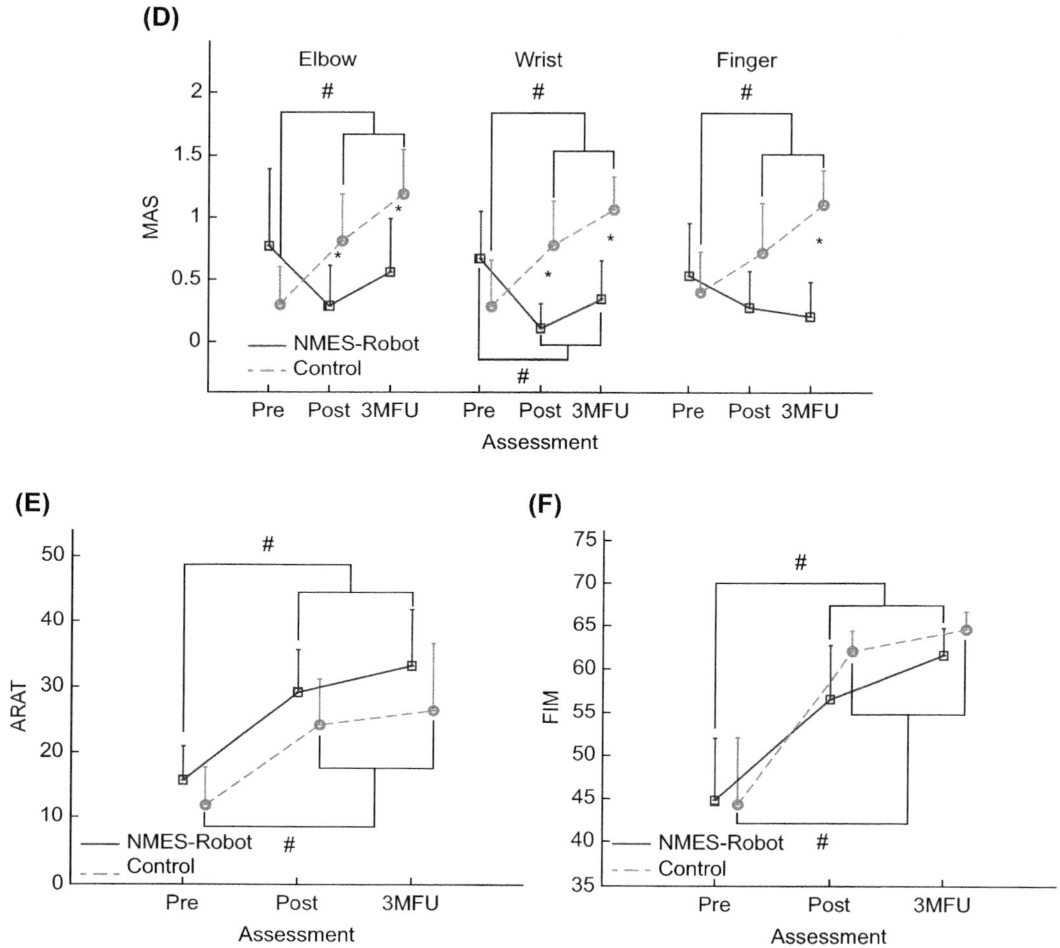


**FIGURE 9.3** Cont'd

increases in FMA scores ($P < .05$, one-way ANOVA) were observed in both groups, suggesting improvement in the entire UE regardless of the type of intervention (Fig. 9.3A). Furthermore, the motor improvements in both groups were maintained for 3 months after training. However, the results also show that after NMES-robotic sleeve training, the 3MFU measurements were markedly higher ($P < .05$, one-way ANCOVA) in the control group. For the FMA shoulder–elbow subscores, the two groups demonstrated similar patterns to those shown in the FMA full scores; that is, there was significant improvement after training, with higher increments in the experimental group, $P < .05$ (Fig. 9.3B). While for the FMA wrist–hand subscore, significant improvement was only obtained in the NMES-robotic sleeve group ($P < .05$, one-way ANOVA) and absent in the control group (Fig. 9.3C). These results illustrate that NMES-robot-assisted training can promote superior motor outcomes in all segments, but especially at the more distal ones. Fig. 9.3D shows MAS scores differ significantly, with the two groups having distinct patterns. The decrease in MAS score in the experimental group at all three parts (i.e., elbow, wrist, and fingers) reflects the release of muscle spasticity around the associated joints, although significant decrease ($P < .05$, one-way ANOVA) was only observed at the wrist joint. For the control group, the MAS increased significantly at all three parts ($P < .05$, one-way ANOVA), indicating exacerbation of muscle spasticity after traditional UE rehabilitation treatments. Compared to the FMA measurement, the increase in ARAT (Fig. 9.3E) also indicates improvement of motor function in the upper limb but improvement was more concentrated in finger and hand movements. After the training, both groups showed significant improvement ($P < .05$, one-way ANOVA) and the outcomes were maintained for 3 months. The changes in FIM in both groups over three evaluation time points ($P < .05$, one-way ANOVA) (Fig. 9.3F) reflect the remarkable recovery of independence in undertaking motor tasks related to daily activities. Significant between-group differences were observed in FMA scores (full score and wrist–hand subscore) and MAS (elbow, wrist, and fingers), indicating that training with the NMES-robotic sleeve promoted distal UE motor function recovery and more effective release of muscle spasticity in the entire UE compared to the traditional rehabilitation treatments.

Besides the clinical scores, the session-by-session evaluation of EMG parameters on BIC, TRI, and FCR muscles and ECU-ED muscle union were monitored to trace the underlying recovery process in the experimental group. An evaluation was conducted in each of the 20 training sessions before the physical training. Fig. 9.4 demonstrates significant changes in EMG parameters, i.e., the EMG activation level of FCR muscle as well as the CI values of FCR-BIC. The EMG activation level was defined as follows [16]:

$$\overline{EMG} = \frac{1}{T}\int_0^T EMG_t\,(t)dt \tag{9 4}$$

where $\overline{EMG}$ is the averaged EMG envelope value of the target muscle *i*, $EMG_t(t)$ represents the normalized EMG values of the muscle during the IMVC evaluation before each training session, and $T$ refers to the length of the signal. The CI value between a pair of muscles $i - j$ was calculated as follows [16]:

$$\mathrm{CI} = \frac{1}{T}\int_0^T A_{ij}(t)dt \tag{9 5}$$

where $A_{ij}(t)$ refers to the overlapping activity of muscles $i$ and $j$ (i.e., minimum [$EMG_i(t)$, $EMG_j(t)$]), and $T$ represents the length of the overlapping region of the two EMG envelopes. The two EMG parameters were further normalized with respect to the maximal values of an individual across his/her whole training sessions. This operation was to minimize the variations among different subjects, and frame the range of values from 0 to 1.

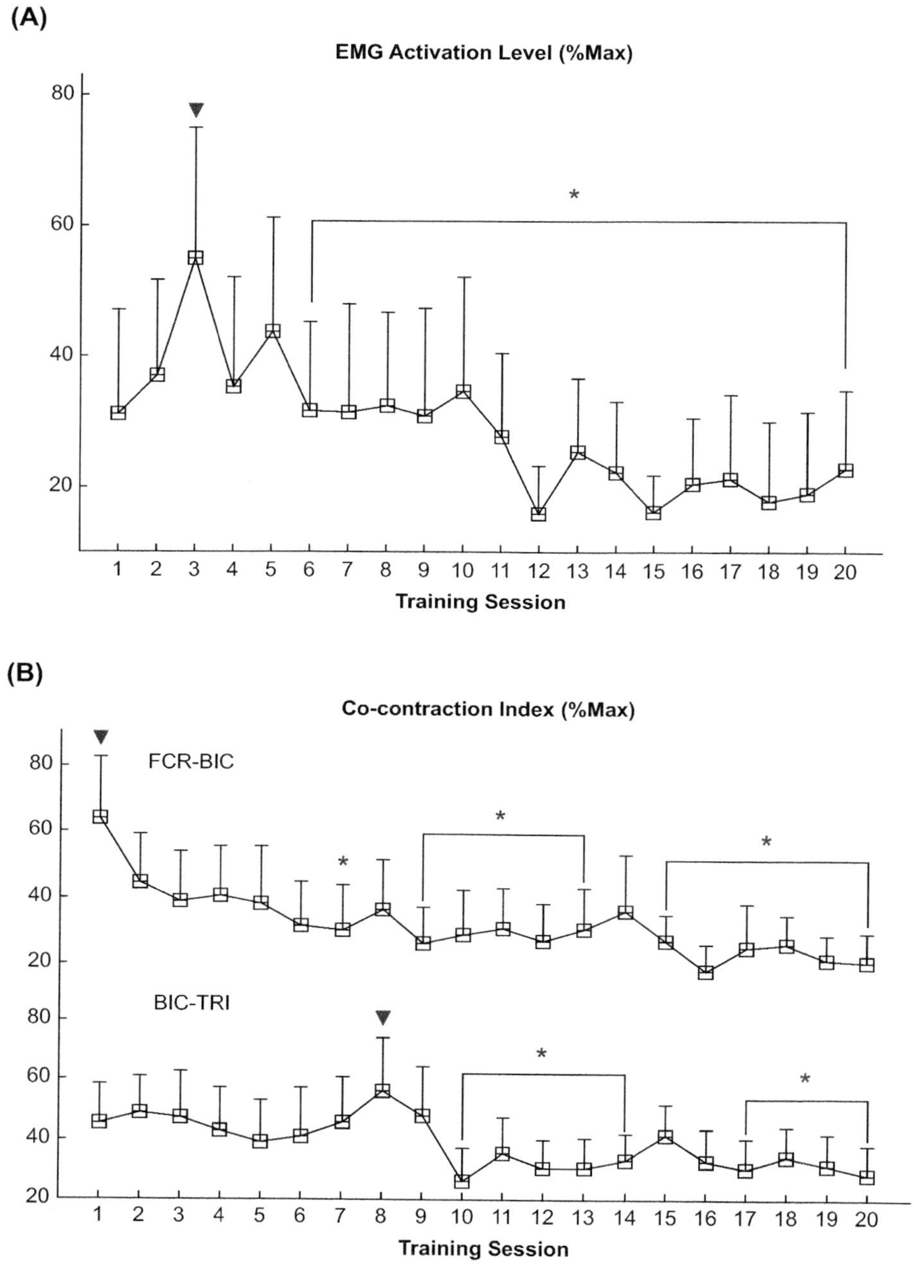


**FIGURE 9.4** The significant variations of the FCR-related EMG parameters in both EMG activation level and cocontraction index values in the early stroke sleeve-assisted UE rehabilitation [16]. The '▼'s indicate the peak values, and the '*'s show the values significantly different from the peak values.

The significant decrease in the EMG activation level in FCR ($P < .05$, one-way ANOVA) reveals that the flexors were released from excessive use during wrist motion tasks (Fig. 9.4A). Meanwhile, the reduction could also have affected the changes in CI values of the FCR–BIC muscle pair, indicating a significant improvement of motion independency between the elbow and wrist joints ($P < .05$, one-way ANOVA, Fig. 9.4B). Another muscle pair, the BIC–TRI, also showed decreasing CIs over the 20 training sessions ($P < .05$, one-way ANOVA). The evidence suggests there is a release of cocontraction patterns between UE extensor and elbow joint-related flexor muscles. The EMG parameters were consistent with the MAS measurements, demonstrating ameliorated spasticity release in the whole arm and improved muscle coordination within single joints or multiple UE joints.

## Application of the EMG-driven NMES-robotic hand in chronic stroke

A large proportion of resources was provided for the early stroke rehabilitation with the concept that motor recovery of the paralyzed limbs should be minimal or plateaued in the chronic phase [22]. However, recent studies have shown the possibility of motor restoration in patients with chronic stroke after high-intensive physical practice. We supposed that the robot-assisted therapy could also work well in long-term stroke rehabilitation and tested the NMES-robotic hand, first through a single-group trial. This trial design was selected as most chronic stroke patients cannot use the hand on the affected side for daily activities and, due to limited resources, conventional rehabilitation strategies rarely provide them with physical practice. Many patients exhibited "learned nonuse" or "learned disuse" with compensatory movements in the entire UE.

The purpose of this work was to investigate the effectiveness of the training of the NMES-robotic system on the heavily affected UE segments, such as the distal finger–hand. Our hope was that motor function gained after the training could contribute to performing daily activities independently.

Fifteen chronic stroke patients with UE impairments were recruited for the 20-session NMES-robot-assisted training. The training comprised three to five sessions per week within seven consecutive weeks. Three separate clinical FMA, MAS, ARAT, and FIM assessments were performed prior to training; further assessments were carried out at training completion and at 3-month follow-up (Fig. 9.5). With the exception of FIM, all clinical scores changed significantly through the three evaluation points. The increase in FMA (full score, shoulder–elbow and wrist–hand subscores) suggests an improvement in voluntary UE motor functions in the related segments. The significant decrease in MAS values after training and 3MFU indicate reduced muscle spasticity in the entire upper limb. The increment in ARAT scores was highly related to improved finger–palm motor function, e.g., pinching, gripping, and grasping. No significant changes in the FIM were observed in this work. Similar results are reported in published literature [23]; they describe the difficulties faced by chronic stroke patients in converting motor recovery into functional improvements in daily activities. In addition to the clinical scores, EMG

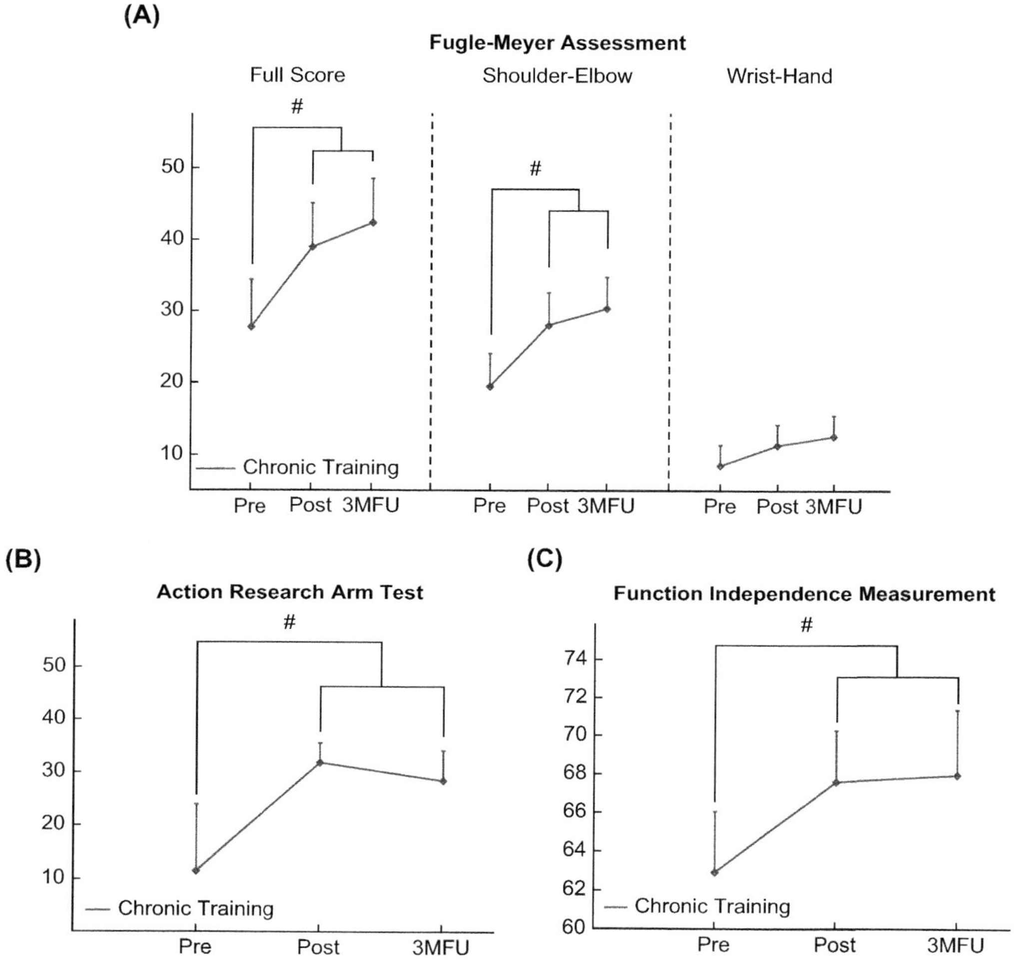


**FIGURE 9.5** Overview of the variations in clinical assessments measured before, immediately after, and 3 months after the NMES-robot-assisted hand training [18]. The '#'s indicate significant intra-group differences.

evaluations were performed before conducting each training session. During the evaluation, four pairs of EMG electrodes were to the bellies of ED, FD, BIC, and TRI muscles; this compares to the two pairs used on ED and FD muscle during practice. Fig. 9.6A shows that EMG activation levels in FD and BIC muscles decreased significantly, indicating a release of excessive movement in the UE flexors during multijoint coordinated motor tasks. Meanwhile, Fig. 9.6B illustrates a significant reduction in CI values of the FD–TRI and BIC–TRI muscle pairs. The former indicates improvement in the activation independence between wrist and elbow joints, while the latter shows a release of muscle cocontraction between the flexor and extensor of the single elbow joint in more proximal UE.

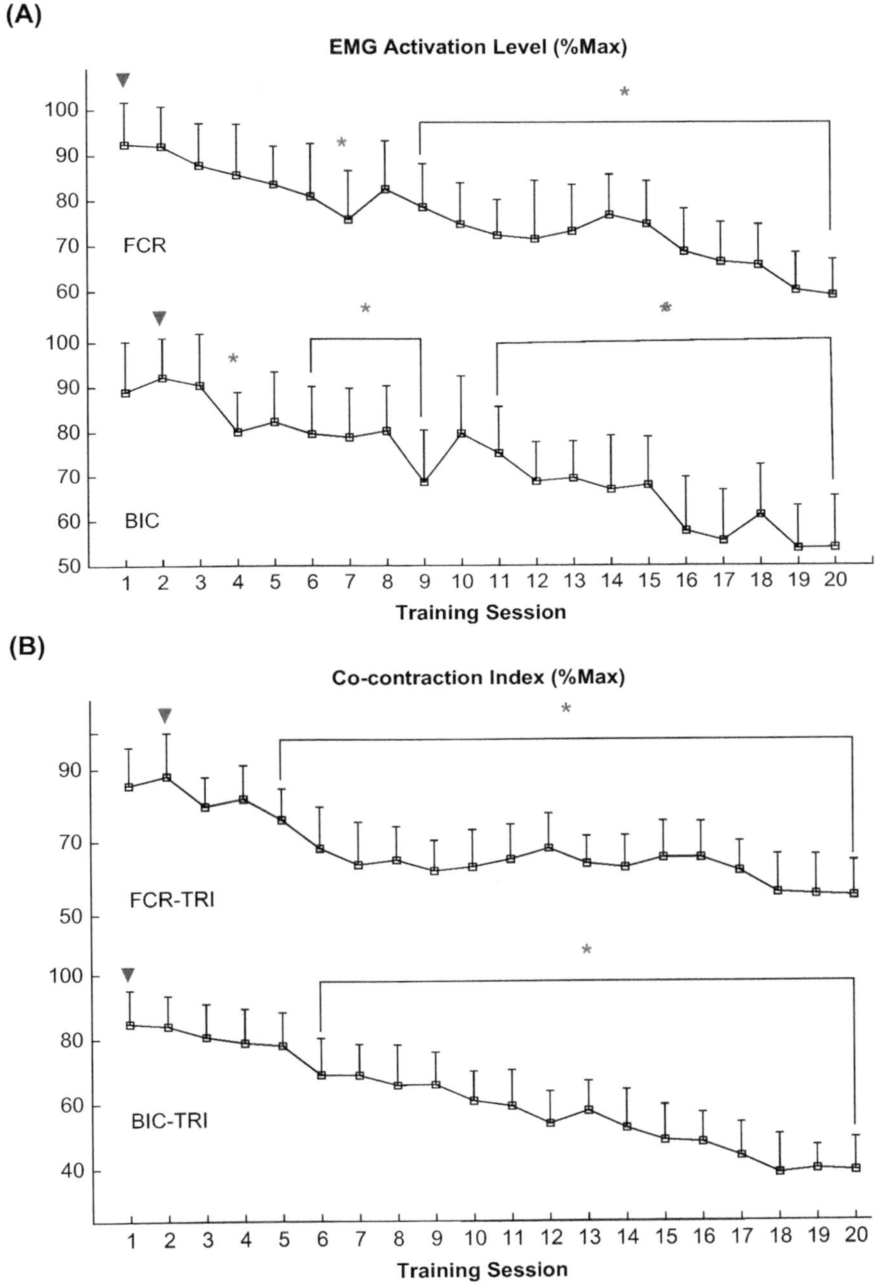


**FIGURE 9.6** The significant variations of the EMG parameters in the chronic stroke robotic-hand-assisted UE trainings [18]. The '▼'s indicate the peak values, and the '*'s show the values significantly different from the peak values.

The EMG-driven NMES-robotic hand managed finger motions while participants practiced multijoint, coordinated upper limb tasks. The robot-assisted training achieved motor improvements at both distal UE, as indicated by FMA wrist–hand, MAS finger, MAS wrist and ARAT, and proximal UE, as illustrated by FMA shoulder–elbow, MAS elbow, and ARAT segments.

## Comparison of different joint-supportive schemes in chronic stroke

Despite the findings of our studies that indicate the NMES-robotic system training is effective in supporting multijoint coordinated physical practice, early robot-assisted therapy studies report inconsistent results [24–27]. Those studies involving voluntary effort reported better training outcomes with assistance from rehabilitation devices, while others adopting CPM mode have negative results. Reports of the sustainability of the training effects over time are also inconsistent. Bovolenta reported the duration of effectiveness to be less than 3 months' after one course of robot-assisted training. Meanwhile Housman found the endurance effect could exceed 6 months. Beside differences between the robot's mechanical structures, a key reason for diverse rehabilitation results could be that the robot-assisted treatments have different supportive strategies. Few studies have investigated this aspect using the same robotic systems or control algorithm. As the patients in this study were usually in a stable state with rare spontaneous recovery, we used both the EMG-driven NMES-robotic hand and the NMES-robotic sleeve to provide different training support for the same motor tasks. We compared the training effectiveness of distal finger–palm support and proximal wrist–elbow support through a randomized controlled clinical trial.

Thirty patients with chronic stroke were recruited into the study and randomly assigned to two group, the NMES-robotic hand group ($n = 15$) and the NMES-robotic sleeve group ($n = 15$). All participants received the same 20-session robot-assisted UE practice, consisting of sequenced and repeated motion tasks to simulate daily activities of multijoint coordinated arm-reaching and arm-withdrawing. Each 70-min training session was comprised of two 30-min training periods with a 10-min break between the two consecutive sections. Participants were provided with five training sessions per week and completed the 20 sessions in 5 weeks.

Fig. 9.7 illustrates the clinical scores of FMA, MAS, and ARAT in both groups. As the results of FIM scores were close to normal for most of the patients, there was no indication of change in their functional independence in the study. The clinical assessments were evaluated three times, as previously described. The values of the FMA full score (Fig. 9.7A) and FMA shoulder–elbow (Fig. 9.7B) were significantly increased in both groups after the treatments ($P < .05$, one-way ANOVA) and the results were maintained for at least 3 months. Although the FMA wrist–hand scores (Fig. 9.7C) were increased in the NMES-robotic hand group ($P < .05$, one-way ANOVA), they were all but absent in the NMES-robotic sleeve group ($P > .05$). The results indicated better motor improvement of distal UE segments with direct support from the NMES-robotic hand. The significant reductions of MAS that were detected in all three parts (i.e., the elbow, the wrist, and the fingers) were also observed in the NMES-robotic hand group ($P < .05$, one-way ANOVA, Fig. 9.7D). There were significant between-group differences in the values of MAS wrist at 3MFU assessment ($P < .05$, one-way ANCOVA) and in the values of MAS finger at post-assessment and 3MFU ($P < .05$, one-way ANCOVA). The results illustrate there was more effective release of muscle spasticity in the entire upper limb with the NMES-robotic support to the distal segments. Furthermore, all the achievements could be maintained for 3 months. Similar patterns were detected for the ARAT scores, with both groups showing significant improvements ($P < .05$, one-way ANOVA) over the three evaluation times; there was no between-group difference ($P > .05$) in this measurement. Improvement in scores is attributed to distal support promoting finger motor functions, such as pinching and gripping; meanwhile proximal support promoted gross joint motor function, for example, lifting the paretic arm up, both of which could affect the results of ARAT scores. The EMG activation level and CI values, which form the two EMG parameters, were also measured in the 20 training sessions (Fig. 9.8). During the before-training EMG evaluation, the electrodes for forearm extensors and forearm flexors were attached to the common area of the FCR–FD muscle pair and ECR–ED muscle

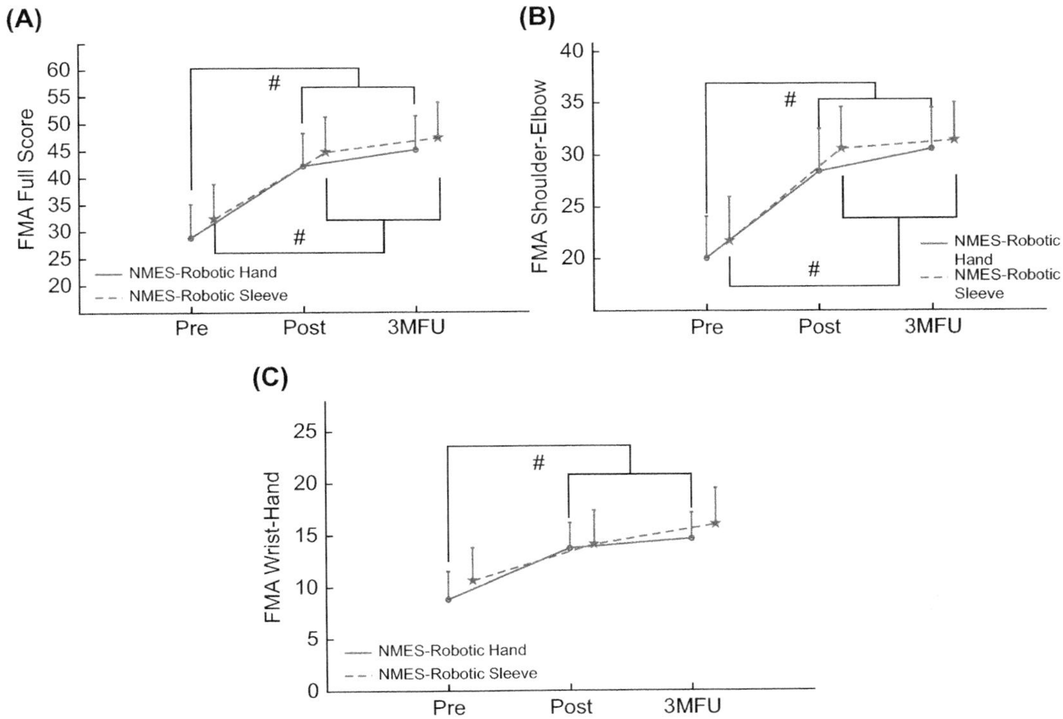


**FIGURE 9.7** Overview of the variations in clinical assessments under different supportive schemes in chronic stroke UE rehabilitation. The '#'s indicate significant intra-group differences and the '*' show the significant inter-group differences.

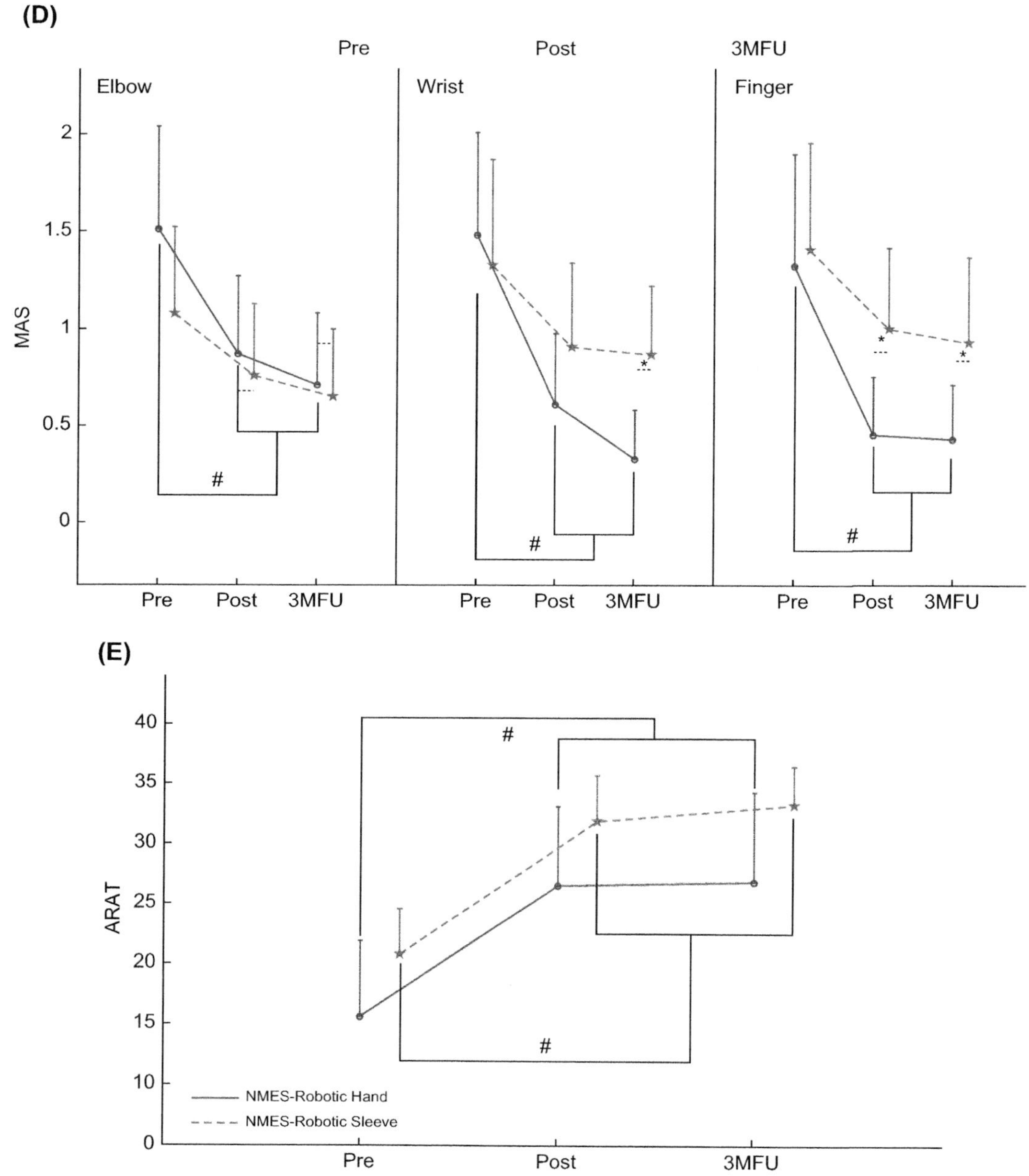


**FIGURE 9.7** Cont'd

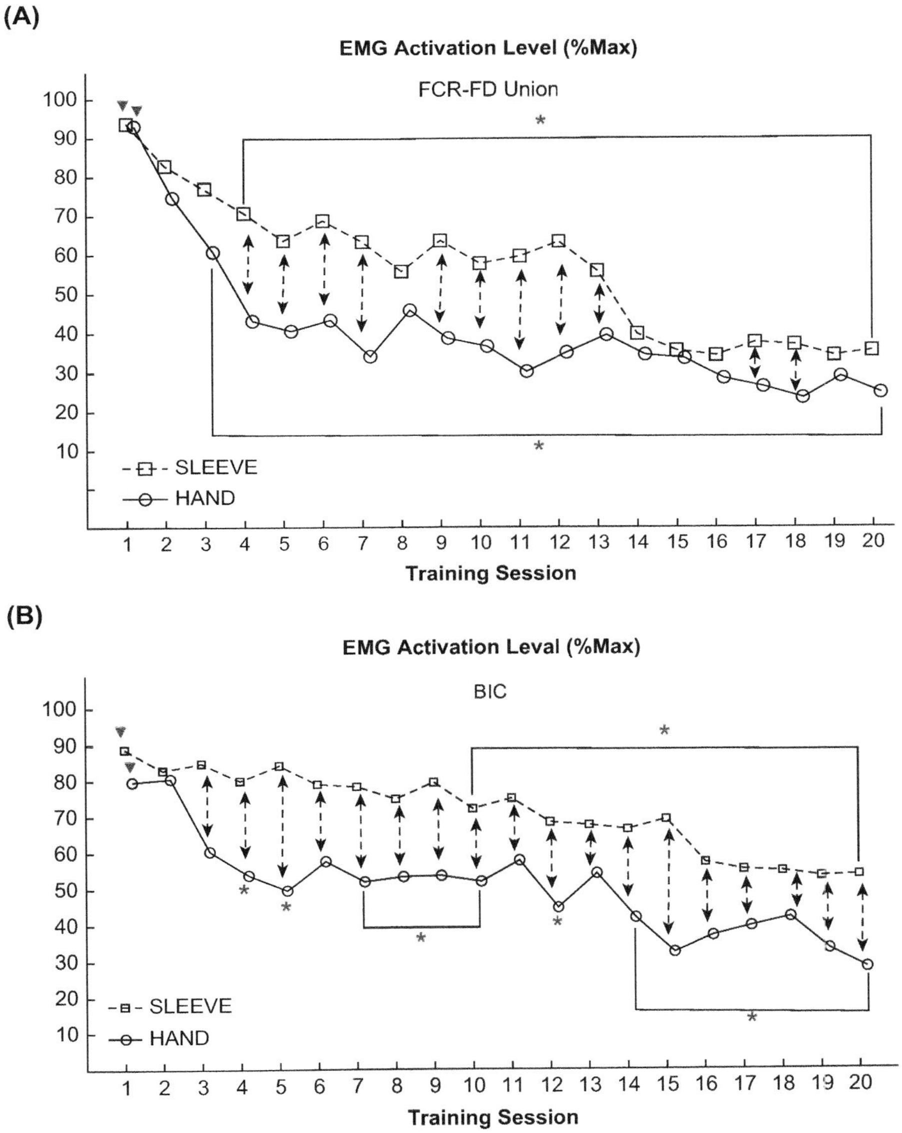


**FIGURE 9.8** The significant variations of the EMG parameters in comparison of both supportive schemes (i.e., finger–hand support in the Hand group and wrist–elbow support in the Sleeve group). Dotted double arrow presents the significant between-group difference (t-test, $P < .05$) in each training session. "▼" illustrates the peak of the EMG parameters through 20 training sessions. "*" demonstrates the significant intragroup differences in each group.

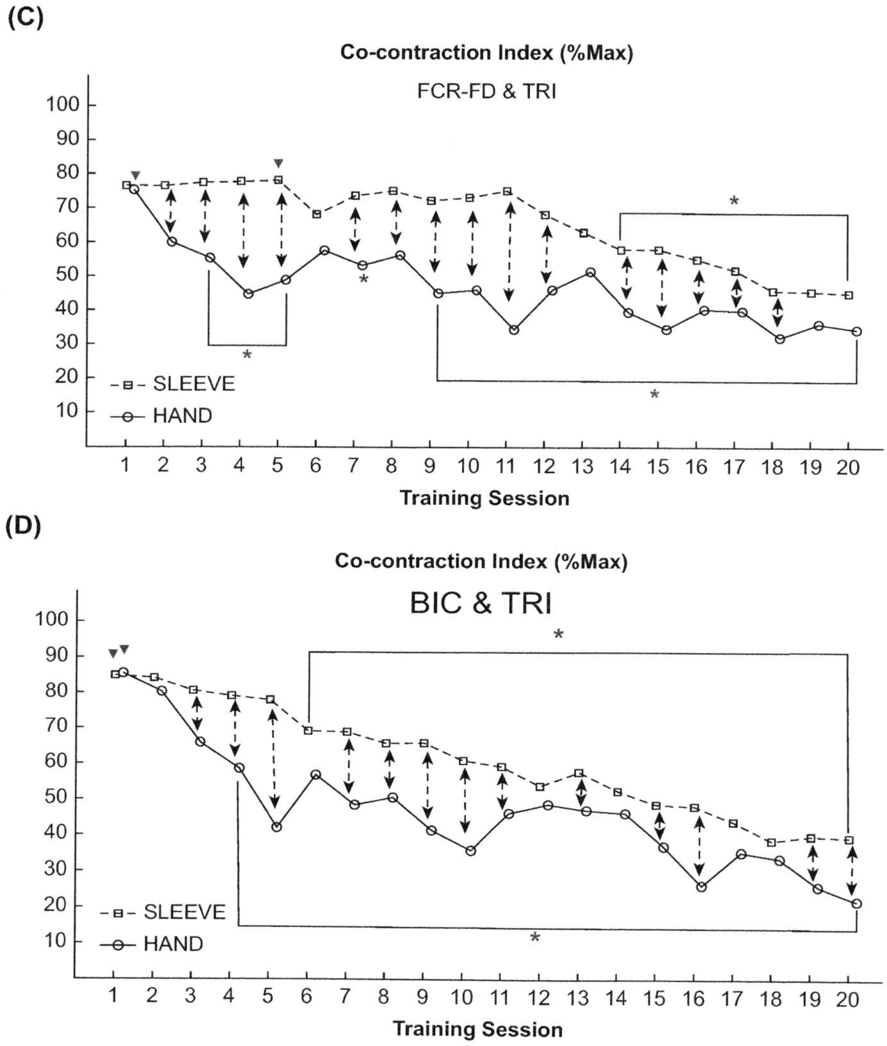


**FIGURE 9.8** Cont'd

pair, respectively. This design was selected because of the close anatomical structures of each pair of muscles and related function in the motion tasks. In this study, the EMG activation levels of FCR muscle and BIC muscle were significantly reduced ($P < .05$, one-way ANOVA) over the 20 sessions, indicating the release of excessive contraction of the UE flexors during the motor tasks. Meanwhile, the CI values of (FCR-FD)–TRI and BIC–TRI muscle pairs decreased significantly during the training process ($P < .05$, one-way ANOVA). This shows a contraction separation of flexor in the distal UE and extensor in the proximal UE and a release of muscle cocontraction patterns of proximal flexor and extensor around the elbow joint. At most training sessions, the values of the EMG activation level and CI values were lower in the NMES-robotic hand group than in the sleeve group ($P < .05$, t-test). These results indicate that direct support to the distal segments could result in better motor outcomes in the UE with respect to improving distal and proximal motor function. Also, the release of muscle spasticity in the whole arm and release of excessive muscle contractions in the UE flexors was ameliorated.

# Conclusion

This chapter introduced the EMG-driven NMES-robots for promoting post-stroke upper limb rehabilitation. The effectiveness of the training system was verified by the three clinical trials for UE motor rehabilitation, in patients with either early stroke or chronic stroke. We also found that providing direct support to the distal fingers was more effective than proximal support to the wrist–elbow joints with the assistance from the NMES-robotic system.

# References

[1] Dobkin BH. Rehabilitation after stroke. New England Journal of Medicine 2005;352(16):1677–84.

[2] Harris J, Eng J. Strength training improves upper-limb function in individuals with stroke: a meta-analysis. Stroke 2010;41(1):136–40.

[3] Hung CS, et al. The effects of combination of robot-assisted therapy with task-specific or impairment-oriented training on motor function and quality of life in chronic stroke. Physical Medicine and Rehabilitation 2016;8(8):721–9.

[4] Woo J, et al. In patient stroke rehabilitation efficiency: influence of organization of service delivery and staff numbers. BMC Health Services Research 2008;8(1):86.

[5] Volpe BT, et al. Robotics and other devices in the treatment of patients recovering from stroke. Current Atherosclerosis Reports 2004;6(4):314–9.

[6] Basteris A, et al. Training modalities in robot-mediated upper limb rehabilitation in stroke: a framework for classification based on a systematic review. Journal of NeuroEngineering and Rehabilitation 2014;11:111.

[7] Hu XL, et al. A comparison between electromyography-driven robot and passive motion device on wrist rehabilitation for chronic stroke. Neurorehabilitation and Neural Repair. Neurorehabilitation and Neural Repair 2009;23(8):837–46.

[8] Hu XL, et al. Variation of muscle coactivation patterns in chronic stroke during robot-assisted elbow training. Archives of Physical Medicine and Rehabilitation 2007;88(8):1022–9.

[9] Hu XL, et al. Quantitative evaluation of motor functional recovery process in chronic stroke patients during robot-assisted wrist training. Journal of Electromyography and Kinesiology 2008;19(4): 639–50.

[10] Hu XL, et al. Post-stroke wrist rehabilitation assisted with an intention-driven Functional Electrical Stimulation (FES)-Robot system. IEEE International Conference on Rehabilitation Robotics 2011; 2011:1–6.

[11] Levin MF, Kleim JA, Wolf SL. What do motor "recovery" and "compensation" mean in patients following stroke? Neurorehabilitation and Neural Repair 2009;23(4):313–9.

[12] Doucet BM, Lam A, Griffinm L. Neuromuscular electrical stimulation for skeletal muscle function. Yale Journal of Biology & Medicine 2012;85:201–15.

[13] Chae J, Sheffler LR, Knutson JS. Neuromuscular electrical stimulation for motor restoration in hemiplegia. Topics in Stroke Rehabilitation 2015;15(5):412–26.

[14] Hu XL, et al. Wrist rehabilitation assisted by an electromyography-driven neuromuscular electrical stimulation (NMES)-Robot after stroke. Neurorehabilitation and Neural Repair 2015;29(8):767–76.

[15] Rong W, et al. A Neuromuscular Electrical Stimulation (NMES) and robot hybrid system for multi-joint coordinated upper limb rehabilitation after stroke. Journal of NeuroEngineering and Rehabilitation 2017;14(34).

[16] Qian QY, et al. Early stroke rehabilitation of the upper limb assisted with an electromyography-driven neuromuscular electrical stimulation-robotic arm. Frontiers in Neuroscience 2017;8:447.

[17] Kwakkel G, Kollen BJ, Lindeman E. Understanding the pattern of functional recovery after stroke: facts and theories. Restorative Neurology and Neuroscience 2004;22(3–5):281–99.

[18] Nam CY, et al. The effects of upper-limb training assisted with an electromyography-driven neuromuscular electrical stimulation robotic hand on chronic stroke. Frontiers in Neurology 2017;8:679.

[19] Rong W, et al. Effects of electromyography-driven robot-aided hand training with neuromuscular electrical stimulation on hand control performance after chronic stroke. Disability and Rehabilitation: Assistive Technology 2015;10(2):149–59.

[20] Good DC, Bettermann K, Reichwein RK. Stroke rehabilitation. Continuum 2011;17(3):545–67.

[21] Langhorne P, Bernhardt J, Kwakkel G. Stroke rehabilitation: stroke care 2. The Lancet 2011; 377(9778):1693–702.

[22] Horn SD, et al. Stroke rehabilitation patients, practice, and outcomes: is earlier and more aggressive therapy better? Archives of Physical Medicine and Rehabilitation 2005;86(12 Suppl. 2):S101–14.

[23] Mehrholz J, et al. Electromechanical and robot-assisted arm training for improving activities of daily living, arm function and arm muscle strength after stroke. Cochrane Database of Systematic Reviews 2015;7(11):CD006876.

[24] Prange GB, et al. Systematic review of the effect of robot-aided therapy on recovery of the hemiparetic arm after stroke. Journal of Rehabilitation Research and Development 2006;43(2):171–84.

[25] Kwakkel G, Kollen BJ, Krebs HI. Effects of robot-assisted therapy on upper limb recovery after stroke: a systematic review. Neurorehabilitation and Neural Repair 2008;22(2):111–21.

[26] Bovolenta F, et al. Robot-based rehabilitation of the upper limb in stroke patients: a longitudinal observational study. Gait & Posture 2011;33(Suppl. 1):S54.

[27] Sale P, et al. Short-term and long-term outcomes of serial robotic training for improving upper limb function in chronic stroke. International Journal of Rehabilitation Research 2014;37(1):67–73.

10

# Soft robotics for hand rehabilitation

Lizhen Wang[1,2], Guangshuai Peng[1,2], Wei Yao[2,4], Stuart Biggar[2], Chaoyi Hu[1], Xiaofei Yin[1], Yubo Fan[1,2,3]

[1]KEY LABORATORY FOR BIOMECHANICS AND MECHANOBIOLOGY OF MINISTRY OF EDUCATION, SCHOOL OF BIOLOGICAL SCIENCE AND MEDICAL ENGINEERING, BEIHANG UNIVERSITY, BEIJING, CHINA; [2]BEIJING ADVANCED INNOVATION CENTER FOR BIOMEDICAL ENGINEERING, BEIHANG UNIVERSITY, BEIJING, CHINA; [3]NATIONAL RESEARCH CENTER FOR REHABILITATION TECHNICAL AIDS, BEIJING, CHINA; [4]DEPARTMENT OF BIOMEDICAL ENGINEERING, UNIVERSITY OF STRATHCLYDE, GLASGOW, UNITED KINGDOM

**Chapter outline**



## Introduction

About 15 million people are affected by stroke annually around the world [1], and more than 70% of stroke survivors have damaged hand function to different degrees [2]. The incidence of stroke has been increasing gradually in recent years. The hands play an important role in performing activities of daily living (ADLs) and living an independent life [3]. Physical therapy is an important treatment to improve nerve function for stroke patients, however the current practice is labor intensive, costly, and requires assistance from physical therapists [4].

Clinical studies indicate that rehabilitation robots give significant improvement in functional gains under intense therapy [5]. There are a number of rehabilitation devices have been developed in this field [6,7]. Currently, the majority of hand rehabilitation and assistive devices consist of multi-degree of freedom linkage structures [8–11], however

**Intelligent Biomechatronics in Neurorehabilitation. https://doi.org/10.1016/B978-0-12-814942-3.00010-6**

these rigid exoskeleton robots are bulky, heavy, costly, and inaccessible outside of a clinical setting [2].

Panagiotis Polygerinos summarized the design requirements of wearable hand rehabilitation devices [12], where the weight of the device mounted on the hand should be less than 0.5 kg. The power, control, and actuators of a device can be mounted on the waist or back, and the weight should not exceed 3 kg. The device also needs to be customizable to the user's hand size. In addition, in order to not resist the motion of the fingers when the device is not powered, the device should be made from soft and compliant material [13].

Rigid exoskeletons are not the only solution for hand rehabilitation. More recently, a list of hand rehabilitation devices combined with soft and compliance materials such as tendon drive or fluidic soft actuators for hand rehabilitation have appeared and been used in hand rehabilitation [14–19]. These devices can provide assistance with ADLs and hand rehabilitation without rigid joints and links. These soft hand rehabilitation robots are more compact, lightweight, accessible, and easy to don and doff.

In this chapter, we introduce a new soft and wearable robotic glove which is based on a cable drive system and pneumatic artificial muscles. The main goal of this design is to develop a soft robotic glove that is both lightweight and low cost.

# Materials and methods

## Actuators design and fabrication

The McKibben pneumatic artificial muscle (PAM) was developed in the 1950s by physician Joseph L. McKibben for actuating orthotics [20,21]. McKibben PAMs have been used for many years due to their unique set of capabilities and features including low cost, compliant behaviour and high power-to-weight ratio. A typical McKibben PAM consists of an inflatable inner tube, an external braided mesh, and a pneumatic fitting. On the basis of previous research achievements [22], the composition material of the PAM used in this chapter is presented in Fig. 10.1A, including a balloon acting as the inner tube, an external braided mesh with Nylon material, and a segment of PU tube used as a pneumatic fitting. Once the inner tube is pressurized, the external braided mesh contracts axially and dilates radially (Fig. 10.1B).

## Actuator characterization

Using the theory of energy conservation, the force produced by a PAM can be described as:

$$F = -p\frac{dV}{dl} \tag{10.1}$$

where $F$ is output force, $p$ is internal air pressure, $V$ is the PAM volume, and $l$ is the PAM length. The relevant parameters are defined in Fig. 10.2.

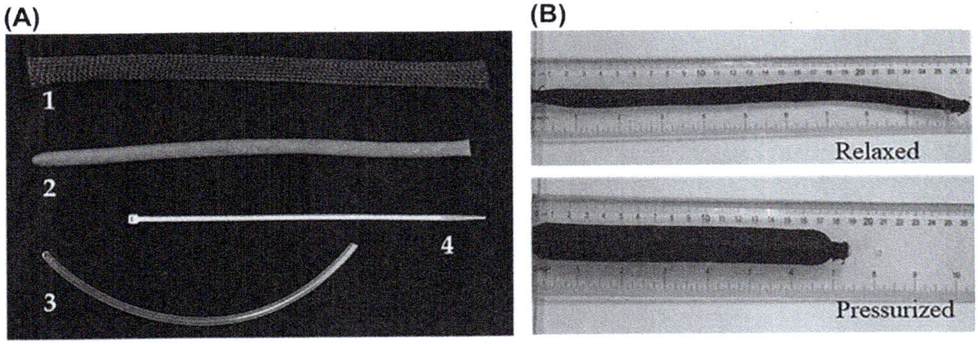


**FIGURE 10.1** Design and fabrication of a PAM: (A) The composition material of a PAM: (1) braided mesh, (2) inflatable inner tube, (3) a segment of PU tube, (4) zip tie. (B) An assembled PAM.

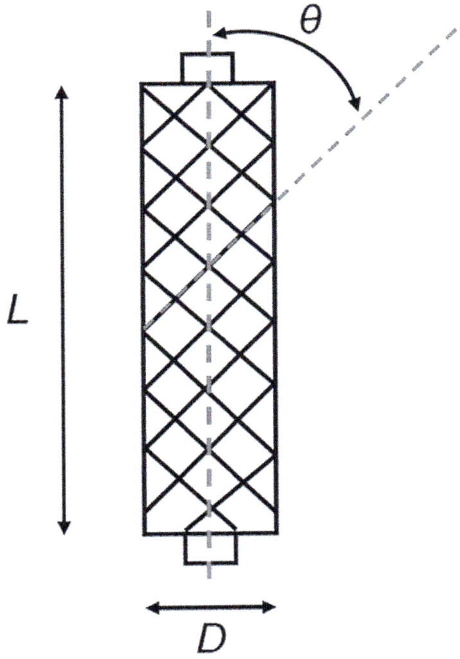


**FIGURE 10.2** Illustration of parameters.

Supposing that there is zero wall thickness and zero friction between the mesh and tube, the force output can be derived:

$$F = p\frac{\pi D_0^2}{4}(3\cos^2\theta - 1) \tag{10.2}$$

where $\theta$ is the braid angle, and $D_0$ is the diameter of the mesh when $\theta$ equal to 90 degrees. The force output is linearly proportional to the pressure and the PAM length. The force output of a PAM is determined by its inner air pressure and its contraction, as shown in Eq. (10.2) [23].

An isometric contraction test and constant pressure contraction test were used to evaluate PAMs [24]. Both tests are carried out using a tensile testing machine (Figs. 10.3A and 10.4A). The force–pressure curve (Fig. 10.3B) shows that the artificial muscle could generate force reaching 100 N in a constant pressure 180 kPa, which should be enough to meet the requirements of 6–10 N in the fingertip for ADLs [22]. The force–contraction curve (Fig. 10.4B) shows that the percent contraction is up to 32% at a pressure of 200 kPa.

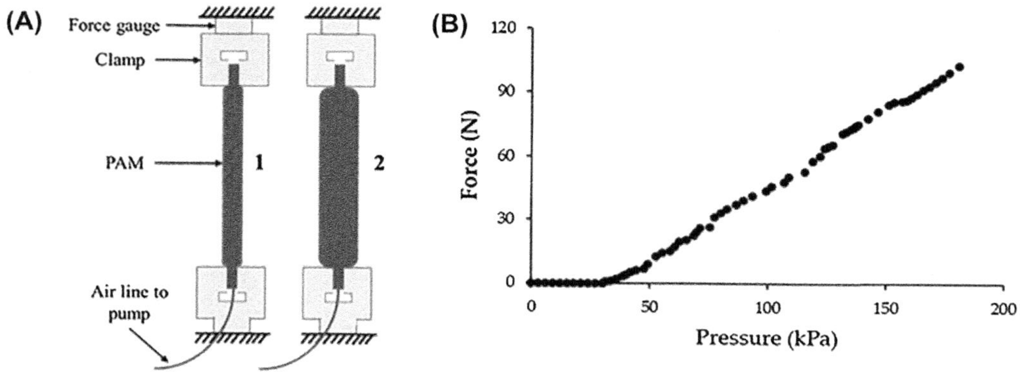


FIGURE 10.3 Isometric contraction test: (A) the schematic of the test procedure; (B) the force–pressure curve.

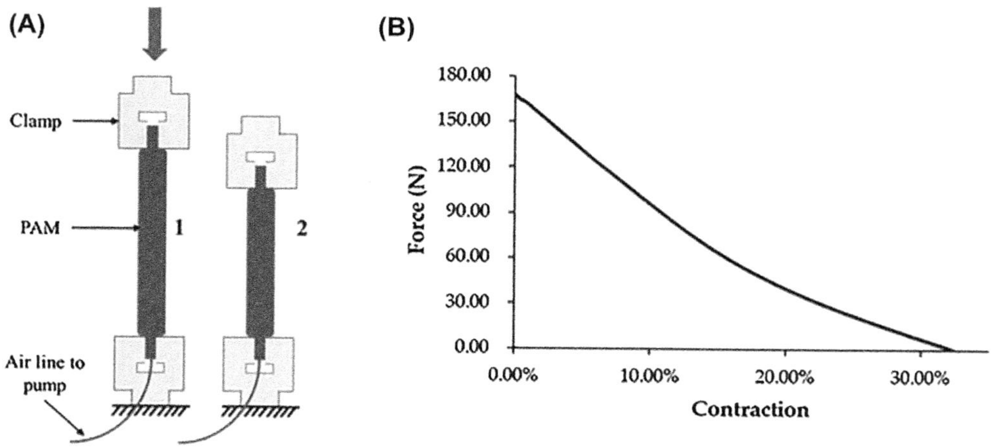


FIGURE 10.4 Constant pressure contraction test: (A) the schematic of the test procedure; (B) the force–contraction curve.

## Cable drive system

In order to reduce the weight of this robotic glove, the user's own skeleton is used to transfer force and torque. The whole structure of the cable drive system consists of a soft glove, cable guides, cables, and PAMs. The PAMs are mounted on the forearm in order to reduce the weight applied to the hand (Fig. 10.5). The total weight of the glove part is about 120 g. The glove part is sufficiently light for stroke patients. Additionally, as the PAMs are worked under air pressure, thus they do not distribute extra weight to the forearm when the actuators are pressurized. The ring finger and little finger were not motorized as they do not play a key role in the grasp motion [4].

An electro-pneumatic control system was developed to allow isolated control of each PAM. The electro-pneumatic control system consisted of a microcontroller (stm32f103zet6, Telesk, China), air pressure sensors (XGZP6847, Risym, China), and miniature solenoid valves (T101U-F, Ousituo, China). Pressure sensors were used to monitor the internal pressure of the PAMs. The controller uses pulse width modulation (PWM) to control the working condition of the valves and pump according to the internal pressure of the PAMs (Fig. 10.6).

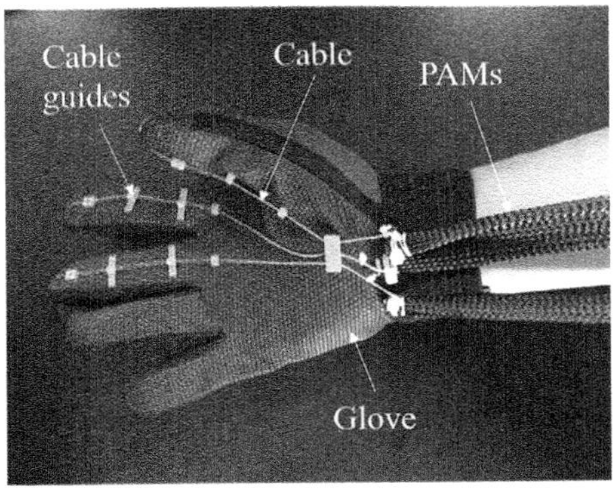


**FIGURE 10.5** The cable drive system based on PAMs.

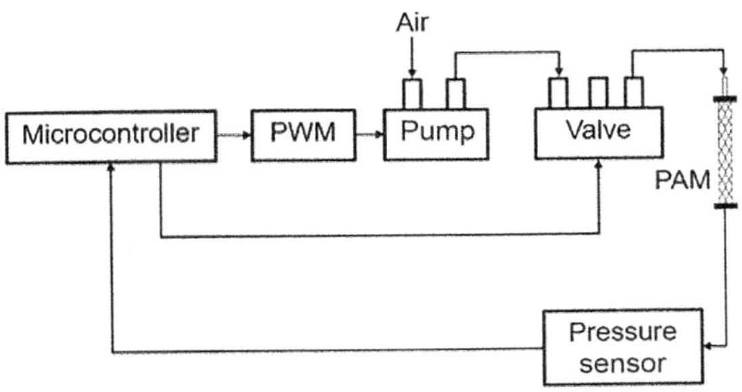


**FIGURE 10.6** The electro-pneumatic control system of the robotic glove.

## User intent detection

User intent of bending and extending fingers can be detected by measuring the sEMG of muscles on the forearm. In this work, we use the healthy hand to control the activities of the robotic glove on the affected hand, where the electrical impulses of the muscle can be used to directly control finger movements. This strategy is suitable for stroke patients with hemiplegia.

In order to measure the activities of the muscles of the forearm, two sEMG sensors were placed on the forearm. The first sEMG sensor was placed at the flexor digitorum superficialis (FDS), which mainly flexes the fingers. The second sEMG sensor was placed at the extensor digitorum communis (EDC), which mainly extends the fingers. Therefore, sEMG was obtained in two signals. The signals were sent to the microcontroller. A new open-loop sEMG logic proposed by Panagiotis Polygerinos et al. [3] has been adopt to judge the user intent. The control of the soft robotic glove can be realized by continuously monitoring the state of the two muscle signals (FDS and EDC) to three predefined conditions: (a) "flex," (b) "extend," and (c) "hold" [25]. The flex condition can be achieved when the sEMG acquired from the FDS muscles is over the flexor threshold, while at the same time, the sEMG acquired from the EDC muscles is below the extensor threshold. In this condition, the PAMs will be pressurized and drive the fingers in flexion. The extend condition will be achieved when the sEMG acquired from the FDS muscles is below the flexor threshold, while at the same time, the sEMG acquired from the EDC muscles exceeds the extensor threshold. In this condition, the PAMs will be depressurized and open the hand. The hold condition can be achieved when neither the flex nor extend conditions are meet. The PAMs inner air pressure will stay unchanged in this condition. The threshold depends on a number of factors and it is different from person to person. The threshold can be adjusted according to the power of the user's sEMG signal.

# Results

We experimentally investigated the performance of the robotic glove by testing the device with one healthy 25-year-old male subject. An air pressure regulator and a force sensor were used to test the force applied by the index finger tip. We measured the force of the healthy male subject's index finger tip while wearing the robotic glove (Fig. 10.7). When the PAM was pressurized, the fingertip applied a vertical force on the force sensor.

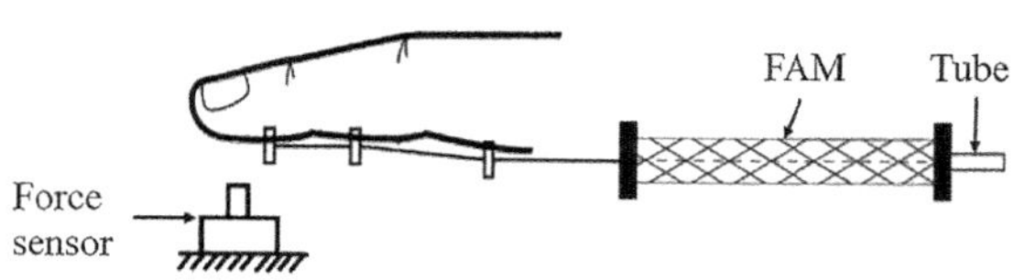


**FIGURE 10.7** The platform for the finger tip force test.

**Table 10.1** The ROM of finger joints when wearing the robotic glove.

| Index finger joint | The range of motion in active motion | The range of motion in passive motion |
|---|---|---|
| MCP | 63.25 degrees | 58.02 degrees |
| PIP | 79.12 degrees | 57.36 degrees |
| DIP | 45.32 degrees | 40.13 degrees |

The maximum fingertip force measured was up to 12 N when the pressure of the PAM was 180 kPa, which is sufficient for performing activities of daily living.

To evaluate the kinematic performance of the soft robotic glove, high-speed cameras were used to film finger motion of the healthy subject while he was wearing the robotic glove. The range of motion (ROM) of index finger joints is shown in Table 10.1, sufficient ROM of the index finger was achieved with the assistance of this device. Although the range of the ROM was smaller than that of voluntary movement when the subject was not wearing the device, it was enough to perform pinch and wrap grasp activities.

With the help of this device, a healthy subject grasped the following target objects: (a) a cube, (b) a plastic bottle containing 300 mL of water, (c) an apple, (d) a smart phone, (e) a pen, and (f) a mouse, to test the ability of the robotic glove to provide assistance in activities of daily living. The subject successfully grasped all the objects for the first time using the robotic glove due to the simple control scheme and good adaption mechanism (Fig. 10.8).

## Conclusions and future trends

In this chapter, a new soft and wearable robotic glove that uses a tendon drive system and PAMs was designed. The weight of the glove part is about 120 g. The weight of the control box, which contains a battery, pump, and microcontroller, is about 2 kg. The robotic glove is compact and lightweight, making it more accessible than rigid hand exoskeleton robots. The myoelectric control strategy was used to control the robotic glove. A new open-loop sEMG logic has been adopted to judge user intent. The maximum fingertip force is about 12 N, which should be enough to provide assistance in ADLs. Finally, with the help of the device subjects were capable of grasping objects of various shapes. The soft and wearable robotic glove can meet the requirements for stroke rehabilitation.

There are some limitations to the cable drive system. The cables fixed between finger tips and PAMs could restrict movement of the wrist and cause injury. In the future, steps to improve the performance and safety of the cable drive system and to develop a more intelligent human–robot interface to detect user intent such as brain-computer interface will be targeted. Also, stroke subjects will be involved in testing in the future.

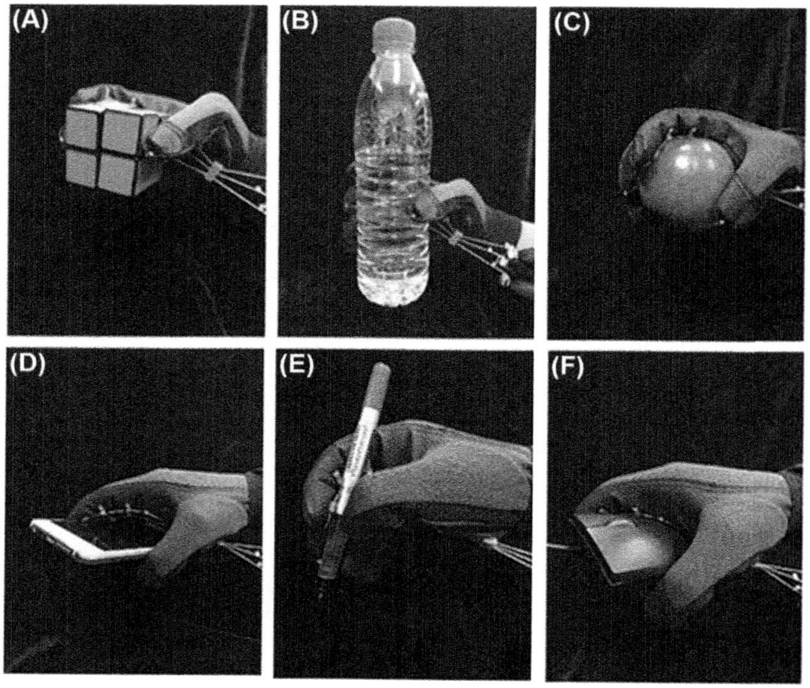


**FIGURE 10.8** Using the robotic glove to pinch and grasp various objects: (A) a cube; (B) a plastic bottle containing 300 mL of water; (C) an apple; (D) a smart phone; (E) a pen; and (F) a mouse.

Soft robots have the advantages of light weight, continuous deformation, and strong adaptability, providing a new way to break through the limitations of traditional rigid robots. Soft robots have broad application prospects in military, medical, and industrial fields.

# References

[1] Takahashi CD, Der-Yeghiaian L, Le V, Motiwala RR, Cramer SC. Robot-based hand motor therapy after stroke. Brain 2008;131:425–37.

[2] Schabowsky CN, Godfrey SB, Holley RJ, Lum PS. Development and pilot testing of hexorr: hand exoskeleton rehabilitation robot. Journal of NeuroEngineering and Rehabilitation 2010;7:36.

[3] Polygerinos P, Galloway KC, Sanan S, Herman M, Walsh CJ. EMG controlled soft robotic glove for assistance during activities of daily living. In: 2015 IEEE international conference on rehabilitation robotics (ICORR). IEEE; 2015.

[4] Biggar S, Yao W. Design and evaluation of a soft and wearable robotic glove for hand rehabilitation. EEE Transactions on Neural Systems and Rehabilitation Engineering 2016;24:1071–80.

[5] Wolf SL, Winstein CJ, Miller JP, Taub E, Uswatte G, Morris D, et al. Effect of constraint-induced movement therapy on upper extremity function 3 to 9 months after stroke: the excite randomized clinical trial. The Journal of the American Medical Association 2006;296:2095–104.

[6] Nef T, Riener R. ARMin - design of a novel arm rehabilitation robot. In: 2005 IEEE international conference on rehabilitation robotics (ICORR). IEEE; 2005. p. 57–60.

[7] Loureiro R, Amirabdollahian F, Topping M, Driessen B, Harwin W. Upper limb robot mediated stroke therapy – GENTLE/s approach. Autonomous Robots 2003;15:35–51.

[8] Heo P, Gwang MG, Lee SJ, Rhee K, Kim J. Current hand exoskeleton technologies for rehabilitation and assistive engineering. International Journal of Precision Engineering and Manufacturing 2012; 13:807–24.

[9] Jones CL, Wang F, Morrison R, Sarkar N, Kamper DG. Design and development of the cable actuated finger exoskeleton for hand rehabilitation following stroke. IEEE 2014;19:131–40.

[10] Lee SW, Landers KA, Park HS. Development of a biomimetic hand exotendon device (biomHED) for restoration of functional hand movement post-stroke. IEEE Transactions on Neural Systems and Rehabilitation Engineering 2014;22:886–98.

[11] Dicicco M, Lucas L, Matsuoka Y. Comparison of control strategies for an EMG controlled orthotic exoskeleton for the hand. In: 2004 IEEE international conference on robotics and automation (ICRA). IEEE; 2004. p. 1622–7.

[12] Panagiotis P, Wang Z, Galloway KC, Wood RJ, Walsh CJ. Soft robotic glove for combined assistance and at-home rehabilitation. Robotics and Autonomous Systems 2014;73:135–43.

[13] Aubin PM, Sallum H, Walsh C, Stirling L, Correia A. A pediatric robotic thumb exoskeleton for at-home rehabilitation: the Isolated Orthosis for Thumb Actuation (IOTA). In: 2013 IEEE 13th international conference on rehabilitation robotics (ICORR). IEEE; 2013.

[14] Radder B, Kottink A, Vaart NVD, Oosting D, Buurke JH, Nijenhuis SM. User-centred input for a wearable soft-robotic glove supporting hand function in daily life. In: 2015 IEEE international conference on rehabilitation robotics (ICORR). IEEE; 2015. p. 502–7.

[15] Zhao HC, Huang RK, Shepherd RF. Curvature control of soft orthotics via low cost solid-state optics. In: 2016 IEEE international conference on robotics and automation (ICRA). IEEE; 2016. p. 4008–13.

[16] Zhao H, Jalving J, Huang R, Knepper R, Ruina A, Shepherd R. A helping hand: soft orthosis with integrated optical strain sensors and EMG control. IEEE Robotics and Automation Magazine 2016; 23:55–64.

[17] In H, Kang BB, Sin MK, Cho KJ. Exo-glove: a wearable robot for the hand with a soft tendon routing system. IEEE Robotics and Automation Magazine 2015;22:97–105.

[18] Nycz CJ, Delph MA, Fischer GS. Modeling and design of a tendon actuated soft robotic exoskeleton for hemiparetic upper limb rehabilitation. In: International conference of the IEEE engineering in medicine and biology society (EMBC). IEEE; 2015. p. 3889–92.

[19] Park S, Bishop L, Post T, Xiao Y, Stein J, Ciocarlie M. On the feasibility of wearable exotendon networks for whole-hand movement patterns in stroke patients. In: 2016 IEEE international conference on robotics and automation (ICRA). IEEE; 2016. p. 3729–35.

[20] Tondu B. Modelling of the mckibben artificial muscle: a review. Journal of Intelligent Material Systems and Structures 2012;23:225–53.

[21] Tondu B, Lopez P. Modeling and control of McKibben artificial muscle robot actuators. IEEE Control Systems Magazine 2000;20:15–38.

[22] Yi J, Shen Z, Song C, Wang Z. A soft robotic glove for hand motion assistance. In: 2016 IEEE international conference on real-time computing and robotics (RCAR). IEEE; 2016. p. 111–6.

[23] Tian Y, Wang K, Yi J, Wang Z, Chen MZQ. Introduction to modeling of the McKibben pneumatic artificial muscle with end constraints. In: 2016 IEEE international conference on information and automation. IEEE; 2016. p. 1624–9.

[24] Obiajulu SC, Roche ET, Pigula FA, Walsh CJ. Soft pneumatic artificial muscles with low threshold pressures for a cardiac compression device. In: ASME international design engineering technical conferences/computers and information in engineering conference (IDETC/CIE). ASME; 2013. p. 1–8.

[25] Hong KY, Ang BWK, Lim JH, Goh JCH, Yeow CH. A fabric-regulated soft robotic glove with user intent detection using EMG and RFID for hand assistive application. In: Proceeding of the IEEE international conference on robotics and automation (ICRA), Stockholm, Sweden, 16–21 May, 2016; 2016. p. 3537–42.

PART

III

# Clinical Applications

11

# Clinical evaluations with robots in rehabilitation

Kevin B. Wilkins, Jun Yao

*PHYSICAL THERAPY MOVEMENT AND HUMAN MOVEMENT SCIENCES DEPARTMENT, NORTHWESTERN UNIVERSITY, USA*

**Chapter outline**



## Introduction

The major goal of rehabilitation is to restore function back to a preinjured state [1]. To date, various robots have been used in rehabilitation for the purpose of either reducing involvement from physicians, increasing the involvement of the stroke user, or enhancing/assisting the motor control of a stroke user. In addition to supplementing rehabilitation, robots can also be used to enhance the quantification of evaluation following interventions, at both the performance and neural levels.

1. At the performance level, robot-quantification acts as a supplement to clinical measures with increased sensitivity; and

Intelligent Biomechatronics in Neurorehabilitation. https://doi.org/10.1016/B978-0-12-814942-3.00011-8

2. At the neural level, robotic simulated environments provide well-controlled conditions for the investigation of neural plasticity related to multijoint movements.

One of the main contributors to upper extremity motor impairments post-stroke is the loss of independent joint control. This loss of independent joint control contributes to the clinically observed flexion synergy in which abnormal coupling between shoulder abductors and elbow/wrist and finger flexion reduces functional reaching distance [2–4] and hand opening [5] when lifting the paretic arm at the shoulder. Importantly, this abnormal coupling between proximal shoulder abductors and distal limb flexors increases as a function of shoulder load, where greater attempt to drive the limb leads to increased expression of the flexion in the distal joints. Although conventional assessments of upper extremity impairments in stroke such as the Fugl–Meyer Assessment (FMA) do include quantifications of this loss of independent joint control in their assessments, it is mixed in with other types of impairments as well. On the other hand, given the dynamic nature of the expression of the flexion synergy and subsequent loss of independent joint control, robotic devices allow us to quantify upper extremity impairments, and thus track intervention-induced changes with high resolution.

The neural mechanisms underlying loss of independent joint control point to an increased reliance on contralesional corticobulbar tracts following damage to ipsilesional corticospinal tract and accompanying corticofugal tracts. For instance, progressively increasing shoulder load while reaching increases reliance on the contralesional hemisphere [6], which is associated with reduced reaching distance and hand opening ability due to the flexion synergy. Evidence from monkeys suggests that this may be due to an increased reliance on contralesional corticobulbar tracts such as the corticoreticulospinal tract [7,8]. These tracts branch more extensively at the level of the cord compared to the corticospinal tract (CST), and thus activation of these tracts leads to muscle activation across multiple joints [9]. Additionally, these tracts preferentially innervate the flexor muscles of the arm, and activation leads to flexion-synergy behavior [10,11]. These findings in monkeys have since been supported in humans, where increased structural integrity in the contralesional reticular formation was associated with greater synergies of the hand and arm [12]. All the above information indicates the limited ability of the contralesional hemisphere to restore function to prestroke levels.

Given the ineffectiveness of contralesional resources, effective interventions are expected and have been shown to increase reliance on the ipsilesional hemisphere, thus restoring neuronal resources, and subsequently lead to true motor recovery [13–19]. However, intervention-induced cortical reorganization has primarily been studied related to the hand or arm in isolation using fMRI or TMS. Unfortunately, these methods do not allow for quantification of neural changes related to the loss of independent joint control or flexion synergy due to practical limitations preventing multijoint movements. The combination of high-density EEG and robotics allows one to examine cortical activity related to dynamic multijoint movements that elicit the flexion synergy in a well-controlled fashion.

## The ACT-3D robotic device

In this chapter we examine examples of application of an Arm Coordination Training 3D Device (ACT-3D) robot (Fig. 11.1, left) for evaluating intervention effectiveness at both performance and neural plasticity levels. The ACT-3D is an admittance-controlled robotic device equipped with a 6° of freedom load cell (JR [3] Inc., Woodland, CA, USA) and an instrumental gimbal at the end effector. It allows unrestrained motion of the upper limb in three dimensions and allows the ability to impose a force in the vertical Z-direction to make the limb feel heavier or lighter.

# Quantifying improvements in shoulder/elbow performance following an intervention

Intervention-induced changes are typically measured by clinical measures such as FMA, Action Research Arm Test, Wolf Motor Function Test, Box and Blocks Test, etc. However, these measures either lack high resolution or can be improved via compensatory strategies. Robotics quantitatively measures changes in performance (e.g., movement patterns) at multiple joints with high resolution, thus providing the opportunity to distinguish motor compensation versus true recovery. In the following example, the ACT-3D robot was used to quantify intervention-induced changes in reaching distance and velocity following a progressive abduction loading reaching intervention [20].

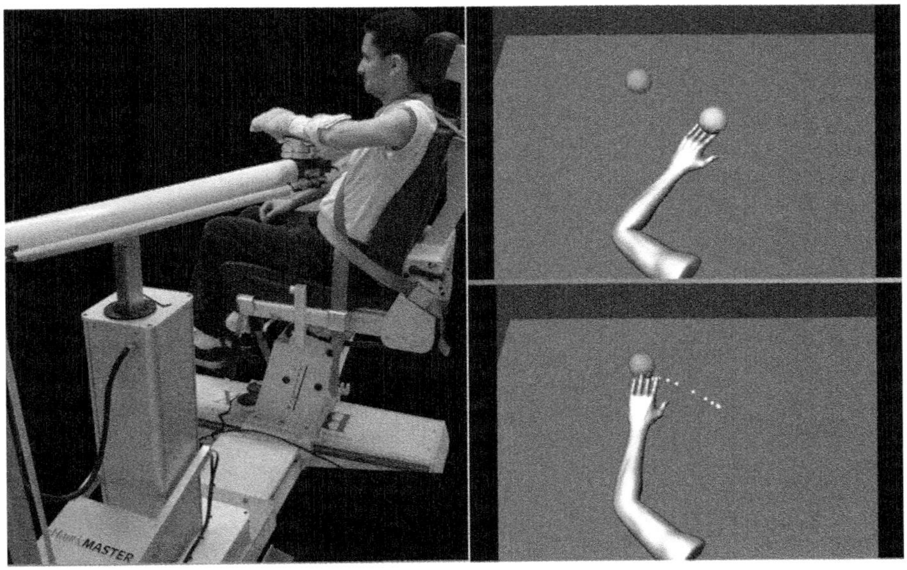

**FIGURE 11.1** Participant set up on an ACT-3D robot (left) and visual display (right) depicting the arm avatar including the home target (*gray*) and reaching target (*red* [dark gray in print version]) with the reaching trajectory shown as *white dots*.

## Progressive abduction loading therapy

Thirty-two individuals with moderate to severe chronic stroke (UEFMA: 16–43) participated in a prospective, single-site, double-blinded, parallel comparison group, randomized clinical trial to determine the therapeutic effects of horizontal-plane viscous resistance in combination with progressive abduction loading therapy on reaching impairment. The intervention lasted 8 weeks with three visits per week. During the visit, subjects sat in an experimental chair with their paretic arm positioned in 40 degrees of horizontal adduction and 110 degrees of elbow flexion. This position is referred to as the "home position." The experimental and comparison interventions were similar in that they both consisted of reaching movements to four standardized target locations while lifting a predetermined percentage of the maximum voluntary shoulder abduction torque replicating the initial preclinical study [21]. These target locations required 100 degrees of elbow extension from the home position, as well as various amounts of horizontal adduction (0 degrees, 20 degrees, 40 degrees, and 60 degrees, respectively). For both groups, the initial level of shoulder abduction loading for each target direction was set as the highest abduction load at which the participant could reach at least 50% of the distance to the target. Participants were trained at this abduction loading level until they could reach at least 80% of the distance to the target in 8 out of 10 repetitions for three out of four sets. The abduction load was then increased to the highest level where, again, participants could reach at least 50% of the distance between the starting position and the target.

For the experimental group, a viscous field was added in the horizontal plane after the abduction loading was determined for all four targets. The viscous field provided a velocity-dependent increase in resistance for the movement in the horizontal plane. Vertical movement was not impacted by the viscous field. Subjects were then trained using the same progression as the comparison group, as stated above. In both groups, shoulder abduction loading was progressed independently for each target. Each intervention session consisted of four sets of 10 repetitions for each of the four target directions totaling 160 repetitions. The time of each session was truncated at 1 h to emulate outpatient rehabilitation. For more information, refer to Reference [21].

## Measuring improvements in reaching distance and velocity on a robotic device

One week before and after the intervention, the ACT-3D robot was used to measure the improvement in both reaching distance and peak endpoint reaching velocity. Participants sat in a Biodex chair with their arm resting in a forearm-hand orthosis attached to the ACT-3D (Fig. 11.1, left). The participant's trunk was immobilized to prevent shoulder girdle movement by a set of straps attached to the experimental chair. The shoulder was positioned with the arm perpendicular to the ground when the arm was resting on a haptically rendered virtual table maintained by the device.

Once positioned and supported by the haptic table, participants were asked to move the tested hand to the home position (gray circle in Fig. 11.1, right) to trigger the data collection. After 1 s, a second circle representing the target (red circle in Fig. 11.1, right) appears on the screen as a cue to begin the movement. The movement target is located requiring an additional 100 degrees of elbow extension and 30 degrees of horizontal adduction from the home position to reach it. Participants were instructed to move as rapidly as possible toward the target and to maintain the final position until the target disappeared (end of data collection). Rapid (ballistic) movements were strongly encouraged through verbal cuing by the experimenter. The avatar of the participant's arm emulated movement in real-time, providing realistic visual feedback of movement performance. During the completion of each target reach, the hand path was displayed to the participant as feedback. The length of data collection was 5 s per trial and 10 consecutive trials were performed for each of the targets.

Reaching distance and velocity were then measured again using the same setup as above, but with ACT-3D modulated abduction loading at 0%, 12.5%, 25%, 37.5%, and 50% of maximum abduction strength. Fig. 11.2 depicts how the ACT-3D provided vertical force while the participant performed the reaching task. It is important to note that the volitional abduction torque required for a given abduction loading condition may be more than the abduction torque required to lift the arm against gravity. In that circumstance, the vertical force provided by the ACT-3D would be directed downwards as depicted in Fig. 11.2. Meanwhile, if the volitional torque required was less than the torque required to lift the limb against gravity, the vertical force would be directed upwards. For all participant assessments, one set of 10 repetitions was performed for each of the five randomized abduction loading conditions in addition to on the table. Therefore, at completion of the primary outcome assessment, participants performed six sets of 10 reaches.

### *Data analysis*

Position data from the ACT-3D were recorded. The reaching distance was calculated as the longest distant of the reaching trajectory that remained within a 30-degree cone centered in the direction of the reach target and while the arm remained off a horizontal haptic surface positioned at shoulder height (Fig. 11.3A). The largest reaching distance was identified from the five reaching trials of the highest abduction loading level that the participate was able to complete during the pretesting session. Maximum reaching distance was normalized to the absolute distance from the home position to the reaching target.

Reaching velocity was calculated as the first peak reaching velocity during each reaching trial from the endpoint position (middle finger tip). The first peak was utilized as it represented the maximum effort to initiate outward movement against synergy from a standardized position and controlled abduction load. Reaching velocity was normalized by maximum reaching velocity during a reach while supported on the haptic surface.

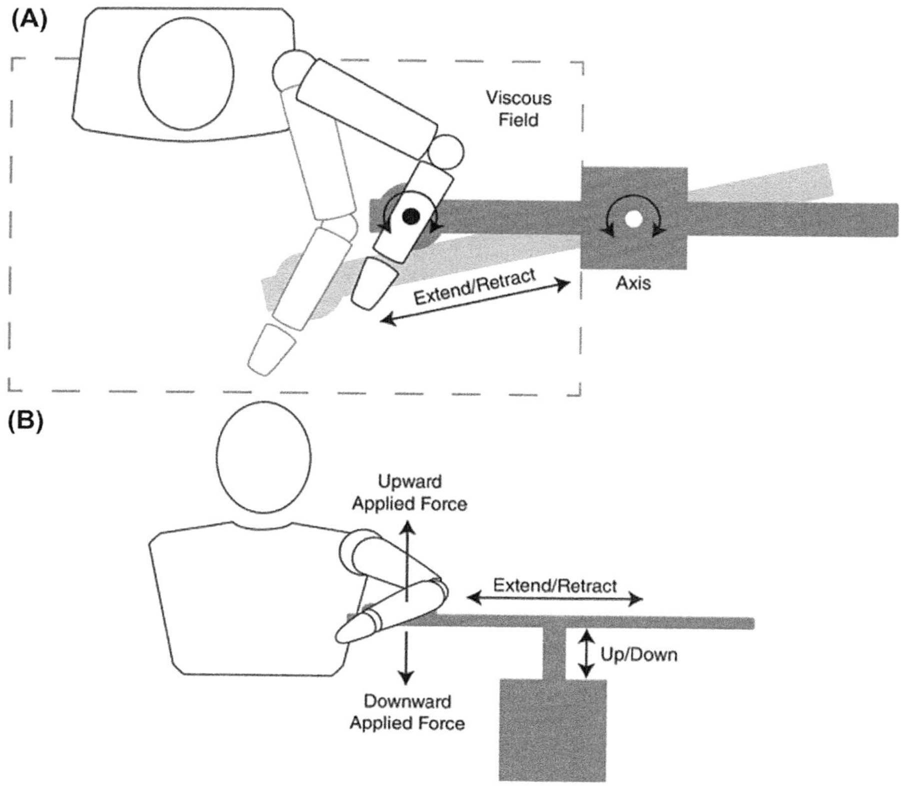


**FIGURE 11.2** Diagram of top (A) and front (B) views illustrating the interface of the participant and device. The top view illustrates both the kinematics (extend/retract and axis rotation motion) and viscous resistance of the device in the horizontal plane during an outward reaching motion involving elbow extension and shoulder horizontal adduction. The front view illustrates both kinematics (extend/retract and up/down motion) and kinetics (upward and downward force) of the device.

### *Results*

Both groups demonstrated significant increases in FMA postintervention (2.21-point increase on average) that was maintained at the 3-month follow-up. However, there was no significant difference between the experimental and comparison groups.

Both groups showed a significant increase in normalized reaching distance postintervention (13.2% increase) that was maintained at the 3-month follow-up, but with no difference between groups (Fig. 11.4 left). Additionally, both groups demonstrated an increase in peak reaching velocity (13.6% and 21% greater for the comparison and experimental group, respectively) that persisted at the 3-month follow-up, but with no significant difference between experimental and comparison groups (Fig. 11.4 right).

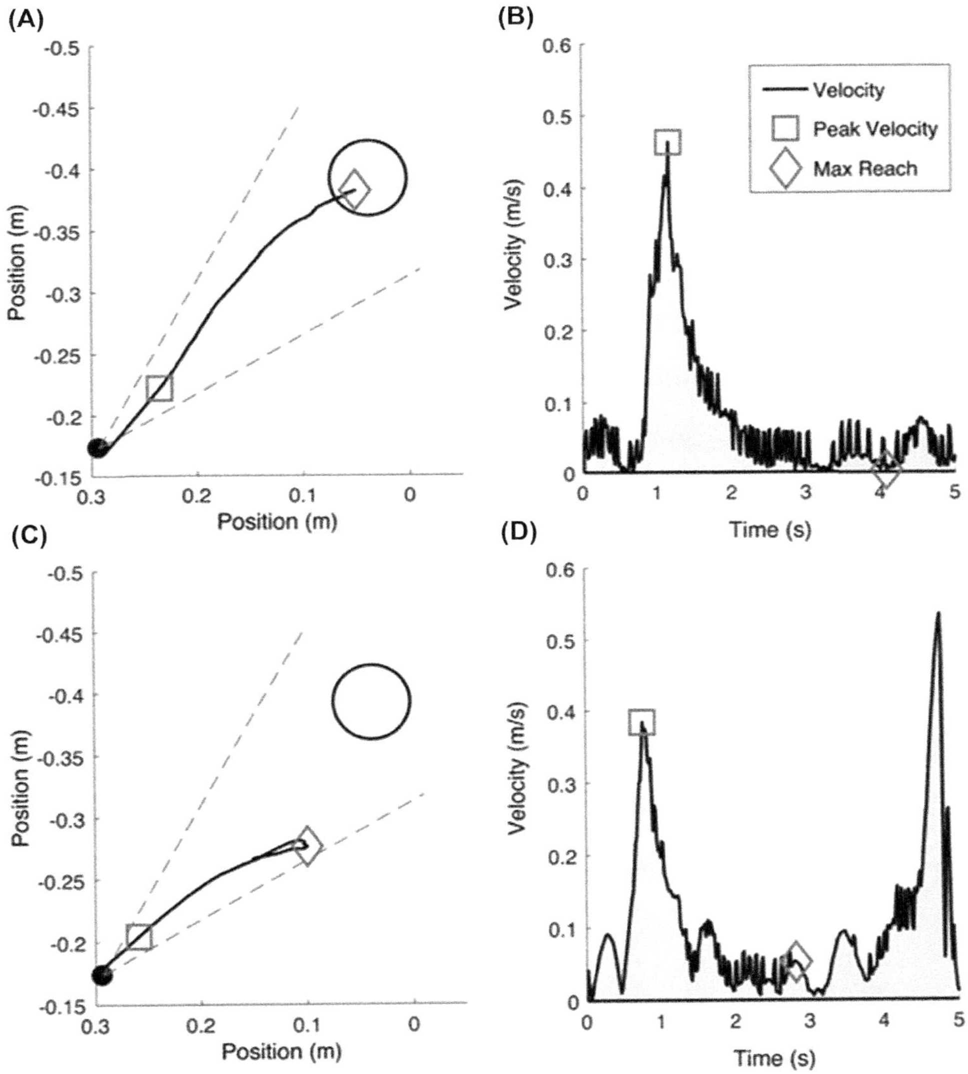


**FIGURE 11.3** Example maximum reaching trajectories (left, indicated by a *diamond*) while supported on a haptic horizontal surface (top) and while abducting off the horizontal surface at 50% of maximum shoulder abduction (bottom). Reaching trajectories are only accepted if within the ±15 degrees cone of tolerance (*gray dotted line*). The endpoint peak reaching velocity (right, indicated by a *square*) illustrates the peak reaching velocity associated with each reaching trajectory.

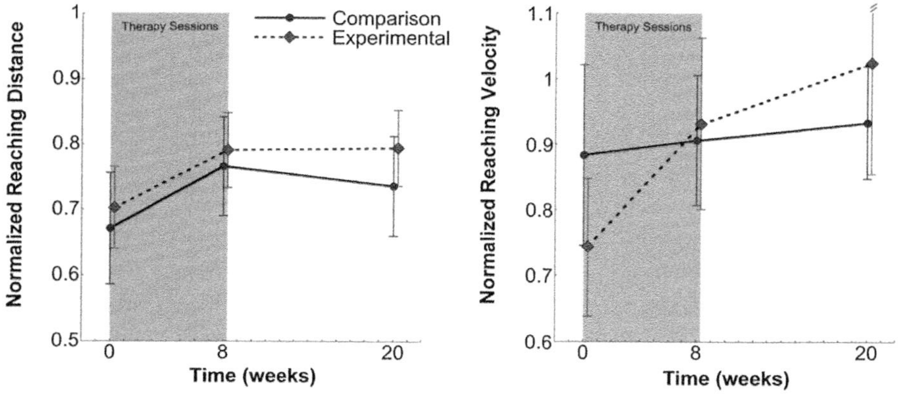


**FIGURE 11.4** Normalized reaching distance (left) and velocity (right) and standard errors for the comparison and experimental groups at pretesting, posttesting, and 3-month follow-up. There was a significant increase in reaching distance and velocity for all participants at posttesting that persisted at 3-month follow-up in all participants.

## Quantifying cortical reorganization related to the hand and arm following an intervention

Cortical reorganization has primarily been measured using either TMS mapping to investigate expansion of cortical maps related to the hand [16,22] or fMRI to investigate changes in activity patterns related to hand opening/closing [17,23,24]. In our work, we expanded this approach to look at cortical changes related to both "hand opening in isolation" and "hand opening while lifting at the shoulder" [25–27]. This is crucial since the addition of lifting further increases reliance on the contralesional hemisphere in stroke [6] and leads to further impairments in hand opening and reaching distance due to the flexion synergy [5,28].

### ReIn-Hand intervention

Eight individuals with moderate to severe chronic stroke (UEFMA: 11–24) participated in a 7-week intervention consisting of three 2-h visits per week. During the visit, subjects completed 20–30 trials of the following sequence of movements: (1) reaching out toward a jar, (2) driving the wrist/finger extensors to open the paretic hand, (3) grabbing the jar, (4) bringing the jar back toward themselves, and (5) releasing the jar. The weight, distance/height, and orientation of the jar relative to the subject were progressively altered to increase the challenge to each subject, as determined by the physical therapist. All subjects started the motor task with the arm supported by the table. Depending on ability, subjects were encouraged to progressively lift the paretic limb actively. During the task, a novel EMG-FES device, called the ReIn-Hand, was used to assist paretic hand opening. This device recorded EMG activities from eight muscles [deltoid, biceps brachii, triceps, extensor communis digitorum, extensor carpi radialis (ECR), flexor digitorum profundus, flexor carpi radialis (FCR), and abductor pollicis]. While the user performed the functional reaching and opening, the ReIn-Hand detected hand opening by extracting EMG features to trigger an Empi transcutaneous electrical neurostimulation device (Vista, CA, USA). The stimulation electrodes were applied to the wrist/finger extensors with the following settings: biphasic waveform, frequency = 50 Hz ± 20%, pulse width = 300 μs, amplitude = sufficient for maximal hand opening without discomfort, and duration = 3 s. The novelty of this device is that even with the increased expression of the flexion synergy at the elbow [4], wrist, and fingers [5,29] during reaching that is prevalent in this population, the device can still detect the hand opening and drive the paretic hand open, thus allowing for a user-driven stimulation to support functional usage of the paretic hand and arm.

For each subject, within 1 week prior to and following the intervention, a licensed physical therapist completed a set of clinical assessments, with the motor-related parts including: Box and Blocks Test (BBT), active range of motion (AROM), and passive range of motion (PROM) averaged over the II and V digits.

## Measuring cortical activity on a robotic device

Participants took part in an EEG experiment within 1 week prior to and following the intervention. Participants sat in a Biodex chair (Biodex Medical Systems, Shirley, NY, USA) with straps crossing the chest and abdomen to limit potential trunk movements. The participant's paretic arm was placed in a forearm-hand orthosis attached to the end effector of the ACT$^{3D}$ robot (JR [3] Inc., Woodland, CA). At the beginning of each trial, participants moved their hand to a home position, with the shoulder at 85 degrees abduction, 40 degrees flexion, and the elbow at 90 degrees flexion angle. The participant received an auditory cue once they reached the home position. Following the cue, the participant relaxed at the home position for 5–7 s and then self-initiated one of two movements: (1) maximum paretic hand opening with the arm resting on a haptic table, or (2) maximum paretic hand opening while lifting against 50% of their maximum shoulder abduction (SABD) force. Participants were instructed to avoid eye movements by focusing on a point and avoid movements of other body parts during the performance of each trial, which was confirmed by electro-oculogram (EOG) traces and visual inspection by the experimenter, respectively. Participants performed 60–70 trials of each condition, which were separated into blocks (one block consisted of 20–30 trials of a particular condition). Blocks were randomized to minimize any order effects. Rest periods varied between 15 and 60 s between trials and 10 min between blocks.

Scalp recordings were made with a 160-channel high-density EEG system using active electrodes (Biosemi, Inc., Active II, Amsterdam, The Netherlands) mounted on a stretchable fabric cap based on a 10/20 system. The impedance was kept below 50 kΩ for the duration of the experiment. Simultaneously, EMGs were recorded from the extensor carpi radialis, flexor carpi radialis, and deltoid of the paretic arm, which were used to detect movement onset for postprocessing purposes. All data were sampled at 2048 Hz.

Additionally, the positions of the EEG electrodes on the participant's scalp were recorded with respect to a coordinate system defined by the nasion and preauricular notches using a Polaris Krios handheld scanner and reflective markers (NDI, Ontario, Canada). This allowed for coregistration of EEG electrodes with each participant's anatomical MRI data.

### *Data analysis for cortical activity*

EEG data were used to estimate the cortical activity during the motor command releasing phase (150–100 ms before the onset of EMG) related to pure hand opening and hand opening while lifting against 50% max SABD. The cortical current density strength ($\mu A/mm^2$) was computed using the standardized low-resolution electromagnetic brain tomography (sLORETA) method (Lp = 1) based on a subject-specific boundary element method model (based on the structural MRI). The inverse results (i.e., estimated activity) were analyzed in specific regions of interest (ROIs) including bilateral primary sensorimotor cortices [primary motor cortex (M1) + primary sensory cortex (S1)] and secondary motor cortices [supplementary motor area (SMA) + premotor area (PM)].

A laterality index ($LI = (I - C)/(I + C)$), where I and C are the current density strengths from the ipsilesional and contralesional sensorimotor cortices (i.e., combined primary sensorimotor and secondary motor cortices) was used to quantify the relative contributions of each cerebral hemisphere to the motor task. A value close to +1 indicates an ipsilesional source distribution and −1 indicates a contralesional source distribution.

Additionally, a cortical activation ratio $CAR = \frac{\sum_{1}^{N} s_n}{\sum_{1}^{M} s_m}$ for each of the four ROIs was calculated, where $s_n$ represents the current density strength of the $n$-th node, and $n$ and $m$ are the indices of nodes in a specific ROI and the whole sensorimotor cortices, respectively. The cortical activity ratio reflects the relative strength from one ROI as normalized by the total combined strength of the four ROIs.

### *Results*

Individuals demonstrated significant increases from pre- to postintervention in BBT, AROM, and PROM. Notably, six out of eight subjects initially scored a 0 on the pre-assessment BBT and showed 0 degrees of AROM due to the severity of their motor impairments at the arm/hand. On average, individuals increased 1.88 blocks in the BBT, 9.9 degrees in AROM, and 11.5 degrees in PROM [27].

These clinical improvements were accompanied by significant cortical reorganization related to the hand and shoulder. Following the intervention, individuals showed a shift toward the ipsilesional hemisphere (indicated by a positive LI) for both hand opening and hand opening while lifting (Fig. 11.5). For hand opening, this ipsilesional shift was driven by a decrease in activity in contralesional primary sensorimotor cortex and a trending increase in ipsilesional secondary motor areas (Fig. 11.6, left). Meanwhile, for

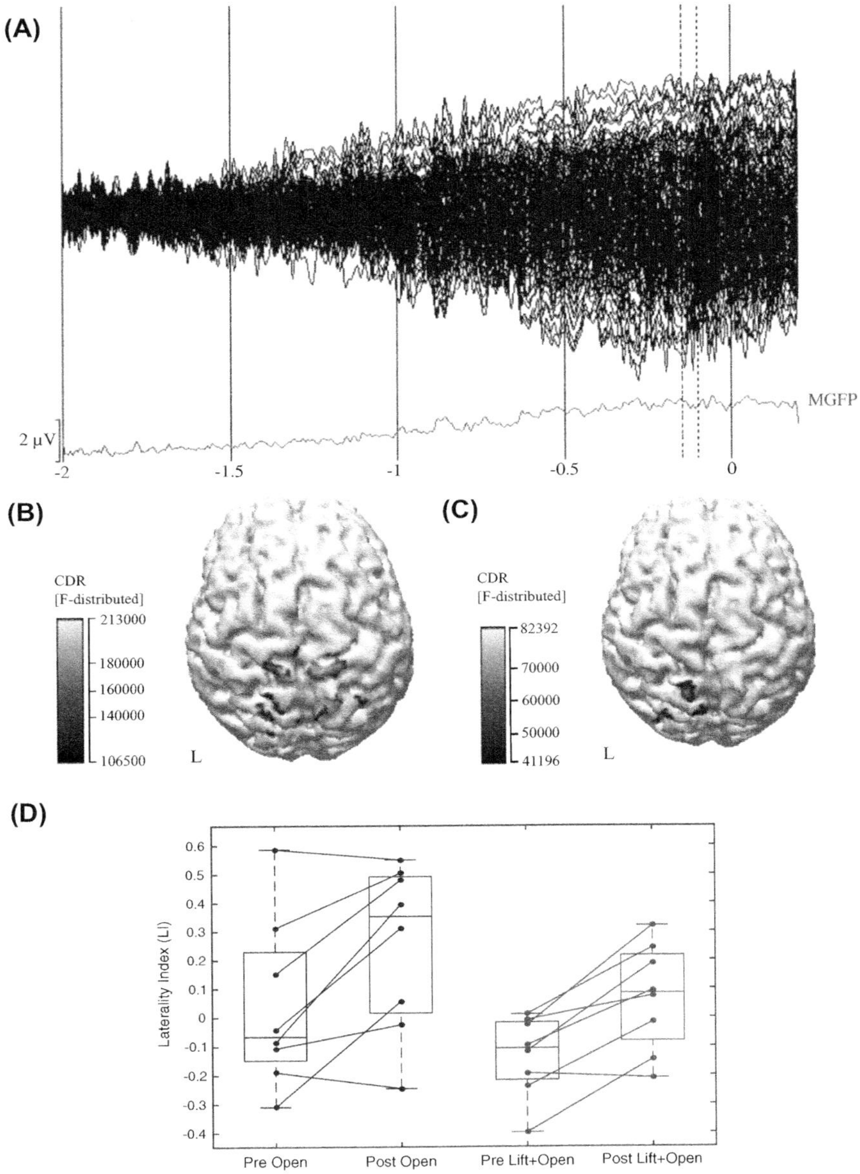


**FIGURE 11.5** (A) Ensemble-averaged EEG of the 160 channels (*blue butterfly plot* [black in print version]) and mean global field power (MGFP; *red line* [gray in print version]) from −2 s to +0.2 s (0 = EMG onset). *Vertical dashed lines* represent the start and end of the window of interest (−150 to −100 ms). A scale bar is included in the lower left; (B) reconstructed cortical activity between −150 and −100 ms prior to movement onset for Subject 1 during hand opening preintervention, and (C) postintervention. Color bars indicate the current density reconstruction (CDR) statistic from sLORETA. Left hemisphere is the lesioned hemisphere. (D) Box plots of laterality index (LI) prior to and following the intervention for paretic hand opening on the table (*blue* [black in print version]) and hand opening while lifting (*red* [gray in print version]). Positive LI indicates predominantly ipsilesional activity. *Indicates $P < .05$.

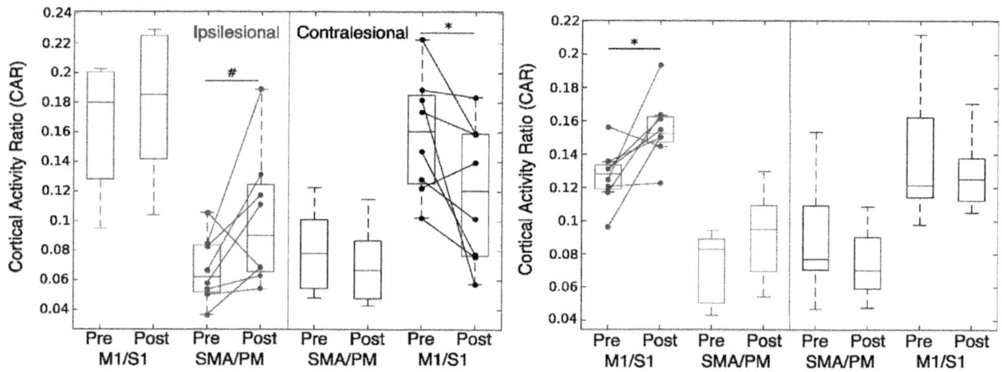


**FIGURE 11.6** Box plots depicting the cortical activity ratio (CAR) prior to and following the intervention for hand opening on the table (left) and hand opening while lifting (right). Regions of interests include M1/S1 and supplementary motor area/premotor area (SMA/PM) for both ipsilesional (blue [black in print version]) and contralesional (red [gray in print version]) hemispheres. *Indicates $P < .05$, #indicates $P = .06$.

simultaneous lifting and opening, this ipsilesional shift was driven by an increase in activity in the ipsilesional primary sensorimotor cortex (Fig. 11.6, right).

# Conclusions

In this chapter, we have demonstrated the utility of a robotic device to supplement clinical measurements by providing both quantitative measurements of performance changes and a well-controlled environment to measure neuronal changes related to multijoint movements following upper extremity interventions. We specifically show applications to movements affiliated with the loss of independent joint control, particularly flexion synergy, but similar principles could also be applied to impairments such as weakness, spasticity, etc.

# References

[1] Levin MF, Kleim JA, Wolf SL. What do motor "recovery" and "compensation" mean in patients following stroke? Neurorehabilitation and Neural Repair 2009;23(4):313–9.

[2] Beer RF, Given JD, Dewald JP. Task-dependent weakness at the elbow in patients with hemiparesis. Archives of Physical Medicine and Rehabilitation 1999;80(7):766–72.

[3] Ellis MD, Acosta AM, Yao J, Dewald JP. Position-dependent torque coupling and associated muscle activation in the hemiparetic upper extremity. Experimental Brain Research 2007;176(4):594–602.

[4] Sukal TM, Ellis MD, Dewald JP. Shoulder abduction-induced reductions in reaching work area following hemiparetic stroke: neuroscientific implications. Experimental Brain Research 2007; 183(2):215–23.

[5] Lan Y, Yao J, Dewald JPA. The impact of shoulder abduction loading on volitional hand opening and grasping in chronic hemiparetic stroke. Neurorehabilitation and Neural Repair 2017;31(6): 521–9.

[6] McPherson JG, Chen A, Ellis MD, Yao J, Heckman CJ, Dewald JPA. Progressive recruitment of contralesional cortico-reticulospinal pathways drives motor impairment post stroke. The Journal of Physiology 2018;596(7):1211–25.

[7] Zaaimi B, Edgley SA, Soteropoulos DS, Baker SN. Changes in descending motor pathway connectivity after corticospinal tract lesion in macaque monkey. Brain: A Journal of Neurology 2012;135(Pt 7):2277–89.

[8] Baker SN, Zaaimi B, Fisher KM, Edgley SA, Soteropoulos DS. Pathways mediating functional recovery. Progress in Brain Research 2015;218:389–412.

[9] Baker SN. The primate reticulospinal tract, hand function and functional recovery. The Journal of Physiology 2011;589(Pt 23):5603–12.

[10] Hirschauer TJ, Buford JA. Bilateral force transients in the upper limbs evoked by single-pulse microstimulation in the pontomedullary reticular formation. Journal of Neurophysiology 2015; 113(7):2592–604.

[11] Davidson AG, Buford JA. Bilateral actions of the reticulospinal tract on arm and shoulder muscles in the monkey: stimulus triggered averaging. Experimental Brain Research 2006;173(1):25–39.

[12] Owen M, Ingo C, Dewald JPA. Upper extremity motor impairments and microstructural changes in bulbospinal pathways in chronic hemiparetic stroke. Frontiers in Neurology 2017;8:257.

[13] Nudo RJ, Wise BM, SiFuentes F, Milliken GW. Neural substrates for the effects of rehabilitative training on motor recovery after ischemic infarct. Science 1996;272(5269):1791–4.

[14] Nishibe M, Urban 3rd ET, Barbay S, Nudo RJ. Rehabilitative training promotes rapid motor recovery but delayed motor map reorganization in a rat cortical ischemic infarct model. Neurorehabilitation and Neural Repair 2015;29(5):472–82.

[15] Jones TA. Motor compensation and its effects on neural reorganization after stroke. Nature Reviews Neuroscience 2017;18:267–80.

[16] Sawaki L, Butler AJ, Leng X, et al. Constraint-induced movement therapy results in increased motor map area in subjects 3 to 9 months after stroke. Neurorehabilitation and Neural Repair 2008;22(5): 505–13.

[17] Takahashi CD, Der-Yeghiaian L, Le V, Motiwala RR, Cramer SC. Robot-based hand motor therapy after stroke. Brain: A Journal of Neurology 2008;131(Pt 2):425–37.

[18] Boake C, Noser EA, Ro T, et al. Constraint-induced movement therapy during early stroke rehabilitation. Neurorehabilitation and Neural Repair 2007;21(1):14–24.

[19] Saleh S, Fluet G, Qiu Q, Merians A, Adamovich SV, Tunik E. Neural patterns of reorganization after intensive robot-assisted virtual reality therapy and repetitive task practice in patients with chronic stroke. Frontiers in Neurology 2017;8:452.

[20] Ellis MD, Carmona C, Drogos J, Dewald JPA. Progressive abduction loading therapy with horizontal-plane viscous resistance targeting weakness and flexion synergy to treat upper limb function in chronic hemiparetic stroke: a randomized clinical trial. Frontiers in Neurology 2018;9:71.

[21] Ellis MD, Sukal-Moulton T, Dewald JP. Progressive shoulder abduction loading is a crucial element of arm rehabilitation in chronic stroke. Neurorehabilitation and Neural Repair 2009;23(8):862–9.

[22] Wittenberg GF, Schaechter JD. The neural basis of constraint-induced movement therapy. Current Opinion in Neurology 2009;22(6):582–8.

[23] Michielsen ME, Selles RW, van der Geest JN, et al. Motor recovery and cortical reorganization after mirror therapy in chronic stroke patients: a phase II randomized controlled trial. Neurorehabilitation and Neural Repair 2011;25(3):223–33.

[24] Hubbard IJ, Carey LM, Budd TW, et al. A randomized controlled trial of the effect of early upper-limb training on stroke recovery and brain activation. Neurorehabilitation and Neural Repair 2015;29(8):703–13.

[25] Wilkins KB, Owen M, Ingo C, Carmona C, Dewald JPA, Yao J. Neural plasticity in moderate to severe chronic stroke following a device-assisted task-specific arm/hand intervention. Frontiers in Neurology 2017;8:284.

[26] Wilkins KB, Dewald JPA, Yao J. Intervention-induced changes in cortical connectivity and activity in severe chronic hemiparetic stroke. NeuroImage: Clinical [submitted for publication] https://doi.org/10.1101/547083.

[27] Camona C, Wilkins KB, Drogos J, Sullivan JE, Dewald JPA, Yao J. Improving hand function of severely impaired chronic hemiparetic stroke individuals using task-specific training with the ReIn-Hand system: a case series. Frontiers in Neurology 2018;9:923.

[28] Beer RF, Ellis MD, Holubar BG, Dewald JP. Impact of gravity loading on post-stroke reaching and its relationship to weakness. Muscle and Nerve 2007;36(2):242–50.

[29] Miller LC, Dewald JP. Involuntary paretic wrist/finger flexion forces and EMG increase with shoulder abduction load in individuals with chronic stroke. Clinical Neurophysiology: Official Journal of the International Federation of Clinical Neurophysiology 2012;123(6):1216–25.

# 12

# Quantitative evaluation

Wai Leung Ambrose Lo, Le Li

*DEPARTMENT OF REHABILITATION MEDICINE, GUANGDONG ENGINEERING AND TECHNOLOGY RESEARCH CENTER FOR REHABILITATION MEDICINE AND TRANSLATION, THE FIRST AFFILIATED HOSPITAL, SUN YAT-SEN UNIVERSITY, GUANGZHOU, GUANGDONG PROVINCE, CHINA*

**Chapter outline**



## Introduction: the need for quantitative outcome measures

Neurorehabilitation covers a broad spectrum of specialties. The aim of any rehabilitation program is to improve a person's quality of life by delivering a set of evidence-based interventions. Intervention effectiveness is often based on functional outcomes. It is perhaps not surprising to find that a large number of outcome measures are function based. While function-based outcome measures are perfectly fit for their purposes, they do suffer from several disadvantages. One of the disadvantages of functional outcome measures is that they only focus on task execution, and do not give sufficient attention to the quality of movement. A very classic example is reach to grasp motion. Reach to grasp motion is a common element in many upper limb function scales. However, biomechanics studies have repeatedly raised doubts as to whether such a task is meaningful since the reach forward and grasp motion could be carried out incorrectly in numerous ways, such as excessive trunk rotation or shoulder flexion. The application of this

Intelligent Biomechatronics in Neurorehabilitation. https://doi.org/10.1016/B978-0-12-814942-3.00012-X

functional scale is also limited to those who have some degree of motor ability to perform the test. For those who have more severe paralysis with little or no movement, alternative forms of evaluation must be established. Another commonly used functional outcome measures is the 6-min walk test. This test involves walking for 6 min with the total walked distance recorded to reflect gait function. The test itself does not take into consideration the quality of movement, such as excessive hip abduction, hip rotations, and foot drop. The identification of the quality of gait pattern is often done by visual inspection, which is subjective and highly dependent on the experience of the clinicians. The well-known Modified Ashworth Scale (MAS) instrument is regularly used by clinicians and researchers to assess muscle spasticity. This test involves the assessor feeling the amount of resistance while moving the joint through the range of a joint passively. The test has repeatedly been cited in the literature as unreliable and lacking validity. Some authors have gone as far as calling for a stop to the use of MAS to evaluate muscle spasticity [1]. Another commonly used clinical test for assessing muscle strength is the Manual Muscle Test (MMT). The MMT requires the clinician to apply a force to a joint and score the amount of resistance on a five-point scale. A validity study on MMT reported moderate to strong Pearson/Spearman correlation coefficients between 0.48 and 0.9 when compared with measurements of strength produced using dynamometers [2]. A review published in 2005 [2] first raised doubt that MMT was insensitive in detecting differences, casting doubt on its suitability as a screening test for strength impairment. Validity studies published 13 years later again suggested that it was not a reliable tool to assess the muscle strength of a single muscle group [3] as only a weak correlation was observed between the maximum voluntary contraction force (measured by a dynamometer) and the MMT classification [4]. Due to these known limitations to existing outcome measures, it is essential to have alternative means of quantitative evaluation to complement existing tools and discover the underlying mechanism that may not be apparent under clinical observation. In this chapter, we introduce some of the emerging evaluation methods used to quantitatively assess muscle function. These evaluation methods are tested and frequently adopted in the clinical research program within our institutes.

## Electrical impedance myography (EIM)

EIM technology is relatively new and is gaining popularity in the assessment of muscle structure due to its convenience. The noninvasive bioimpedance-based technique is generally used to detect and quantify muscle health by sending a high-frequency, low-intensity current into a discrete region of muscle tissue. The consequent voltage within the region is then measured [5]. The technique relies on the principle of conductivity (the quality to pass electrical current) and permittivity (the capability to store electrical charge) of soft tissues. The characteristics of conductivity and permittivity are altered in muscle dysfunction, causing changes in the voltage being generated. Two basic EIM parameters are recorded: resistance, which represents the resistivity to current flow through the intra- and extracellular fluids; and reactance, which indicates how the current flow is affected by cell membranes and by the various fascia of the body. Anisotropy is a key phenomenon of EIM. It refers to the characteristic of varying impedance values when electrical current flows along versus across a skeletal muscle. Early research indicated that it was possible to identify various types of neuromuscular disorders [6], and differentiate between type I and type II muscle fibers [7] with EIM measurements. The majority of the existing literature was conducted in rat models or in healthy individuals. The application of EIM in neurorehabilitation to identify muscle dysfunction is a relatively new area and needs to be continuously refined. In addition, most of the current literature has focused on measuring the muscle impedance while the muscle was in a relaxed state. There is minimum information available on how muscle impedance is changed at different levels of muscle contraction.

## EIM measurements during muscle contraction

A study published by our research group previously investigated the feasibility of using EIM to assess the architecture of muscles and how it changed at different rates of contraction. In the study, 19 healthy participants were recruited and biceps brachii muscle function of the dominant arm was assessed by an hand-held electrical array impedance measurement system (EIM1103; Skulpt Inc., Boston, MA, USA) (Fig. 12.1). The arrangement of the electrodes is shown in Fig. 12.2. Participants were seated in a comfortable position with the tested arm securely resting on a bespoke elbow torque-measuring device. The elbow of the tested arm was maintained at 90 degrees flexion, shoulder at 45 degrees abduction, with the wrist joint supported in a neutral position. The temperature of the laboratory was maintained at 22°C during data collection.

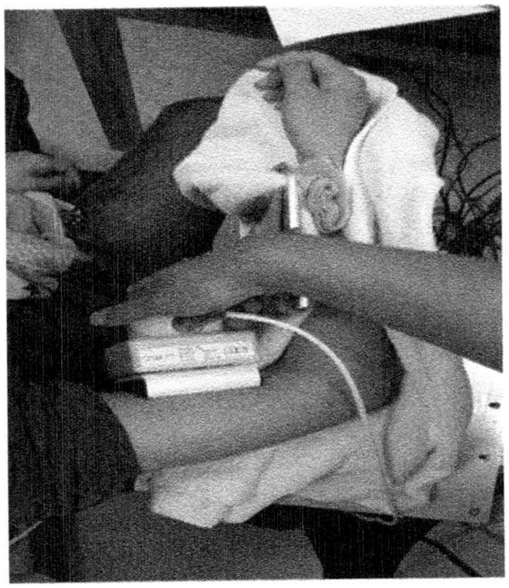

**FIGURE 12.1** Hand-held electrical array impedance measurement system.

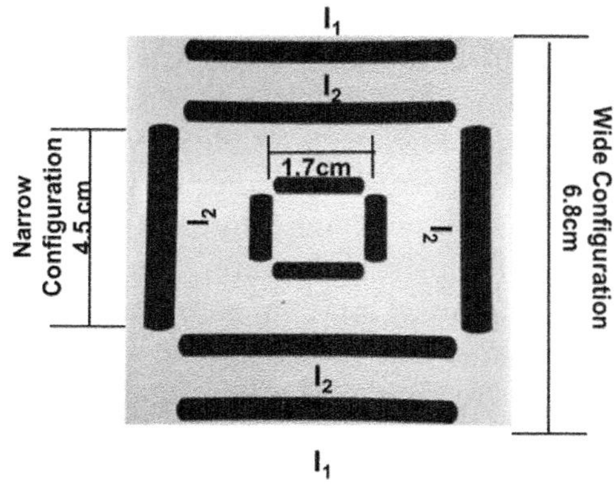


**FIGURE 12.2** Setup for the electrodes of a hand-held electrical array.

Impedance measurements were recorded during; (1) rest (baseline); (2) maximum voluntary contraction (MVC); (3) 20% (MVC); and (4) 60% MVC. An additional fatigue contraction test was conducted where the participants were required to maintain isometric contraction at 60% MVC. Task failure was defined as being unable to maintain within 10% threshold of the 60% MVC value for longer than 3 s. EIM measurements for fatigue test were recorded at (1) baseline (at rest prior to start of the task); (2) start of contraction; (3) end of contraction (before task failure); and (4) 1 min postfatigue during the relaxed state. The time to task failure was also noted. Our results indicated that resistance increased significantly during a high level of contractions and decreased with muscle fatigue. These findings indicated that the measurement of resistance reflects a net resistance from the metabolic and architectural changes during muscle contraction. The parameter of reactance, however, did not have any significant changes during each level of isometric contraction. A possible reason for this was that muscle contractions did not affect the muscle fiber thickness, but more of the muscle length. A previous study by Li et al. [8] found that membrane capacitance had the greatest impact on reactance value and that muscle fiber size reduces the charge storage capacity of the tissue. Thus a physical reduction in muscle fiber size would be more likely to affect reactance reading than a change of muscle fiber length during muscle contraction.

Muscle fatigue is a complicated area to study since it involves both a subjective feeling of weakness or muscle pain, with or without the presence of muscle force reduction [9]. Therefore, quantifying muscle fatigue and its associated physiological markers are essential for clinical practice and research. From the results of a fatigue contraction test, a significant decrease in resistance was observed as the muscle fatigued. The resistance recorded before task failure was lower in comparison to the measurement recorded at the beginning of the contraction. A positive significant correlation between the change to resistance and the contraction time was observed. The review performed by Allen et al. [10] proposed that the accumulation of metabolites during prolonged isometric contraction is a contributing factor to muscle fatigue. Several other studies also indicated that sustained isometric contraction contributed to muscle ischemia. The reduction in resistance was likely related to the build-up of metabolites and intracellular fluid during sustained contraction, which in turn increased muscle conductivity and reduced resistance. Unlike resistance, the reactance parameter did not differ significantly during contraction and fatigue. One of the contributing factors to the lack of difference was that reactance may return quickly to baseline after contraction despite the muscle still being in a fatigue state. The reactance of cell membranes depends on the frequency of the injecting current, thus an increase in metabolites and intracellular fluid during sustained contraction was unlikely to affect the measurement of reactance.

To date, there is some evidence to demonstrate that the EIM is a valid method to assess the state of muscles. The results from this research indicated that the parameters of resistance may detect changes in muscle architecture and physiological factors and potentially be used to detect the fatigue state of muscles. These data provided an insight into assessing muscle function, in particular in the aspect of force generation in patients with neurological conditions.

## Application of EIM in spinal cord injury

Scientific studies published over the past several years have investigated whether EIM could identify muscle structure alterations in people with neurological deficits. One of the areas that may benefit from EIM application is the assessment of hand and upper limb muscles. Upper limb functional recovery after neurological insults has been a long-standing issue [11]. There is an urgent need to assess the architectural changes to muscle quantitatively, either as outcome measures or to provide a new insight into alternative rehabilitation techniques. Two separate studies were conducted by our research group to investigate the feasibility of using EIM in people with spinal cord injury. In one of the studies, 16 people with incomplete spinal cord injury were assessed (injured levels between C2 to C6, American Spinal Injury Association impairment levels range between B to D). EIM measurements of bilateral thenar, hypothenar muscle groups and the first dorsal interosseous muscles were recorded and compared to healthy individuals. The setup of EIM was the same as that used in a previous study (Fig. 12.2). Measurements were recorded at rest with the arm supported by the chair arm. The results indicated there was minimal difference in resistance between the stronger and weaker side of the body within the SCI group, and between the spinal cord injury group and the healthy group. The lack of difference is somewhat difficult to explain due to contradictory theories proposed in the literature on the architectural and physiological alterations in muscles post spinal cord injury. One of the theories was that as muscle atrophy occurs, fatty tissues and connective tissues may replace muscle fibers [12], which in turn increase resistance. The early literature indicated that spinal cord injury survivors have increased amounts of type II muscle fibers 1 year after injury [13]. Type II muscle fibers have less fatty content in comparison with type I fibers [14], which would have reduced the resistance. As the sample population included people with spinal cord injury duration between 1 and 30 years, they may have been in different stages of physiological changes within the muscle fibers. It was therefore difficult to draw firm conclusions from the results. The parameter of reactance, which was associated with the oscillating charges across the membranes of muscle cells, was significantly lower in the spinal cord injury group than the healthy group in all tested muscles. This finding may reflect the integrity of cell membrane after injury since reactance is mainly affected by the sarcolemmal surface area, which is particularly sensitive to myocyte atrophy that is associated with denervation disorders. This interpretation is given support by a study that investigated bioimpedance analysis in athletes with muscle injury [15]. The study reported that muscle reactance measurement decreased with increasing muscle injury severity, suggesting disruption to cellular membrane integrity and injury.

In another study, we assessed the architecture of bicep brachii [16] in people with spinal cord injury. Seventeen people with neurological injury between levels C2 to C6 were recruited. The mean duration of injury was 8.7 years. A further 23 healthy individuals were recruited as controls. EIM measurements were performed on bilateral biceps brachii in the spinal cord injury group and on the dominant arm of healthy individuals. Participants were seated in a height-adjustable chair. The involved arm was supported by the chair arm or by the examiner with the elbow joint placed at 90 degrees. The parameters of resistance and reactance were recorded. The findings were similar to our previous study in that no significant difference in muscle resistance was observed between healthy participants and participants with spinal cord injury. As discussed previously, resistance was proposed to be mainly affected by muscle size. A previous study revealed that the wrist extensor muscles were 20%–35% those in people with tetraplegia as compared with healthy individuals. While reduced muscle size is expected to reduce resistance, however, as previously mentioned, there are multiple physiological changes that may contribute to different results.

The result for reactance measurement was consistent with our previous study that a significant difference was observed between healthy participants and participants with spinal cord injury. Several authors indicated that the reduction in reactance may be related to deterioration of the cell membrane. The deterioration is not limited only to physical deterioration but also refers to the ability to control ionic flow across the cell membrane or decreased ability to store electrical charge due to reduced size of individual myocyte sarcolemma.

These findings substantiate that EIM may be a feasible way to assess cellular membrane integrity or as an outcome measure to evaluate target therapies.

In summary, existing evidence indicated that EIM is a feasible way to provide quantification of muscle fiber size, structure, and overall composition. There are several aspects of EIM that need to be refined. One of these aspects includes the impact of the subcutaneous fat layer on EIM measurement. There has been a continued discussion on this topic since subcutaneous soft tissue has a direct influence on the passing current. In addition, there is a different setup of the electrodes that may also have an impact of the measurements. One of our studies demonstrated that a longitudinal configuration with wide distance between electrodes may reduce the effects of subcutaneous fat when compared with the narrow distance setup [17]. While this study shed essential insight on how to reduce the impact of the subcutaneous fat layer on EIM readings, it remains unclear as to whether these parameters are clinically meaningful or have a strong correlation to clinical status in different neurological conditions.

## Muscle spasticity

Muscle possesses both viscous and elastic properties. The viscoelastic property of muscle is referred to as muscle tone. From a biomechanical point of view, the mechanical stress that is present in a relaxed muscle may be regarded as muscle tone. It can also be regarded as a manifestation of stretch reflex neural motor control [18]. A degree of resting muscle tone is necessary for the background active muscle work to maintain posture and body position between different trunk and limb segments. The tone in all neuromuscular systems is considered to be provided through the central nervous system (neural component), and through cellular tone (nonneural component). Increased muscle tone may increase the pressure on blood vessels and nervous tissue, resulting in pain and suboptimal muscle contraction. Low muscle tone may indicate flaccidity or weak muscles. There are many muscular dysfunctions that are associated with altered muscle tone such as muscle contracture and chronic pain. Altered muscle tone is suggested to be involved in the pain–spasm–pain vicious circle [19], which may be a contributor to chronic pain conditions. Therefore, muscle tone is one of the key considerations in the field of neural rehabilitation and objectively quantifying muscle tone is highly desirable in research and clinical setting. As mentioned previously, muscle tone is often clinically assessed by the Modified Ashworth Scale. This scale has received criticism over recent years over a lack of reliability and validity. Surface electromyography (sEMG) has traditionally been used to assess muscle electrical activity to reflect muscle tone. Some authors have proposed that sEMG may not fully reflect muscle tone, since it mostly concerns electrical activity in the muscle cells instead of the mechanical properties of tone that is present in the absence of electrical activity. One of the methods that was proposed approximately 40 years ago to quantify muscle tone is the soft-tissue indentation method. The principle of the indentation method is based on applying a known amount of mechanical force to the muscle and generating oscillation within the muscle fiber, the mechanical muscle tone can be quantitatively assessed. This method appears to be widely popular among researchers, as evidenced in the sharp increase in the number of publications on myotonometry utilizing the indentation method over recent years.

## Myotonometer—validity

The validity of the technology used for myotonometry has been investigated in several studies [20–22]. A recently published studies validated the myotonometric measurement of elbow flexors against the stretch technique that measures the changes in resistance torque during repeated joint rotations controlled by a servomotor [20]. The results indicated myotonometry and conventional passive stretch techniques could identify substantial changes in the spastic muscles and both measurements indicated a significant linear relation between the two measurements. Frohlich-Zwahlen et al. [21] investigated the validity of the lower limb muscles of chronic stroke patients with limited hypertonia by comparing the parameters to muscle thickness and muscle strength measured by ultrasound. The results indicated that muscle strength was positively correlated with biceps femoris stiffness and tone. Muscle thickness was positively correlated with tibialis anterior stiffness and tone. The tone and stiffness of hand muscles were also shown to be significantly correlated with hand strength and upper limb motor function [22]. In addition to the validation studies, statistically significant differences in muscle mechanical proprieties between different age groups in women [23] and in a mixed population [24], between patients with Parkinson's disease and healthy individuals, and between patients with chronic stroke and healthy individuals [25] were observed in published clinical trials. Existing evidence suggests that myotonometry is a valid technology to record muscles' mechanical properties. Despite these evidences and the fact that the technology has been around for over a decade, there appears to be ongoing concern regarding its validity. The ongoing argument against the indentation method is that it somewhat omits the contribution from all the soft-tissue structures that are above the muscle fibers (e.g., skin, subcutaneous soft tissue, and myofascial that surrounds the muscle fibers). It is currently unclear if the fixed amount of force delivered by the device is sufficient to be transmitted through all the soft-tissue structures and reach the desired muscle to induce the oscillation within the fiber. If the indentation force was not transmitted to the desired muscle structure, then the measurement may lack validity. This issue may be resolved by limiting the use of myotonometry to individuals with a "normal" body mass index (BMI). However, the BMI is an indicator of whole body fat ratio and does not take into consideration the variation of subcutaneous soft-tissue thickness of local muscles. Our research group identified two studies that reported the relations between subcutaneous fat and muscle tone. One of the studies measured subcutaneous fat thickness of rectus femoris with ultrasound imaging and correlated the findings with myotonometer-measured muscle tone. Fröhlich-Zwahlen et al. [21] investigated whether subcutaneous fat layers influenced myotonometer measurements in 20 participants with stroke and healthy participants of different ages and genders. Ultrasound imaging was used to measure subcutaneous fat thickness over the rectus femoris, vastus lateralis, bicep femoris, and tibialis anterior. The correlation coefficients between muscle tone and subcutaneous fat thickness were −0.56, −0.02, −0.74 (statistically significant), and −0.19 (nonstatistically significant), respectively. Although these correlation values were statistically significant, the low to moderate correlation indicated that subcutaneous fat may not affect the measurement of muscle tone. This study has the limitation of pooling the data from different age groups and genders. This did not allow for the effects of different factors, such as age and gender, that are known to vary, on the relationship with subcutaneous fat to be taken into account. A pooled correlation analysis may exaggerate the relationship between muscle tone and subcutaneous fat. Agyapong-Badu [24] conducted a study to further investigate the relationship between muscle tone and subcutaneous fat by subgrouping healthy people of different ages. The study recruited young (18–35 years old) and elderly individuals (65–90 years old) [24]. The sample cohort had normal mean BMI. The results indicated a significant moderate correlation in young female, elderly male, and female groups, and a low nonstatistical significant correlation in the young male group. The findings from both studies were inconclusive but at least indicated that myotonometer-measured muscle tone value may not be significantly affected by subcutaneous fat layers, at least in people of certain age groups.

### *Myotonometer—reliability*

The reliability of myotonometry has been investigated extensively in several pieces of research concerning people with neurological conditions and other neuromuscular pathologies. For the stroke population, the reliability of myotonometry has been investigated in both laboratory and clinical settings and in people at different stroke stages. Our research group was among the first to investigate the reliability of myotonometry in people with acute stroke in a clinical setting. The primary goal of the study was to investigate whether using the device in a clinical environment was feasible and to establish the likely range of error in muscle tone measurement in this population group. Muscle tones of biceps brachii, brachioradialis, rectus femoris, and tibialis anterior muscles were recorded. Measurements were recorded by one assessor on two consecutive days (between-days intrarater reliability) [26], and by two assessors within the same day (within-day interrater reliability) [27]. The results indicated adequate reliability for both between-days (ICC $>0.75$) and within-day (ICC 0.63–0.97) muscle tone measurement. While the overall ICC value appears to be acceptable, there were concerns about the variability within the dataset. The lower bound of the ICC for some of the measurements was below 0.75. According to some literature, the wide confidence interval implies low statistical power, leading to inconclusive conclusions. Some authors argued that a poor lower bound value is indicative of poor precision of the instrument. Depending on the requirements for the measurement, a wide confidence interval may be deemed unacceptable, despite the acceptable ICC value. One of the key findings from our studies was that different muscle groups appeared to have different levels of reliability, with some muscle groups yielding higher reliability indices than others. We believe that although myotonometry is a more useful instrument for numerically quantifying muscle tone than subjective scales, the interpretation of the data must be treated with caution due to the varying level of consistency.

### Interpretation of myotonometric data

Despite myotonometry having been around for many years, the interpretation of myotonometric muscle tone measurement continues to be unclear. There is little information regarding the normative values that clinicians could use as a reference, limiting its application in the area of diagnostics. This is not uncommon for many other outcome measures, such as sEMG or dynamometry, where their main application is to quantify changes. Thus, a myotonometer should not be used alone to diagnose muscle spasticity condition, at least the current body of literature does not support such practice. Instead myotonometric measurements should be used to compare between body sites to identify differences in muscle condition. Some literature has proposed that a between-side difference of below 5% may be considered normal [28]. However, the study that was conducted within our research group in the acute stroke population revealed that not all tested muscles on the paraplegic side had over a 5% difference. We observed the muscle tone of brachioradialis was 12% lower on the affected side than the unaffected side. Muscle tone of biceps brachii, rectus femoris, and tibialis anterior have between-sides differences of between 10.12% and 2.8%. Our findings therefore did not support that the myotonometer-measured muscle tone may be able to distinguish between the affected and unaffected sides for all muscle groups. This finding should be interpreted with caution as we only recorded muscle tone in people who scored between 0 and 2 on the Modified Ashworth Scale. Thus, it is unclear whether the sensitivity would be improved if people with high muscle tone were included. One of the more promising applications would be used as an outcome measure since it only concerns the difference before and after the intervention. The weight of evidence suggests that the myotonometer device may be sufficiently reliable to be used as an outcome measure tool. There are data available within the body of literature to identify the error range for difference muscle groups with a range of neurological disorders. A summary of the data from reliability studies is provided in Table 12.1. These data may form the reference for clinicians when using such a device. To date, it is unclear if the reported error range is clinically acceptable as there are insufficient data to determine the amount of change that can be expected from intervention.

## Ultrasound imaging

Ultrasound is a form of noninvasive medical imaging that allows in vivo measurement of muscle architecture. It is commonly cited in the literature that kinematic analysis and clinical scales are frequently used in the clinical setting to evaluate neuromuscular and functional outcomes after intervention [29]. Kinematic analysis has the main focus on motor execution and clinical scales rely [30] on the clinical experiences of the assessor [31]. These parameters do not assess changes at the level of muscle structure. Therefore, muscle architectural changes post stroke and post intervention assessed by ultrasound imaging may assist in evaluation of the mechanism underlying the pathophysiology and

**Table 12.1** A summary of the data from reliability studies for myotonometer-measured muscle tone.

| Study | Condition | N (type of reliability) | Muscles | ICC | SEM (Hz) | SRD (Hz) |
|---|---|---|---|---|---|---|
| Chuang et al. [22] | Subacute and | 12 | Biceps | 0.96 (0.86–0.99) | 0.34 | 0.79 |
| | Chronic stroke | Within-day | Triceps | 0.90 (0.68–0.97) | 0.7 | 1.63 |
| Chuang et al. [36] | Chronic stroke | 61 | Deltoid | 0.92 (0.87–0.95) | 1.16 | 3.21 |
| | | Within-day | Biceps brachii | 0.94 (0.89–0.96) | 0.5 | 1.38 |
| | | | Triceps brachii | 0.87 (0.78–0.92) | 0.78 | 2.16 |
| | | | Extensor digitorum | 0.92 (0.86–0.95) | 0.72 | 1.99 |
| | | | Flexor carpi radialis | 0.95 (0.92–0.97) | 0.66 | 1.82 |
| | | | Flexor carpi ulnaris | 0.93 (0.88–0.96) | 0.65 | 1.8 |
| Lo et al. [26,27] | Acute stroke | 29 | Biceps brachii | 0.83 (0.63–0.92) | 0.49 | 1.14 |
| | | Within day | Brachioradialis | 0.93 (0.86–0.97) | 0.48 | 1.12 |
| | | | Rectus femoris | 0.86 (0.71–0.93) | 0.73 | 1.71 |
| | | | Tibialis anterior | 0.86 (0.71–0.93) | 0.63 | 1.47 |
| Lo et al. [26,27] | Acute stroke | 28 | Biceps brachii | 0.75 (0.49–0.89) | 0.76 | 2.41 |
| | | Between days | Brachioradialis | 0.82 (0.47–0.93) | 0.35 | 1.96 |
| | | | Rectus femoris | 0.81 (0.58–0.91) | 0.83 | 2.52 |
| | | | Tibialis anterior | 0.81 (0.37–0.86) | 1.24 | 3.08 |
| Ko et al. [37] | SCI | 13 | Right medial gastrocnemius | 0.91 (0.62–0.97) | 0.84 | |
| | | Within-day | Left medial gastrocnemius | 0.90 (0.65–0.97) | 1.06 | |
| | | | Right lateral gastrocnemius | 0.90 (0.66–0.97) | 1.02 | |
| | | | Left lateral gastrocnemius | 0.37 (−1.05–0.81) | 3.84 | |
| | | | Right tibialis anterior | 0.6 (−0.41– 0.88) | 1.51 | |
| | | | Left tibialis anterior | 0.78 (0.25–0.93) | 0.91 | |

rehabilitation program. Beyond basic morphology of muscle thickness and contraction rate, fascicle length is probably the most functionally important aspect of skeletal muscle [32] since the functional range is affected by the length of the fibers. There is a large amount of literature reporting the method to obtain the muscle pennation angle from ultrasound images and the estimation of fascicle length. Readers are encouraged to further explore the literature on the detailed calculation of these parameters. A study published in 2007 [32] validated the reliability of fascicle length measurements of brachialis muscle in people with stroke. The study also reported the difference in muscle architecture changes between the affected and nonaffected sides. In the study, seven adults with stroke-related hemiparesis that lasted for at least 2 years were recruited.

Muscle architect was assessed by B-mode ultrasonography scanner. Ultrasound images were taken at rest when the elbow was in a range of different angles (range between 10 and 90 degrees of flexion using 10-degree increments), and at different contraction levels (20%, 40%, 60%, and 80% of MVC) when the elbow joint was fixated at 90 degrees flexion. The results supported that the measurement of muscle pennation angle and fascial length were reliable parameters (ICC 0.869 and 0.85, respectively) when used in the stroke population group. The study observed that the pennation angle increased and fascicle length decreased as the muscle contraction level increased. This finding was consistent with another study that also observed these changes during isometric contraction of the gastrocnemius in healthy individuals [33]. The inverse change of muscle pennation angle with the fascicle length in muscle contraction was related to a decrease in the fascicle length. The fascicle length of brachialis on the affected side was shorter than the unaffected side and was further compromised when the elbow joint moved into a flexed position. The shortening of fascicle length may be related to spastic hypertonia since individuals with stroke have a tendency to hold the arm in a flexed position or have limited movement. This may contribute to shorter fascicle length, which in turn increases the pennation angle. The alteration of muscle architecture is related to the ability to generate force. The changes of pennation angle and fascicle length have a direct influence of the physiology cross-section area [34]. It is reported that the larger the physiological cross-section area of the muscle, the larger the force generation potential [34]. A study published by our research group investigated the possibility to estimate muscle force by means of biomechanical modeling [35]. Ultrasound measurements were recorded on the elbow extensor from five individuals who had stroke-related hemiparesis for over 1 year, and healthy individuals. The results indicated that the calculated maximum isometric muscle stress by the biomechanical model of individuals with stroke was lower than for healthy individuals, suggesting weakness in potential muscle force generation.

## Muscle architecture changes induced by intervention

In one of our published studies, we attempted to demonstrate that the changes to lower limb motor recovery post intervention may be related to changes of muscle architecture in a randomized controlled trial. Fifteen participants with subacute stroke and eight age-matched healthy individuals were recruited to the study. The mean age of the subacute stroke cohort was 60.5 years old (range: 51–73 years old). Participants were randomly allocated to routine rehabilitation care and body weight support treadmill training with routine rehabilitation. Ultrasound measurements were taken at baseline and post intervention when the muscles were in relaxed and contracted states. Muscle architecture parameters of pennation angle and fascicle length were recorded from tibialis anterior and medial gastrocnemius. Our results demonstrated that the pennation angle and muscle thickness of tibialis anterior increased post intervention. The fascicle length of medial gastrocnemius and strength were increased. A significant positive correlation was observed between tibialis anterior pennation angle and muscle strength. Muscle pennation angle is related to the number of muscle fibers and the contractile unit attached to the tendon but not the efficiency to transmit force from muscle fibers to tendon. Findings from our study indicated that treadmill training facilitated more muscle fiber growth in the cross-sectional area.

## Conclusion

This chapter has critically evaluated the evidence for several quantitative evaluation methods. These methods complement existing evaluation methods, assisting in the identification of change within muscles fibers and subtle changes that might not be obvious. These quantitative evaluations need to be continuously refined to gain further understanding of their applications in quantitative evaluation of the neural rehabilitation field.

## References

[1] Pandyan AD, et al. A review of the properties and limitations of the Ashworth and modified Ashworth Scales as measures of spasticity. Clinical Rehabilitation 1999;13(5):373–83.

[2] Bohannon RW. Manual muscle testing: does it meet the standards of an adequate screening test? Clinical Rehabilitation 2005;19(6):662–7.

[3] Baschung Pfister P, et al. Manual muscle testing and hand-held dynamometry in people with inflammatory myopathy: an intra- and interrater reliability and validity study. PLoS One 2018;13(3): e0194531.

[4] Manikowska F, et al. Validation of Manual Muscle Testing (MMT) in children and adolescents with cerebral palsy. NeuroRehabilitation 2018;42(1):1–7.

[5] Rutkove SB. Electrical impedance myography: background, current state, and future directions. Muscle & Nerve 2009;40(6):936–46.

[6] Li J, et al. Distinguishing neuromuscular disorders based on the passive electrical material properties of muscle. Muscle & Nerve 2015;51(1):49–55.

[7] Sanchez B, et al. Differentiation of the intracellular structure of slow- versus fast-twitch muscle fibers through evaluation of the dielectric properties of tissue. Physics in Medicine and Biology 2014;59(10):2369–80.

[8] Li J, et al. Electrical impedance alterations in the rat hind limb with unloading. Journal of Musculoskeletal and Neuronal Interactions 2013;13(1):37–44.

[9] Romani A. The treatment of fatigue. Neurological Sciences 2008;29(Suppl. 2):S247–9.

[10] Allen DG, Lamb GD, Westerblad H. Skeletal muscle fatigue: cellular mechanisms. Physiological Reviews 2008;88(1):287–332.

[11] Chen L, et al. Assessing the relationship between motor anticipation and cortical excitability in subacute stroke patients with movement-related potentials. Frontiers in Neurology 2018;9:881.

[12] Gorgey AS, Dudley GA. Skeletal muscle atrophy and increased intramuscular fat after incomplete spinal cord injury. Spinal Cord 2007;45(4):304–9.

[13] Burnham R, et al. Skeletal muscle fibre type transformation following spinal cord injury. Spinal Cord 1997;35(2):86–91.

[14] Izumiya Y, et al. Fast/Glycolytic muscle fiber growth reduces fat mass and improves metabolic parameters in obese mice. Cell Metabolism 2008;7(2):159–72.

[15] Nescolarde L, et al. Localized bioimpedance to assess muscle injury. Physiological Measurement 2013;34(2):237–45.

[16] Li L, et al. Alterations in localized electrical impedance myography of biceps brachii muscles paralyzed by spinal cord injury. Frontiers in Neurology 2017;8:253.

[17] Li L, et al. The effect of subcutaneous fat on electrical impedance myography: electrode configuration and multi-frequency analyses. PLoS One 2016;11(5):e0156154.

[18] Simons DG, Mense S. Understanding and measurement of muscle tone as related to clinical muscle pain. Pain 1998;75(1):1–17.

[19] Mense S, Masi AT. Increased muscle tone as a cause of muscle pain. Springer Berlin Heidelberg; 2010. p. 207–49.

[20] Li X, et al. Assessing muscle spasticity with myotonometric and passive stretch measurements: validity of the myotonometer. Scientific Reports 2017;7:44022.

[21] Frohlich-Zwahlen AK, et al. Validity of resting myotonometric assessment of lower extremity muscles in chronic stroke patients with limited hypertonia: a preliminary study. Journal of Electromyography and Kinesiology 2014;24(5):762–9.

[22] Chuang LL, Wu CY, Lin KC. Reliability, validity, and responsiveness of myotonometric measurement of muscle tone, elasticity, and stiffness in patients with stroke. Archives of Physical Medicine and Rehabilitation 2012;93(3):532–40.

[23] Kocur P, et al. Effects of aging on mechanical properties of sternocleidomastoid and trapezius muscles during transition from lying to sitting position-A cross-sectional study. Archives of Gerontology and Geriatrics 2017;70:14–8.

[24] Agyapong-Badu S, et al. Measurement of ageing effects on muscle tone and mechanical properties of rectus femoris and biceps brachii in healthy males and females using a novel hand-held myometric device. Archives of Gerontology and Geriatrics 2016;62:59–67.

[25] Rydahl SJ, Brouwer BJ. Ankle stiffness and tissue compliance in stroke survivors: a validation of myotonometer measurements. Archives of Physical Medicine and Rehabilitation 2004;85(10):1631–7.

[26] Lo WLA, et al. Between-days intra-rater reliability with a hand held myotonometer to quantify muscle tone in the acute stroke population. Scientific Reports 2017;7(1):14173.

[27] Lo WLA, et al. Relative and absolute interrater reliabilities of a hand-held myotonometer to quantify mechanical muscle properties in patients with acute stroke in an inpatient ward. BioMed Research International 2017;2017:4294028.

[28] Gapeyeva H, Vain A. Methodological guide: principles of applying Myoton in physical medicine and rehabilitation. Tartu, Estonia: Muomeetria Ltd; 2008.

[29] Ju MS, et al. Performance of elbow tracking under constant torque disturbance in normotonic stroke patients and normal subjects. Clinical Biomechanics 2002;17(9):640–9.

[30] Yeh C-Y, Chen J-JJ, Tsai K-H. Quantitative analysis of ankle hypertonia after prolonged stretch in subjects with stroke. Journal of Neuroscience Methods 2004;137(2):305–14.

[31] Pizzi A, et al. Application of a volar static splint in poststroke spasticity of the upper limb. Archives of Physical Medicine and Rehabilitation 2005;86(9):1855–9.

[32] Li L, Tong KY, Hu X. The effect of poststroke impairments on brachialis muscle architecture as measured by ultrasound. Archives of Physical Medicine and Rehabilitation 2007;88(2):243–50.

[33] Narici MV, et al. In vivo human gastrocnemius architecture with changing joint angle at rest and during graded isometric contraction. Journal of Physiology 1996;496(Pt 1):287–97.

[34] Narici M. Human skeletal muscle architecture studied in vivo by non-invasive imaging techniques: functional significance and applications. Journal of Electromyography and Kinesiology 1999;9(2): 97–103.

[35] Li L, Tong RK. Combined ultrasound imaging and biomechanical modeling to estimate triceps brachii musculotendon changes in stroke survivors. BioMed Research International 2016;2016. 5275768.

[36] Chuang LL, Lin KC, Wu CY, Chang CW, Chen HC, Yin HP, Wang L. Relative and absolute reliabilities of the myotonometric measurements of hemiparetic arms in patients with stroke. Archives of Physical Medicine and Rehabilitation 2013;94(3):459–66.

[37] Ko Chang-Yong, Choi Hyuk-Jae, Ryu Jeicheong, Kim Gyoosuk. Between-day reliability of MyotonPRO for the non-invasive measurement of muscle material properties in the lower extremities of patients with a chronic spinal cord injury. Journal of Biomechanics May 2018;73:60–5. https://doi.org/10.1016/j.jbiomech.2018.03.026.

13

# Automation in neurorehabilitation: needs addressed by clinicians

Ka Leung Marko Chan[1,2,3]

[1]HEALTH CARE (HEALTH TECHNOLOGY) (HKPOLYU); [2]STROKE AND CLINICAL NEUROSCIENCE (CUHK); [3]BIOMEDICAL ENGINEERING (CUHK)

## Chapter outline



Acquired brain injury (ABI) is damage to the brain caused by events after birth. It is not related to genetic or congenital disorders. The injury could result in the impairment of different aspects including cognitive, physical, sensory, perceptual, or psychosocial areas that affect daily function. ABI could be further categorized as traumatic brain injury (e.g., head injury due to accidents or assaults, neurosurgery) and nontraumatic injury (e.g., stroke, brain tumors, poisoning, hypoxic brain damage, encephalopathy).

Clients with acquired brain injury will commonly have cognitive impairment. The impairment may be global and affect various cognitive and perceptual components which include attention, memory, apraxia, executive function, and language. These

Intelligent Biomechatronics in Neurorehabilitation. https://doi.org/10.1016/B978-0-12-814942-3.00013-1

impairments limit the client's function in daily activities and assistance is needed, resulting in increased burden for the care taker.

In view of the great demand on healthcare services to tackle the aging problem and longer life expectancy, the application of rehabilitation technology and automation equipment has develop rapidly in recent decades. It not only improves the treatment effectiveness of the healthcare professional, but also enhances manpower as the therapist can focus on interventions which are human-dependent. The automation system can also lessen the variability between different clinical settings. As such, the global rehabilitation service quality can be improved.

In this chapter, we introduce the conventional approach for the occupational therapist to engage in neurological rehabilitation with a focus on cognitive aspects. In addition, the latest developments in advanced technology and automation systems are discussed.

## Conventional approach in cognitive rehabilitation

### Cognitive functional evaluation

Cognitive function is important for daily performance and is one of the main areas of concern for occupational therapists after injury. Therapists would like to know how the impairment affects the client's life and limits their participation in activities. Therapists also explore the residual strength of the client and then formulate an effective treatment plan to enhance their function.

There are different intervention models for the occupational therapy practitioner to follow. One of the most commonly adopted models in the last decade is cognitive functional evaluation (CFE). This was developed by three renowned occupational therapy professors: Hartman-Maeir et al. [1]. There are six stages to the CFE which guide therapists in conducting a comprehensive evaluation of cognitive aspect and understanding of the impact of the injury on clients' performance. They include the following:

1. Conduct interview and collect background information including an occupational history;
2. Perform cognitive screening and baseline status tests;
3. Conduct general measures of cognition in occupations;
4. Conduct cognitive tests for specific domains;
5. Measure specific cognitive domains in occupations;
6. Environmental assessment.

In the first stage of the interview, the main aim is to collect the clients' detail, which includes premorbid function, social environment, and caretaker support. The therapist could know the client's premorbid performance and understand the disease impact. Throughout the interview with the client and her caretaker, the therapist can discover the self-awareness level of the client and the comments of family members which can predict the compliance and motivation for the intervention program. After conducting the interview, the therapist would conduct a screening test to measure the global cognitive function of the client. Examples of commonly use assessments in local occupational therapy departments include Abbreviated Mental Test (AMT), Mini Mental States Examination (MMSE), and Hong Kong version of Montreal Cognitive Assessment (HK-MoCA). According to the results of those screening tests, the therapist obtains a global picture of the client's cognition and identifies which aspects may need further investigation. Moreover, the occupational therapist will observe the clients' cognitive function in the task. One of the most common assessments is the kettle test. This is a brief performance measure which is based on a complex everyday task and is designed to assess cognitive processes [2]. The client has to prepare tea for the guests and their performance is rated in terms of working memory, sequencing, initiation, and problem-solving.

After we have a brief picture of the client's cognitive function and performance in daily tasks, we can go through a more detailed assessment in specific cognitive areas and assess their impact on daily life. For example, regarding memory, the therapist may conduct the Rivermead Behavioral Memory Test which assesses everyday memory and the therapist will gain an understanding of whether the visual or auditory memory is more impaired through this test. If the therapist suspects the client suffers from a specific attention problem, they may conduct a test of everyday attention to confirm the deficit. Some functional tests also can be adopted to measure how the cognition deficit affects daily performance. For example, for those subjects suffering from unilateral neglect, which is a common deficit after right parietal lobe injury after stroke, the therapist could adopt a behavioral inattention test to determine the severity of the symptoms which are reflected in a real-life situation. This test consists of several function tasks like telephone dialing, menu/article reading, map navigation, coin sorting, etc. It provides rich information for the therapist to design the treatment program. Finally, the environmental assessment [Safety Assessment of Function and Environment for Rehabilitation (SAFER) Tool] will be conducted. This assesses an individual's abilities to safely manage functional activities within their home and covers 97 daily life items. An example of a cognitive assessment flow is given in Fig. 13.1. The long form of assessments are listed in Table 13.1.

## Neurofunctional approach

The neurofunctional approach (NFA) is designed for clients with severe deficits following traumatic brain injury (TBI) [3]. It is indicated for those individuals who had limitations in their problem-solving ability or generalized skills across different tasks and environments. In addition, it is indicated for those who lack insight and are limited in their engagement in the rehabilitative process. The approach emphasizes a client-centered and goal-driven approach and combines the principles of skill learning to promote the day-to-day and ability development of the actual activities required for everyday life.

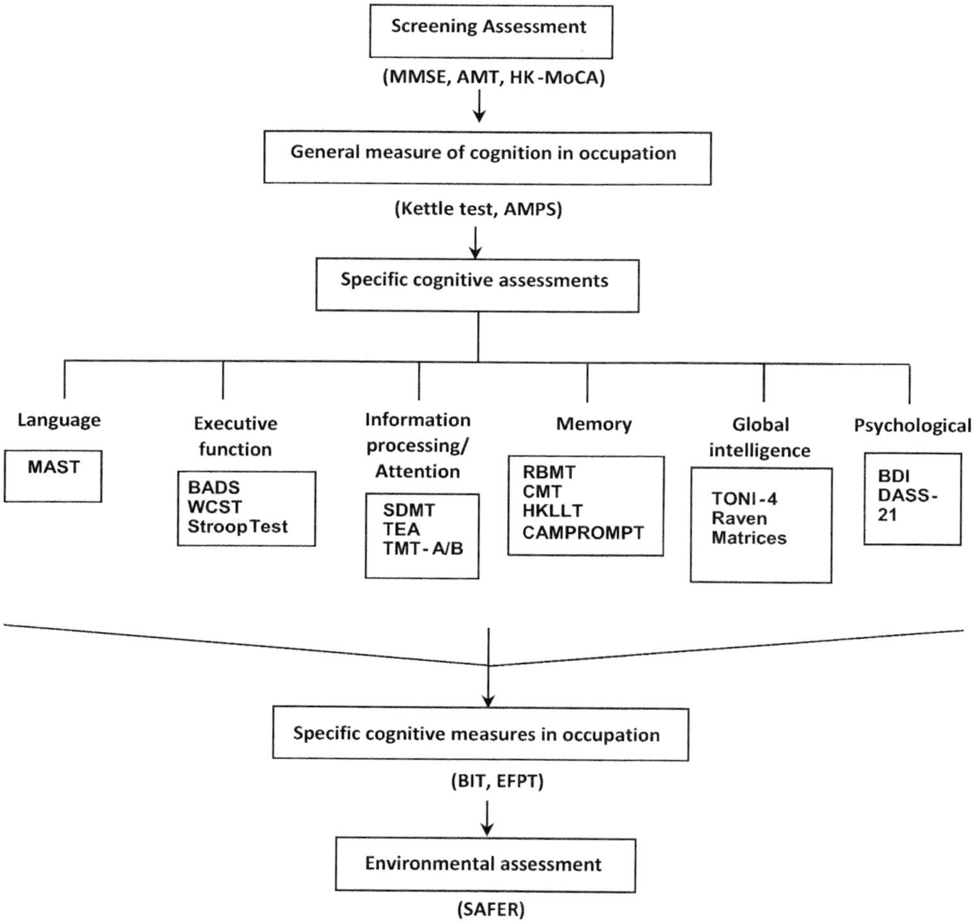


**FIGURE 13.1** An example of cognitive assessment flow according to cognitive functional evaluation (CFE).

**Table 13.1** Abbreviation of assessments.

| | | | |
|---|---|---|---|
| MMSE | Mini Mental State Examination | CMT | Contextual Memory Test |
| AMT | Abbreviated Mental Test | HKLLT | Hong Kong List Learning Test |
| HK-MoCA | Montreal Cognitive Assessment Hong Kong Version | CAMPORMPT | Cambridge Prospective Memory Test |
| AMPS | Assessment of Motor and Process Skills | TONI-4 | Test of Nonverbal Intelligence |
| MAST | Mississippi Aphasia Screening Test | BDI | Beck Depression Inventory |
| BADS | Behavioral Assessment of the Dysexecutive Syndrome | DASS-21 | Depression Anxiety Stress Scale |
| WCST | Wisconsin Card Sorting Test | BIT | Behavioral Inattention Test |
| SDMT | Symbol Digit Modalities Test | EFPT | Executive Function Performance Test |
| TEA | Test of Everyday Attention | SAFER | Home Safety and Accessibility Assessments |
| TMT | Trail Making Test | RBMT | Rivermead Behavioral Memory Test |

*CMT*, contextual memory test.

There are eight stages in NFA:

- Development of a positive therapeutic alliance with a client-centered approach;
- Gathering and assimilating relevant information in order to understand the client's functioning in their environment and anticipating their responses to intervention;
- Observation of performance in everyday situations which included client, task, and environment characteristics;
- Explore the resources and constraints which may facilitate the client achieving the treatment goal;
- Define specific operational performance goals for the retraining programs;
- Creation of skill-retraining programs which may incorporate adaptive aids or environmental supports developed from an analysis of the person, activity and environment;
- Generalize the skills in the client's everyday life;
- Provide feedback to the client to encourage further improvement.

Since the therapeutic relationship is the key component in cognitive rehabilitation processes, development of the alliance between the therapist and client is crucial as it will affect the compliance and motivation of the client in both assessment and treatment. In addition, a more realistic goal could be set with trust. The therapist could gather more information about the client's performance, task requirements, and environment through real-life situation observation and assessment. The therapist may adopt a systematic observation or standardized assessment to collect the client's result. Throughout the process the therapist can identify the strengths and weaknesses of the client. With those assessment results, the therapist could work with the client to setup goals and design a training program which facilitates the client to achieve their target. Within the treatment processes, the therapist may adopt different strategies (e.g., goal management training for executive function, memory strategies like association, chunking) and aids (e.g., notebook, digital device) to improve the client's performance. The client could also benefit from repeated drilling of specific skills. During the final stage of the intervention, the therapist will ask the client to perform the skills in her own context which enhances generalization. During the process, the therapist may have to modify the task or reduce the complexity to facilitate the client transferring their skills. The therapist monitors the client's performance during training and provides appropriate feedback which improves the client's self-awareness. It could act as a positive reinforcement for the client, improving their compliance and performance. It also prompts the client to sustain acquired cognitive skills.

In conclusion, NFA is a systematic approach which could guide the therapist to provide quality cognitive rehabilitation training to the clients. It is a holistic approach to know the client's performance and her environment which could boost the treatment effect.

## Human elements in the cognitive rehabilitation

Although advancement of the technology and automation system bring advantages over the manpower restraint, there are still some areas which should not be replaced, especially for those clients with a self-awareness problem. Compared with physical rehabilitation, the client with cognitive problems may be less aware of the deficits, resulting in lower compliance to the treatment. Even though some computer rehabilitation may draw the client's attention by attractive design and multiple feedback, some clients are still reluctant to accept training and it is commonly seen for the elderly or clients who are less familiar with the digital technology. In addition, for those clients who suffer from frontal lobe syndrome, which results in low frustration tolerance and difficulty in expressing emotions, a more specific and tailor-made approach is indicated. In these scenarios, therapists' expertise to adjust the intervention is needed. The following intervention (EDO-GFS) is recommended to improve patient compliance and improve treatment outcomes in cognitive rehabilitation programs.

According to this intervention, we focus on improving the client's awareness toward their disease impact, how the cognitive deficits affect their daily life, strength and weakness which facilitate the rehabilitation, and their interaction with the physical and social environment. After reviewing all these factors, the client can perform goal setting with the therapist to make a consensus and draft a treatment goal. The goal attainment scale (GAS) or Canadian occupational performance measures (COPM) are commonly adopted at this stage. The specific goal should meet the SMART principle (specific, measurable, achievable, realistic, timely) so that it can be quantified. As a result, the training will be more focused and the missing cognitive components or skills can be drilled. During the training process, the therapist has a role to feedback the client's performance and adjust the strategies to ensure the client can generalize the improvements in daily activities. Moreover, the therapist should also work with the client to sustain the performance across different scenarios. The flow of the treatment is summarized in Fig. 13.2.

## Types of cognitive assessments

Cognitive assessments adopted by occupational therapists commonly consist of standardized cognitive assessment, which focuses on the specific cognitive domain and ecological assessments which reflect cognitive performance in daily activities. It may conducted in visual or auditory aspects to enable the detection of abnormalities in a specific modality. In addition, the time domain is another important element to assess the client's processing speed and reaction time. As the floor effect and ceiling effect may happen as the clients have various abilities level, an assessment battery which

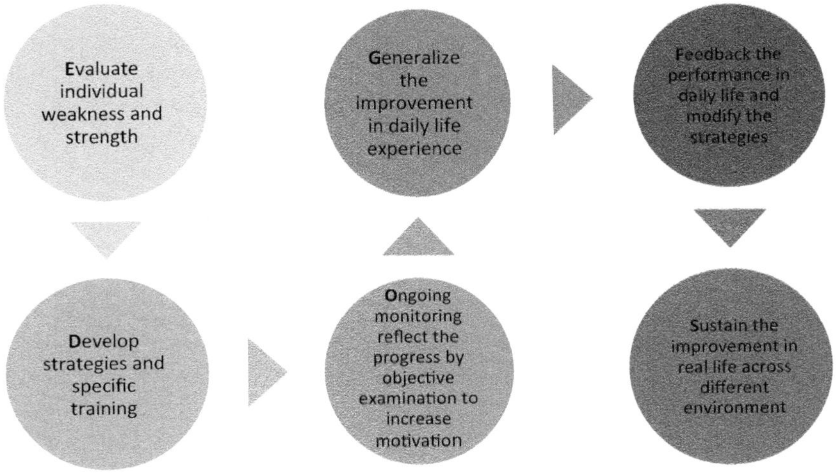


**FIGURE 13.2** Cognitive training intervention protocol (EDO-GFS).

measures the same cognitive area with different complexity is indicated. Moreover, the therapist may also adopt the dynamic assessment approach to assess the client. Toglia [4] describes the importance of dynamic assessments as conventional standardized cognitive tests are static in nature and they examine individual performance in specific context for the purposes of identifying and quantifying cognitive deficits. Dynamic assessment is an interactive process that systematically and objectively measures the improvement that occurs in response to cues, strategies, feedback, or task conditions which are introduced during testing. Moreover, dynamic assessment focuses on individual variations and changes instead of direct comparison with the norm. The goal is to measure how and to what extent performance can improve with guidance [5].

## Computerized cognitive assessment

The mode of conventional neurocognitive assessment is mainly delivered by face-to-face, pencil-and-paper tests, or behavioral observations [6]. Since the conventional assessment is labor-intensive, the therapist may need to spend 45 min to an hour to do a one-to-one assessment with a client. Man et al. [7] suggested these modes of implementation have limitations for clients with stroke as it can be time-consuming to arrange the assessment space and schedule, as well as the time taken for result interpretation. On the contrary, for the computerized cognitive assessment, the clients may be more willing to respond to sensitive questions as they are less affected by a social desirability bias [7]. In addition, the client could attend the computerized assessment by his own with minimal supervision from a healthcare professional. This would minimize the manpower burden as the assessments should be reviewed regularly to monitor the client's progress. Both the American Academy of Clinical Neuropsychology and the National Academy of Neuropsychology agree that the computerized cognitive assessment has several benefits in clinical care [8]. Their joint position paper suggests computerized assessment has improved (1) efficiency (huge capacity to test multiple individuals quickly; shortened assessment times through the use of adaptive testing protocols), (2) ease of access (ready availability of assessment services without advance notice; administering assessments in different languages; increased accessibility to patients in areas or settings in which professional neuropsychological services are inadequate), (3) accuracy (the ability to measure performance on time-sensitive tasks, such as reaction time more accurate), (4) costing (reduced costs relating to test administration and scoring), and (5) automation and research development (automated data exporting for research purposes; the ability to integrate and automate interpretative algorithms such as decision rules for determining impairment or statistically reliable change). Although computerized assessments bring several advantages, there are also some limitations. Unlike the therapist-conducted assessment, the computerized assessment mainly reflects the client's performance on specific components. Behavioral observations, such as the motivation and attention of the client, fatigue level, or the mood status, are lacking. For those with lower education or older age, they may not be able to use a computer for evaluation.

## Examples of computerized assessment

One of computerized cognitive assessment examples is the Wisconsin Card Sorting Test: Computer Version 4—Research Edition. The Wisconsin card sorting test was developed in 1948 by Grant and Berg, and Milner adopted it for brain injury assessment in the 1960s to assess prefrontal cortex executive function. It involved a lot of therapist effort to administer the test as the therapist would provide feedback and record the response of the client. Errors often occur during operations and scoring in a multitasking environment. Thus, the computerized version was developed on the basis of reliability research [9]. Their results show that there are nil significant differences between the manual and computerized versions for the outcome of perseverative responses, perseverative errors, and set breaks. Their results also suggest the computer version is more reliable in the scoring and administration procedures, as reflected by lower variance. Another group of researchers [10] compared different ways of administering the computerized version of the test like keyboard, mouse click, mouse auto and touch screen, and their results showed significant differences between the manual version and the computerized versions. They concluded that the assessor should be cautious when interpreting the results of the computer version. They could not simply assume the result of the computer test would be the same as a manual test.

There is another local assessment example which adopts a computerized version (Rivermead behavioral memory test – Chinese version). The Rivermead behavioral memory test (RBMT) is a functional memory assessment evaluating an individual's ability to use memory function for the performance of daily tasks and attempts to correlate the memory function and daily performance. The test consists of 11 common daily tasks, such as immediate and delayed recall details of a short story, immediate and delayed route recall, memorizing a name, a personal belonging and a prospective appointment, and face and picture recognition. Man et al. [7] designed an online version (OL-RBMT) and suggested the use of an audio or video presentation in computer-assisted assessment might be advantageous for patients who may be illiterate or have verbal communication problems. They proposed an online version of the Rivermead behavioral memory test, which has three advantages for stroke patients compared to traditional methods. These are: (1) the multimedia format could be more adaptable to match with clients' cognitive abilities; (2) the assessment could be more cost-effective and flexible; and (3) the online version integrated visual, auditory, and data collected, which enhanced the efficiency and quality of assessments. Moreover, their overall results suggest that online RBMT seems to provide an alternative way to present the same memory tasks in a multimedia format and successfully attract the attention of respondents (e.g., use of videos and 3D rendering animations). However, a few subtests show weaker correlations between the online and the face-to-face version. They recommend briefing and pretesting for stroke patients to familiarize themselves with the assessment process and ensure that patients are able to follow the test.

## Technology-enhanced cognitive assessment using a physiological signal (quantitative electroencephalography and eye tracking)

Electroencephalography (EEG) is a measure of the electrical pattern that reflects cortical activity on the surface of the brain. EEG has the unique feature that it has high temporal resolution and test–retest reliability. QEEG analysis refers to the mathematical processing of the EEG and includes a vast array of tools and techniques [11]. Quantitative EEG refers to parameters calculated for a brief epoch of EEG data and then plotted versus time. Since cognitive, perceptual, and motor processes are fast, most cognitive processes occur within tens to hundreds of milliseconds and the temporal sequence that may span hundreds of milliseconds to a few seconds. The high temporal resolution EEG is a suitable method to capture those processes. It is 2–3 times faster than other methods which measure the hemodynamic response.

The 10-20 System for the electrode position is commonly adopted and it is the only method officially recommended by the International Federation of Clinical Neurophysiology. There are different commercially available EEG systems which range from 4 to 256 channels. According to the American Society of Clinical Neurophysiology, it is recommended that 16 simultaneous recording channels be used as the minimum number required to display areas that produce normal or abnormal EEG patterns [12].

Measurements will be carried out under different conditions, such as eye closed, eye open, or during effortful activities. Researchers adopt the quantitative electroencephalography (QEEG) to measure the cognitive function of the client. They are looking for different QEEG parameters like alpha peak [13], alpha delta ratio [14], and theta coherence along the midline electrodes [15] which may indicate improvement over specific brain function like memory or motor control. The clinician could correlate the QEEG parameters with the standardized paper and pencil test score or daily life performance.

## Automation in QEEG

Although adoption of QEEG is a new approach for assessment and intervention in acquired brain injury rehabilitation, researchers have started to develop the automation system for analysis. Wickering et al. [16] used an automated detection system to detect delayed cerebral ischemia (DCI). They measured the EEG signal according to the 10–20 system and using ADR (alpha delta ratio) and RAV (relative alpha variability) trending to predict DCI occurrence. The EEG data are grouped into six regions: right frontal (F4–C4), left frontal (F3–C3), right rolandic (C4-T4), left rolandic (C3-T3), right posterior (P4–O2), and left posterior (P3–O1), which roughly correspond to territories of the anterior cerebral artery (ACA), middle cerebral artery (MCA), and posterior cerebral artery (PCA), respectively. Their overall findings show that both alpha–delta ratio-based and relative alpha variability-based trending method only achieved fair sensitivity (65%–80%) and specificity (27%–43%), which is well below that of the nonautomated method. They concluded that an automated EEG-based early DCI detection system is not yet ready for clinical use and the detection algorithm requires modification.

## Eye tracking

As the low-cost eye tracker device has become more popular and reliable [17], studies in neuroscience started using this new technology as an assessment means. Eye tracking measures provide noninvasive and rich indices of brain function and cognition [18]. They can provide additional information in addition to accuracy and response times due to their high temporal resolution, making it possible to measure how people respond to task demands every moment. Gaze analysis reveals the client's attention and cognitive strategies during a task. There are different parameters in eye tracking, which has attracted the attention of researchers and clinicians. Those parameters include fixation, gaze points, areas of interest, time to first fixation, fixation sequence, first fixation duration, and average fixation duration, etc. In addition, saccades, scan path length, and heatmap are also frequently reported in gaze analysis research. Those parameters can reveal which parts of the displayed information or stimulus are most salient, and to what extent and in what order information is processed within a multiple stimulus environment.

The researchers are using eye movement to study cognitive function as it is a biomarker which is natural and frequently occurs during human behavior. People will look at something when they acquire information and the fixation location/duration corresponds to the duration of cognitive processing of the material located at fixation [19]. Visual problems and impaired cognition are common after stroke and affect daily performance. Thus, assessment for those areas and understanding their relationship are important for rehabilitation.

There are some studies which measure the relationship between gaze parameter and cognitive function. Chan et al. [20] recruited 15 subjects with stroke (aged 39–63, educational level ranged from 9 to 18 years) for the correlation study. They found that there was a significant negative correlation between the fixation duration mean (ms) and HK-MoCA score. In addition, there was a positive correlation between the path velocity and HK-MoCA score. This provides evidence to support the relationship between gaze parameters and cognitive function.

Rather than just using eye tracking to detect cognitive function, some researchers adopt it with conventional assessment as an add-on to enhance its effectiveness. Lagun et al. [21] adopted visual paired comparison (VPC) to detect the memory decline in people with mild cognitive impairment. They used noninvasive eye tracking to identify how subjects view novel and repeated visual stimuli. The healthy control subjects would spend more time looking at the novel image, which is the repeated stimuli. Their machine learning could detect clients who suffer from mild cognitive impairment while their performance in fixations, saccades, and refixations during the VPC task is different from their healthy counterparts. With the gaze parameter, this combined approach has better accuracy (67%–87%), sensitivity (60%–97%), and specificity (73%–77%).

A similar study was done by Bott el al [22] who applied a web camera instead of the expensive eye tracker to capture eye movement. Their results showed the human scoring of a VPC decisional task using a built-in laptop web camera correlated strongly with automated scoring of the same task using a standard high frame rate eye tracker camera. Since built-in web cameras are commonly seen in smart devices, they suggest that it may be an effective mean to track eye movements on decisional tasks with adequate accuracy and minimal cost.

## Virtual reality

Virtual reality (VR) allows the user to experience and interact with a computer-generated environment. It can motivate the client to participate in the rehabilitation program. There are different types of virtual reality. Both immersive and nonimmersive VR have been adopted in cognitive rehabilitation. As immersive VR may be more realistic and facilitate the generalization to the real-life performance, a considerable number of patients with brain damage may have cyber sickness and withdraw from training. On the contrary, nonimmersive VR could be also beneficial as the client could learn the sequence step by step. Fong et al. [23] conducted research to study the usability of nonimmersive VR to assess and train people with stroke to use the an automatic teller machine for cash withdrawal and transfer function. The results show that there is satisfactory correlation between performance in the virtual reality and the real-life environments. The VR system also shows good specificity and sensitivity. Moreover, their study also compared the effectiveness of the conventional educational approach with PowerPoint and VR training. The results showed that the VR training group showed better performance in a more complex task (e.g., money transfer) but there was no significant difference in the simpler task (e.g., money withdrawal). This suggests use of virtual reality is more useful to teach clients for tasks involving multiple steps.

There was another study done by Yip and Man [24] using a virtual reality-based prospective memory training program for clients with acquired brain injury. They designed a nonimmersive, high-resolution convenience store scene to train client's event base and time base prospective memory. The participants performed several real-life tasks like calling back home when a specific target was seen, recalling to take the food out of the oven after a specific time period, and performing shopping according to a given list. The program consisted of prospective memory, retrospective memory, and inhibition components. They reported after 12 sessions training that the VR group showed significantly greater improvement in real-life event-based prospective memory test and color trail test. They suggested the improvements are not only found within the virtual environment, but also in basic cognitive components and real-life tasks. They emphasized the relevance between virtual reality and real-life environments which helps to promote generalization to real situations. This gives us an insight into the development of the virtual reality programs in cognitive rehabilitation.

## Automation in functional home-based rehabilitation

Measurement of the clients' performance or activity levels at home may reflect their health condition. The latest research uses smart home sensor data to develop automated functional assessments [25]. This is called the clinical assessment using activity behavior (CAAB) approach, which predicts the cognitive and mobility scores of smart home residents by monitoring a set of basic and instrumental activities of daily living. They captured the raw data from different smart home sensors (motion, light, temperate, door, and bed sensors) and built the model to detect the activities of daily living. Their activity recognition algorithm could label the activity by sensor events (date, time, sensor identifier, and sensor value). They also correlated the data with the conventional clinical test score—Repeatable Battery for the Assessment of Neuropsychological Status measure of cognitive status (RBANS) and time up and go test. Their study invited 18 elderly clients to participate in the study and their results found that there was a statistically significant correlation between CAAB-predicted and clinician-provided cognitive assessment scores and a statistically significant correlation between CAAB-predicted and clinician-provided mobility scores. Moreover, CAAB has a classification accuracy of 72% while classifying cognitive assessment scores and 76% while classifying mobility scores.

They concluded that it is feasible to predict standard clinical scores using smart home sensor data and learning-based data analysis has the potential to be a tool for clinician use.

## The way forward—developing the automation system for a cognitive rehabilitation program

People suffering from acquired brain injury persist with chronic and progressive symptoms. Moreover, their life expectancy has become longer. Rehabilitation technology and automation systems have become an essential part of rehabilitation development. Oña et al. [26] suggest an "automated rehabilitation cycle" which highlights several processes in the rehabilitation cycle that could be automated. These are automatic assessment system (AAS), decision support system (DSS), and the end effector-robotic rehabilitation systems (RRS). Although the cycle originated for the automated process for upper limb rehabilitation, it could also apply to the cognitive rehabilitation process. According to their suggested cycle, the result of automatic assessment should correlate with the gold standard clinical test. It minimizes the therapists' time to conduct the assessment and interpretation. The results obtained with the AAS then automatically update in the client's profile. These results serve as input parameters to the decision support system (DSS). The DSS aims at designing the most optimal treatment protocol for the client and determining the intervention plan. It may be based on the normative data collected for the specific age and educational range. In addition, artificial intelligence may be applied in this system which could simulate the judgment of the healthcare professions. Finally, the end effector in the cognitive rehabilitation is the training protocol. The computer remediation program may be the feasible solution which concentrates on the area identified by the decision support system. The automated system could determine the treatment dosage, mode of training, and selection of cognitive strategies. This cycle provides a framework to develop the system.

There are several aspects in cognitive rehabilitation which could be enhanced by automation. First of all, in the assessment aspect, it is expected that the automation system could do a global screening and identify which aspect needs further investigation. There is a local example where Wong et al. [27] developed a self-administered computerized test which was called "Computerized Cognitive Screen (CoCoSc) – Hong Kong version" It aims at quick screening of individuals with cognitive impairment (CI) in community settings. The computerized cognitive assessment is conducted in a touch screen format. It assesses memory, executive functions, orientation, attention/working memory, and prospective memory. It is supposed to be finished within 15 min so that more clients are able to access it. Their reliability study results showed that total scores of the CoCoSc and MoCA are significantly correlated and it is a promising computerized cognitive screen for self-administration in a community setting. Moreover, the automation system could take the concept of dynamic assessment. That means the cues will be provided whenever the clients fail to answer and see whether the provided cue can improve the performance. It not only measures the clients' strength but also motivates the client to participate.

In order to enhance the automation in the cognitive rehabilitation, we may strike the balance between generality and specificity. The design of the cognitive assessments and intervention should be universal enough so that it can be applied without the constraint of language or context. It should be automatically adjusted to fit with different educational and impairment levels without a floor or ceiling effect. Moreover, with the concept of tele-rehabilitation, the automated cognitive assessments and treatments could be done in the client's home. It not only reflects the performance over the entire day but is also able to increase the dosage as the participants could practice at home.

For specificity, the automation system should be able to meet individual needs as each patient may have unique features. Although each client may have similar deficits, such as memory problems, they may have different attributes which may compensate for their deficit areas (e.g., the client may use notebook daily). In addition, their environment may be different and they may present with different impact severity (e.g., a working age young brain-injury patient vs. a retired elderly person).

In conclusion, the design and use of a personalized and automated cognitive rehabilitation system is essential to tackle the huge patient demand. The multidisciplinary team consisting of a therapist, psychologist, and engineer is indicated to develop a holistic and effective automated program.

# References


[1] Hartman-Maeir A, Katz N, Baum CM. Cognitive functional evaluation (CFE) process for individuals with suspected cognitive disabilities. Occupational Therapy in Health Care January 1, 2009;23(1): 1–23.

[2] Hartman-Maeir A, Harel H, Katz N. Kettle test—a brief measure of cognitive functional performance: reliability and validity in stroke rehabilitation. American Journal of Occupational Therapy September 1, 2009;63(5):592–9.

[3] Clark-Wilson J, Giles GM, Baxter DM. Revisiting the neurofunctional approach: conceptualizing the core components for the rehabilitation of everyday living skills. Brain Injury December 1, 2014; 28(13–14):1646–56.

[4] Toglia JP. A dynamic interactional approach to cognitive rehabilitation. Cognition and occupation across the life span: models for intervention in occupational therapy, vol. 2; 2005. p. 29–72.

[5] Haywood HC, Lidz CS. Dynamic assessment in practice: clinical and educational applications. Cambridge University Press; December 18, 2006.

[6] Caetano C, Christensen AL. The design of neuropsychological rehabilitation: the role of neuropsychological assessment. In: Neuropsychological rehabilitation: fundamentals, innovations and directions; 1997. p. 63–72.

[7] Man DW, Chung JC, Mak MK. Development and validation of the online Rivermead behavioral memory test (OL-RBMT) for people with stroke. NeuroRehabilitation January 1, 2009;24(3):231–6.

[8] Bauer RM, Iverson GL, Cernich AN, Binder LM, Ruff RM, Naugle RI. Computerized neuropsychological assessment devices: joint position paper of the American Academy of Clinical Neuropsychology and the National Academy of Neuropsychology. The Clinical Neuropsychologist February 1, 2012;26(2):177–96.

[9] Tien AY, Spevack TV, Jones DW, Pearlson GD, Schlaepfer TE, Strauss ME. Computerized Wisconsin card sorting test: comparison with manual administration. The Kaohsiung Journal of Medical Sciences August 1996;12(8):479–85.

[10] Feldstein SN, Keller FR, Portman RE, Durham RL, Klebe KJ, Davis HP. A comparison of computerized and standard versions of the Wisconsin Card Sorting Test. The Clinical Neuropsychologist August 1, 1999;13(3):303–13.

[11] Sinha SR. Quantitative EEG analysis: basics. In: Continuous EEG monitoring. Cham: Springer; 2017. p. 173–89.

[12] Sinha SR, Sullivan LR, Sabau D, Orta DS, Dombrowski KE, Halford JJ, Hani AJ, Drislane FW, Stecker MM. American Clinical Neurophysiology Society guideline 1: minimum technical requirements for performing clinical electroencephalography. The Neurodiagnostic Journal October 1, 2016;56(4):235–44.

[13] Angelakis E, Lubar JF, Stathopoulou S, Kounios J. Peak alpha frequency: an electroencephalographic measure of cognitive preparedness. Clinical Neurophysiology April 1, 2004;115(4):887–97.

[14] Schleiger E, Sheikh N, Rowland T, Wong A, Read S, Finnigan S. Frontal EEG delta/alpha ratio and screening for post-stroke cognitive deficits: the power of four electrodes. International Journal of Psychophysiology October 1, 2014;94(1):19–24.

[15] Anguera JA, Boccanfuso J, Rintoul JL, Al-Hashimi O, Faraji F, Janowich J, Kong E, Larraburo Y, Rolle C, Johnston E, Gazzaley A. Video game training enhances cognitive control in older adults. Nature September 2013;501(7465):97.

[16] Wickering E, Gaspard N, Zafar S, Moura VJ, Biswal S, Bechek S, O'Connor K, Rosenthal ES, Westover MB. Automation of classical QEEG trending methods for early detection of delayed cerebral ischemia: more work to do. Journal of Clinical Neurophysiology: Official Publication of the American Electroencephalographic Society June 2016;33(3):227.

[17] Ooms K, Dupont L, Lapon L, Popelka S. Accuracy and precision of fixation locations recorded with the low-cost Eye Tribe tracker in different experimental setups. Journal of Eye Movement Research 2015;8(1).

[18] Eckstein MK, Guerra-Carrillo B, Singley AT, Bunge SA. Beyond eye gaze: what else can eyetracking reveal about cognition and cognitive development? Developmental Cognitive Neuroscience June 1, 2017;25:69–91.

[19] Irwin DE. Fixation location and fixation duration as indices of cognitive processing. In: The interface of language, vision, and action: eye movements and the visual world, vol. 217; July 1, 2004. p. 105–33.

[20] Chan KL, Tong KY, Chung YK, Ng CW, Ng SW. The correlation between gaze parameter and cognitive function for people with stroke. In: Brain 2018 – the 15th multidisciplinary neuroscience conference, CUHK, HKSAR, jan 2018; 2018.

[21] Lagun D, Manzanares C, Zola SM, Buffalo EA, Agichtein E. Detecting cognitive impairment by eye movement analysis using automatic classification algorithms. Journal of Neuroscience Methods September 30, 2011;201(1):196–203.

[22] Bott NT, Lange A, Rentz D, Buffalo E, Clopton P, Zola S. Web camera based eye tracking to assess visual memory on a visual paired comparison task. Frontiers in Neuroscience June 28, 2017;11:370.

[23] Fong KN, Chow KY, Chan BC, Lam KC, Lee JC, Li TH, Yan EW, Wong AT. Usability of a virtual reality environment simulating an automated teller machine for assessing and training persons with acquired brain injury. Journal of Neuroengineering and Rehabilitation December 2010;7(1):19.

[24] Yip BC, Man DW. Virtual reality-based prospective memory training program for people with acquired brain injury. NeuroRehabilitation January 1, 2013;32(1):103–15.

[25] Dawadi PN, Cook DJ, Schmitter-Edgecombe M. Automated cognitive health assessment from smart home-based behavior data. IEEE Journal of Biomedical and Health Informatics July 2016;20(4): 1188–94.

[26] Oña ED, Cano-de la Cuerda R, Sánchez-Herrera P, Balaguer C, Jardón A. A review of robotics in neurorehabilitation: towards an automated process for upper limb. Journal of Healthcare Engineering 2018;2018.

[27] Wong A, Fong CH, Mok VC, Leung KT, Tong RK. Computerized cognitive screen (CoCoSc): a self-administered computerized test for screening for cognitive impairment in community social centers. Journal of Alzheimer's Disease January 1, 2017;59(4):1299–306.

# 14

# Translation of robot-assisted rehabilitation to clinical service in upper limb rehabilitation

Yanhuan Huang[1], Will Poyan Lai[2], Qiuyang Qian[1], Xiaoling Hu[1], Eric W.C. Tam[1,2], Yongping Zheng[1]

[1]DEPARTMENT OF BIOMEDICAL ENGINEERING, THE HONG KONG POLYTECHNIC UNIVERSITY, KOWLOON, HONG KONG; [2]JOCKEY CLUB REHABILITATION ENGINEERING CLINIC, DEPARTMENT OF BIOMEDICAL ENGINEERING, THE HONG KONG POLYTECHNIC UNIVERSITY, KOWLOON, HONG KONG

**Chapter outline**



## Background

Stroke remains the leading cause of adult permanent disability [1]. Due to the rapidly expanding stroke population and insufficient professional manpower, various rehabilitation robots have been proposed for human therapists when conducting labor-demanding physical training [2–5]. Among these, the voluntary intention-driven (e.g., electromyography, EMG) rehabilitation robots can provide highly intensive and

Intelligent Biomechatronics in Neurorehabilitation. https://doi.org/10.1016/B978-0-12-814942-3.00014-3

repetitive training, and have proven to be effective and could be a cost-effective compensation for the conventional rehabilitation service [6–9]. Indeed, evidence that positively supports robot-assisted rehabilitation was mainly gathered through research-oriented clinical trial studies, while the effectiveness of robot-assisted rehabilitation gathered in a real clinical service configuration has been seldom reported. This is because during the translation of robot-assisted rehabilitation from clinical trials to clinical service settings, it is commonly assumed that the satisfactory effectiveness of the robotic devices obtained in the laboratory setting would naturally continue in the same manner in real service situations following commercialization. However, this reality is not always achieved, and differences or even discounts of training outcomes may occur when translating these efforts from well-controlled research studies to more flexible services. A number of studies have cast doubt on whether robotic devices have a useful role to play within the clinical service environment, noting that it is not easy to oversee the quality of the trial throughout a long-term service [10–14]. Difficulties of the head-to-head investigation regarding training effectiveness between clinical services and clinical trials are seen across three aspects. First, the schedule of a rehabilitation program for the participants in the clinical trials is usually restricted because it is free of charge, and sometimes participants may even be paid for their participation in a trial. However, rehabilitation schedules in clinical services are relatively flexible when sustainability payments are required. Second, the variation is large and random in motor impairments of the clients who come for the clinical service, compared with the subject recruitment in a clinical trial for research purpose only. Also, it was seen that inclusion criteria within a laboratory setting were not always suitable for implementation in a clinical service (especially in the private sector). Third, participants in clinical trials are usually not allowed to attend other physical treatments during the investigation period, to avoid any possible interference, while clients of a clinical service may attend other treatments they believe to be helpful. In our previous study, in order to assess the rehabilitation effects of an EMG-driven robotic hand, a single group clinical trial was conducted [15]. Following commercialization, EMG-driven robotic hands were manufactured and employed to offer clinical services to local communities in the private sector, in the context of a self-financed university clinic from 2011. This study aims to examine the rehabilitation benefits of an EMG-driven robot hand-assisted upper limb training program in the context of a research trial in a laboratory configuration compared to real clinical practice in a private clinic, making sure there was as little disruption to routine clinical management and service provided to the clients as possible.

## The EMG-driven robotic hand

Fig. 14.1A presents the EMG-driven robotic hand system used in this work. The system can assist with finger extension and flexion of the paretic limb for patients after stroke. In this study, real-time voluntary EMG detected from the abductor pollicis brevis (APB) and

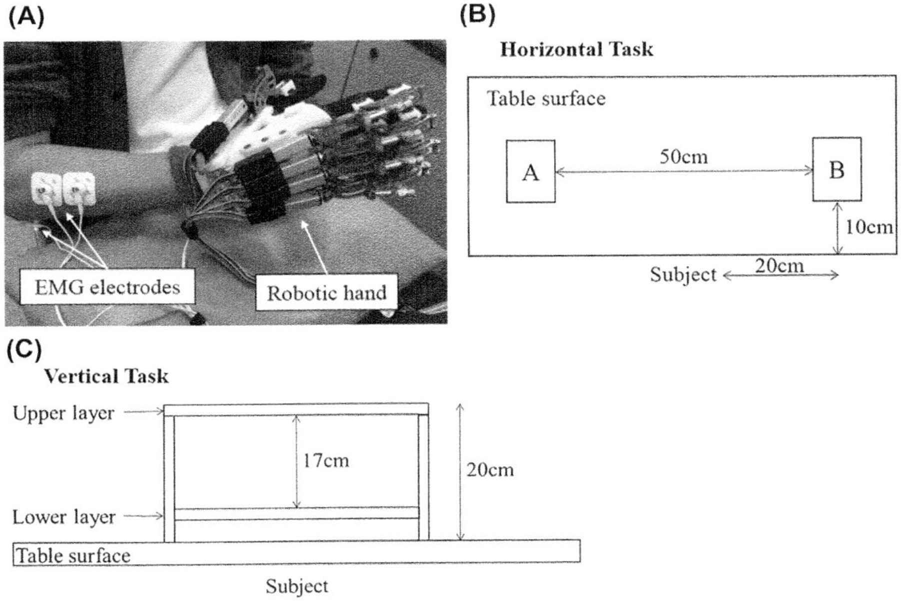


**FIGURE 14.1** The electromyography (EMG)-driven robotic hand system: (A) The wearable system consisting of a mechanical exoskeleton of the robotic hand and EMG electrodes; (B) the table configuration during the horizontal task for the training; (C) the table configuration during the vertical task for the training.

extensor digitorum (ED) muscles was used to control the respective hand closing and open movements. A detailed working principle of the robotic hand was introduced in Chapter 9, and the difference between the robot hand in this chapter and that in Chapter 9 is that there was no neuromuscular electrical stimulation applied.

# Clinic versus laboratory

This work is a controlled trial without randomization within two different settings, including a clinical service setting in a business environment and a laboratory setting.

## The clinical setting

The clinical service was hosted in the Jockey Club Rehabilitation Engineering Clinic (JCREClinic), which is found within the campus of the Hong Kong Polytechnic University. The JCREClinic aims to serve the local communities and provide them holistic and professional clinical services such as rehabilitation therapies, orthoses, and prostheses. The interior configuration of the JCREClinic is shown in Fig. 14.2 [16], which mainly consists of a main entrance, a reception counter, corridor, waiting area for guests, and several treatment rooms. Consistent with other private clinics, all consultations and treatments in the JCREClinic could be scheduled via walk-in, phone, email, or WhatsApp message. For rehabilitation services through robotic hand training, a client needs to

FIGURE 14.2 The interior configuration and training setup of the robotic hand training in Jockey Club Rehabilitation Engineering Clinic: (A) Entrance, (B) corridor, (C) waiting area for guests and reception counter, (D) treatment room with estimated area presented in square meters, and (E,F) the training setup of the robotic hand rehabilitation system assisted by a physical therapist.

make an appointment first and then consult with the physiotherapist, a charge of the service. When consulting with the client, the physical therapist reviews their medical and rehabilitation background, before assessing the functional recovery of the affected limb with clinical scores. Following this, the client completes a trial of robotic hand training, with the assistance of the physical therapist, evaluating the fit and size, and the ability to use the voluntary EMG signal to control the robotic hand system. Potential rehabilitation benefits as seen in earlier trials would also be explained to the client by the physical therapist [15]. Once the consultation is completed, a training schedule of 20 sessions would be organized by the clinic, if the client wants to go ahead with the robotic hand training. These sessions would be scheduled based on the availabilities of both the client and the physiotherapist, ideally resulting in three to five sessions a week. In most cases, no more than four sessions per week were allocated to a client, however, the client can reschedule because of commitments that may occur at a later time. At the end of each session, a 400 Hong Kong Dollar service charge is applied, and clients could leave at any juncture during the service period, without being penalized.

## The laboratory setting

The EMG-driven robotic hand upper limb training was carried out in a neurorehabilitation laboratory at Hong Kong Polytechnic University (Fig. 14.3), which comprised of a physical training area (where the robotic hand training took place), a cognitive training area, and an office area [16]. Participants in the robotic hand treatment in the laboratory setting received their treatment for free [15]. Detailed comparisons of the clinical setting and the lab setting can be found in Table 14.1 [16].

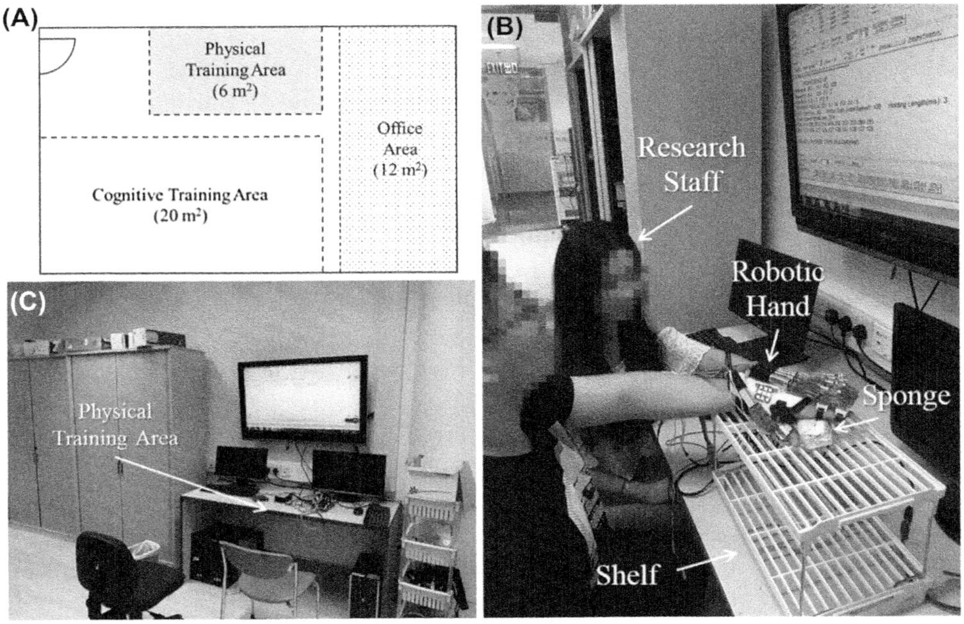


**FIGURE 14.3** The interior configuration and training setup of the robotic hand training in a neurorehabilitation laboratory: (A) Lab planar graph with estimated area presented in square meters, (B) physical training area, and (C) the training setup of the robotic hand rehabilitation system assisted by a research staff.

**Table 14.1** Clinic versus laboratory.

| | Clinic | Laboratory |
|---|---|---|
| Interior configuration | | |
| Entrance | √ | √ |
| Reception counter | √ | × |
| Corridor | √ | × |
| Waiting area | √ | × |
| Treatment room/area | √ | √ |
| Appointment | | |
| Walk-in appointment | √ | × |
| Scheduled appointment | √ | √ |
| Schedule | | |
| Mutual agreement | √ | √ |
| Fixed training intensity | × | √ |
| Accept reschedule | √ | × |
| Contact person | | |
| Reception assistant | √ | × |
| Research staff | × | √ |
| Trainer | | |
| Physical therapist | √ | × |
| Research staff | × | √ |
| Fee | √ | × |
| Withdrawal | √ | √ |

## Participants

This study was given ethical approval by the Human Participants Ethics Sub-Committee of Hong Kong Polytechnic University. In total, the study had 32 participants, who were divided equally into a lab group (since their training was given in the laboratory) and the clinic group (who received their training in the clinical service setting). Recruitment processes varied according to the group in question. The lab group were chosen from local districts, according to clear inclusion criteria [16], namely: (1) no less than 6 months had elapsed since the onset of the singular and unilateral brain lesion caused by a stroke; (2) the metacarpophalangeal (MCP) and proximal interphalangeal (PIP) joints were both capable of being extended to 180 degrees passively; (3) the spasticity during extension at the finger joints and the wrist joint was less than, or equal to 3, as measured by the Modified Ashworth Scale (MAS) [17]; (4) participants needed to exhibit measurable voluntary EMG signals from the target muscles in their paretic side, that is, the amplitude of the signal should be more than 3 SD above the baseline mean; and (5) the participants were capable of understanding and following instructions throughout the experiment, as evaluated by the Mini-Mental State Examination (MMSE >21). They were also told not to receive other upper limb physical training during the robotic hand training as part of the consent in this study, otherwise, they would be excluded. Those taking part in the clinic group were recruited from a wider group of clients who had planned to receive the robotic hand training at the JCREClinic. These individuals were screened and examined against the inclusion criteria used for the lab group, to find possible suitable candidates. Clients with an interest in participating in this project, and who consented to not receiving other upper limb treatment during the training period, were recruited in this study. All recruited participants gave their written consent before the training.

## Training protocol

All study participants took part in 20 sessions of robotic hand-assisted upper limb training. Each session lasted for 90 min, during which participants were taught to carry out repetitive upper limb movements—for example, hand grasp and release motions, and lateral and vertical task training. For the lateral task, participants with left hemiplegia were asked to hold onto a target object, move it 50 cm horizontally from point A on the left side of a table to point B on the right side of a table (as seen in Fig. 14.1B), let it go, grasp it again, and take it back to its original position at the point A. On the other hand, participants with right hemiplegia undertook the same target object transporting cycle as the left hemiplegia group, with the key difference being that they began the task at point B and moved it to point A, then back to point B. For arm transportation, the testing hand was raised from the table by 2–5 cm from the lowest point of the hand, to the table surface. The vertical task involved each participant grasping the target object on the midline of the lower layer of a shelf, raising it across a vertical distance of 17 cm, positioning it at the midline of the upper layer of the shelf, picking it back up, and replacing it at the starting point (as seen in Fig. 14.1C). The reaching movements involved a movement range of 40 degrees of flexion to 180 degrees of extension for the elbow joint [15].

There were certain key differences noted between the lab group and the clinic group. First, the training frequency per week was not equal. In the clinic group, clients received robotic hand training in a quiet treatment room in the JCREClinic assisted by a physiotherapist in a one-to-one manner, with a negotiable training frequency (maximum of four sessions/week). The final averaged training frequency in the clinic group was 2.26 sessions/week (ranging from one to three sessions/week) owing to the clients' requests for rescheduling training sessions. The lab group all received robotic hand training in the laboratory as arranged by a research assistant with a fixed training frequency of four times per week, to continue over a 5-week period. Second, the training pace was relatively flexible in the clinic group compared with the lab group. For instance, participants of the clinic group could stop the practice for a rest of 5 min whenever needed, in order to avoid significant muscle fatigue, while the participants in the lab group would have a 10-min break every 20 min of training with a 60-min accumulated practicing time as in the previous trial [15]. It was noted that the clinic could gradually increase the total practicing time, from less than 45 min to over 60 min on average, across the treatment process. Additionally, various target objects were used in the upper limb motions. For the lab group, a sponge was provided as the target object to grasp during the upper limb motions. For the clinic group, various target objects providing different tactile perception were prepared for the training task, which might include a sponge, tennis ball, alloy tube, or toy carrot. Verbal communication took place between the physiotherapist and the participant, and the therapist encouraged the clinic group throughout.

## Rehabilitation outcome

### Outcome evaluations and statistics

The robotic hand training outcomes were assessed by pre- and post-evaluation of clinical scores, including Fugl-Meyer assessment [18] (FMA) with the full score of 66 for the upper limb assessment further divided into shoulder/elbow (42/66) and wrist/hand (24/66)), Modified Ashworth Scale (MAS) [17] on the flexors related to the fingers, wrist and elbow, Action Research Arm Test (ARAT) [19] and Functional Independence Measure (FIM) [20]. Fugl-Meyer assessment evaluates the motor function impairment in voluntary limb movements, and an increased FMA score indicates improvement in the related motor functions. MAS measures the resistance during passive muscle stretching and indicates muscular spasticity, mainly in the flexors. A decrease in the MAS suggests released muscle spasticity in the relevant muscle. ARAT assesses upper limb voluntary function with a focus on finger activities, and the increased scores in the ARAT instrument may reflect improved upper limb motor functions, especially in the hands and fingers. FIM was used to rate the basic quality of activities of daily living (ADLs) for patients with stroke. A greater FIM score puts forward the notion that there is a greater level of independence when it comes to ADLs. All the clinical scores mentioned above were also used in Chapter 9.

In order to compare the intergroup variations of the post-training clinical assessments, the pre-assessment was used as a covariate and then one-way covariance analysis (ANCOVA) was carried out. Next, a paired t-test was utilized to evaluate the two groups' intragroup differences at a number of time points (i.e., before and after training). Then, an independent t-test assessed the intergroup comparisons between the improvements of every clinical assessment after the treatments. This study's levels of statistical significance were set at 0.05.

## Functional achievement after training

Fig. 14.4 shows a comparison between clinical scores before and after training for the two groups. In the clinic group, improvements with statistical significance ($P < .05$, paired t-test) were seen with FMA full score, FMA shoulder/elbow, FMA wrist/hand, ARAT, and FIM, while significant decreases ($P < .05$, paired t-test) were obtained in the

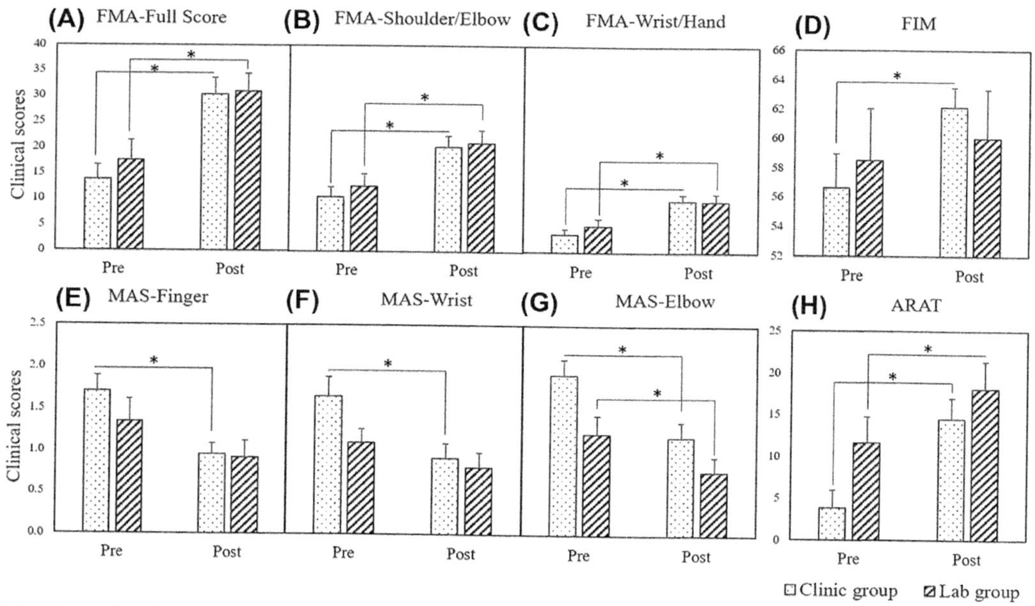


**FIGURE 14.4** The clinical scores (evaluated before the first and after the 20th training sessions) of the participants in both the clinic and lab groups: (A) Fugl-Meyer Assessment (FMA) full scores, (B) FMA shoulder/elbow scores, (C) FMA wrist/hand scores, (D) Functional Independence Measure (FIM) scores, (E) Modified Ashworth Scale (MAS) scores at the fingers, (F) MAS scores at the wrist, (G) MAS scores at the elbow, and (H) Action Research Arm Test (ARAT) scores, presented as mean values and SE (error bar) in each evaluation session. The significant intragroup difference is indicated by "*" ($P < .05$, paired t-test).

MAS finger, MAS wrist, and MAS elbow in the post-training tests. In the lab group, significant increases were obtained in the scores of the FMA full score, FMA shoulder/elbow, FMA wrist/hand, and ARAT ($P < .05$, paired t-test), and a significantly decreased MAS elbow score was observed ($P < .05$, paired t-test). The significantly increased FMA score suggested a better voluntary performance in the entire upper limb in both groups, and the improved ARAT score indicated an improvement in finger coordination for fine precision grasping and joint stability of the fingers. The significantly decreased MAS scores at the elbow, wrist, and fingers for the participants in the clinic group implied that muscle spasticity was reduced after training, while muscle spasticity for the participants in the lab group was only released at the elbow joint. In addition, the major improvement reflected in the FIM scores of the clinic group showed that the EMG-driven robotic hand had a significant positive impact on chronic stroke patients' ability to be independent in their daily lives—whereas the significant improvement in FIM score was not observed for the lab group, after training. Furthermore, no significant intergroup differences were found in the post-clinical scores ($P > .05$, one-way ANCOVA), which indicated that the rehabilitation outcomes after the robotic training were comparable between the two groups. However, one-way ANCOVA could not be used to compare the group differences for the post-FIM scores due to the significant interaction between the group and the preclinical scores of the FIM scores. Therefore, there was a need to further evaluate the group differences across the variations of the clinical scores, particularly for the FIM scores. Fig. 14.5 demonstrates improvements for each clinical assessment for the two groups after robotic hand training. Significantly higher variations in the clinic group were pointed out on the FIM scores compared with the lab group ($P < .05$,

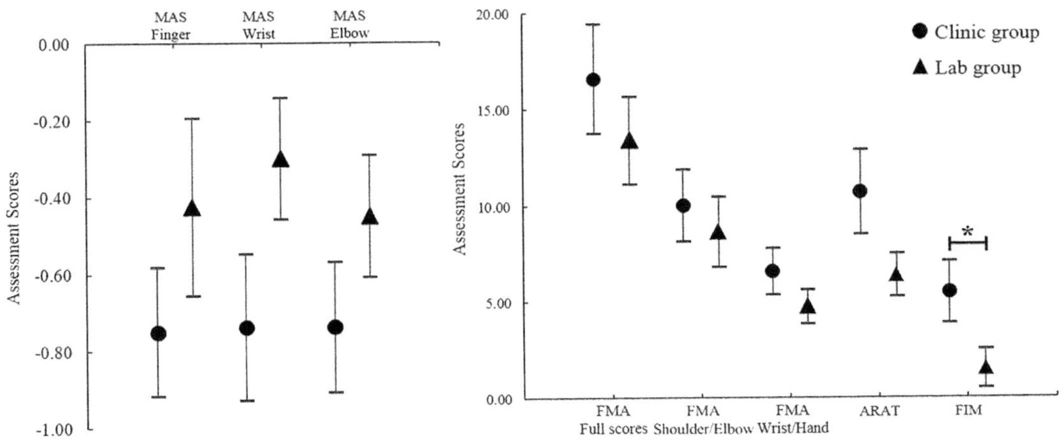


**FIGURE 14.5** The changes to each clinical assessment after the treatments in both the clinic and lab groups: Modified Ashworth Scale (MAS) scores at the fingers, wrist, and elbow, Fugl-Meyer Assessment (FMA) full scores, FMA shoulder/elbow, FMA wrist/hand, Action Research Arm Test (ARAT), Functional Independence Measure (FIM) and, presented as mean value and SE (error bar) in each evaluation session. The significant difference is indicated by "*" ($P < .05$, independent t-test).

independent t-test), which implied that the clinic group achieved more improvement in the ADLs compared with the lab group.

## Discussion

Based on the results of this study, equivalent motor improvement could be obtained after training for the clinic service setting and the lab setting. It should be noted that the clinical service group showed greater improvement in daily independency, despite lower training frequency, when compared to the lab group. Several reasons could explain the better rehabilitation outcomes achieved in the clinic group after the translation of the robotic hand training. One possible reason was that the training pace was flexible for the clients. The clients in a clinical service could more actively participate in the regulation of their training pace under the supervision of the therapist, as they could voluntarily ask for a rest whenever they needed, or continuously perform the training without any rest, once the therapist evaluated that continuous practice was safe. Each session's average accumulated practicing time varied from 45 to 60 min through the entire training program. In the beginning, most participants asked for a rest every few minutes of training, in their early training sessions. Once the clients were more familiar and experienced with the training program, they increased their practice time to roughly 60 min per session, and some participants stated they could go on for more than 60 min per session if there was no time limit involved. On the other hand, the participants in the lab group had a fixed training pace (a 10-min break for every 20 min of training) with a 60-min accumulated practice time per session. This suggested that voluntary exercise could achieve more significant training outcomes than fixed training could, and similar findings have been found in the post-stroke mice model [21,22]. For example, Ke et al. [21] compared the training effectiveness of the voluntary exercise of wheel running and the fixed exercise of treadmill running on post-stroke rats, and the results showed that the post-stroke rats engaged in voluntary exercise could achieve more motor recovery compared with the fixed exercise group. It is possible that voluntary exercise might also be a more effective training style when it comes to facilitating motor recovery in human beings.

In this study, one feature of the clinical service was that richer somatosensory stimulation was applied in the clinic group, and that might be effective for improving motor function after stroke [23,24]. Sensory deficiency after stroke will reduce sensory input to the brain, which is particularly important for the brain to plan and execute voluntary movements and provides access to the external world of physical objects [25,26]. In light of recent neurophysiological research, it is reported that sensory stimulation may assist in enhancing sensory input for stroke patients, which can facilitate motor movements and further improve motor functions [27]. In addition, Gallien et al. [28] reported poorer rehabilitation outcomes for stroke patients when there was insufficient sensory stimulation, while Huang et al. [29] suggested that improvements in neurological scores can be obtained when increasing activation of the primary motor cortex (M1) and the primary somatosensory cortex (S1) by somatosensory stimulation for both acute and chronic stroke patients. As a result, it is considered that sensory stimulation is a crucial component for motor recovery. In this study, various target objects were prepared in the clinic group to provide different sensory stimulations to the paretic hand. For instance, the sponge provides soft textile perception with a very light weight, while the alloy tube provides a feeling of hardness and coldness. Meanwhile, the tennis ball provides a perception of fluffiness and roughness, while the toy carrot provides a smooth tactile sensation. However, only the sponge could be used by the participants in the lab setting, with the absence of variety in the sensory stimulation compared with the clinic group.

Another possible reason for the better rehabilitation outcomes obtained by the clinic group could be the additional daily self-practice done by the participants. The physiotherapists encouraged stroke patients to apply the learned motor skill into their daily functions, such as practicing the hand grasp and release motion and arm reaching motions every day. Following these professional suggestions, participants in the clinic group performed ADLs actively including dressing, bathing, and self-feeding with their affected limb. However, no such suggestions were given to the participants in the lab group. It is possible that the noteworthy improvement in FIM scores in the clinic group could demonstrate that self-practice within day-to-day routines plays an important role in effecting change.

In this study, higher motivation for robotic hand training was noted for stroke patients in the clinic group rather than those in the lab group. It has been reported that motivation plays an important role in post-stroke rehabilitation, and patients with high motivation can obtain greater improvement than those with less enthusiasm [30,31]. Motivation is dependent on multiple aspects including personal characteristics (e.g., socioeconomic status, age, personality traits), social factors, and rehabilitation environment [32,33]. It was noted that the stroke patients in the clinic group with higher socioeconomic status had greater motivation compared with participants in the lab group. For instance, four stroke patients in the clinic group remained working, and expressed strong motivation to regain their independence in ADLs. However, all participants in the lab group had quit their jobs and therefore might have lower motivation to regain motor recovery. Social factors played a part in patient motivation [34], in particular, the abilities of the practitioner and the relationship between the patient and practitioner. It was considered that a positive effect could be seen through practitioner confidence and communication, while negative impacts would be experienced if the practitioner appeared neutral or unconfident [35,36]. As a result, the physiotherapist in the clinic group offered strict, professional rehabilitation advice, which can subconsciously boost patient motivation (and subsequent outcomes) in the clinic group. In addition, patients' motivation toward the treatment they received could be affected by a different rehabilitation environment, and a stimulating rehabilitation environment with a well-maintained treatment room is a crucial factor and could result in higher motivation [33]. Therefore, strategies could be determined to raise the motivation of patients during rehabilitation training by providing an encouraging environment and increasing active interaction between the patients and the therapists.

Although, in this study, no specific robotic system was applied either to the joints of the shoulder or the elbow, after robotic hand training, FMA shoulder/elbow scores nevertheless went up. This could be because the training tasks may benefit the entire upper limb, since they focus on a number of joints [37]. When participants were undertaking their vertical and lateral task training, these upper limb motions may also have engaged the shoulder and elbow muscles. Second, it is possible that the adjacent proximal joint would be improved at the same time as the muscle around the joint was trained [38,39]. Previous studies have found that wrist training can often result in an improvement in elbow function [39] and, in turn, elbow training may result in shoulder improvement [38]. It is therefore clear that whole upper limb training, which is task-oriented, could well be more effective than an approach which concentrates on joint-by-joint rehabilitation, when the proximal to distal gradient of motor deficit is not present [40,41].

## Conclusion

This chapter has argued and provided evidence for the contention that robotic hand-assisted upper limb therapy for post-stroke patients is both practicable and effective, within a clinical service. The study results showed that the EMG-driven robotic hand training was helpful and valuable in a clinical service, and that its results were comparable to the rehabilitation effects which were noted in a research-based clinical trial. When compared with the results of the lab group, the clinic group demonstrated a rise in independence in their daily lives and the activities they undertook, and a more marked and effectual release of muscle spasticity. It is possible that the better outcomes found in the clinic group may be the result of flexible training, self-exercise, sensory stimulation, and a higher degree of motivation on the part of the participants. In the future, studies in this area will focus on large-scale clinical trials, with a variety of randomized groups located in a number of centers, to produce valid, generalizable evidence of the positive impact of device-assisted post-stroke rehabilitation training.

## Acknowledgments

The study was supported in part by PolyU Central Fund (1-ZE4R and G-YBRS). The authors would also like to thank the subjects who participated in this study.

## References

[1] Donnan GA, et al. Stroke. The Lancet 2008;371(9624):1612–23.

[2] Maciejasz P, et al. A survey on robotic devices for upper limb rehabilitation. Journal of Neuroengineering and Rehabilitation 2014;11(1):3.

[3] Takahashi K, et al. Efficacy of upper extremity robotic therapy in subacute poststroke hemiplegia an exploratory randomized trial. Stroke 2016;47(5):1385–8.

[4] Kwakkel G, Kollen BJ, Krebs HI. Effects of robot-assisted therapy on upper limb recovery after stroke: a systematic review. Neurorehabilitation and Neural Repair 2008;22(2):111–21.

[5] Norouzi-Gheidari N, Archambault PS, Fung J. Effects of robot-assisted therapy on stroke rehabilitation in upper limbs: systematic review and meta-analysis of the literature. Journal of Rehabilitation Research and Development 2012;49(4):479.

[6] Volpe BT, et al. Robotics and other devices in the treatment of patients recovering from stroke. Current Atherosclerosis Reports 2004;6(4):314–9.

[7] Song R, et al. Assistive control system using continuous myoelectric signal in robot-aided arm training for patients after stroke. IEEE Transactions on Neural Systems and Rehabilitation Engineering 2008;16(4):371–9.

[8] Hu XL, et al. A comparison between electromyography-driven robot and passive motion device on wrist rehabilitation for chronic stroke. Neurorehabilitation and Neural Repair 2009;23(8):837–46.

[9] Lattari E, et al. Corticomuscular coherence behavior in fine motor control of force: a critical review. Revue Neurologique 2010;51(10):610–23.

[10] Patton J, Small SL, Zev Rymer W. Functional restoration for the stroke survivor: informing the efforts of engineers. Topics in Stroke Rehabilitation 2008;15(6):521–41.

[11] Bishop L, Stein J. Three upper limb robotic devices for stroke rehabilitation: a review and clinical perspective. NeuroRehabilitation 2013;33(1):3–11.

[12] Demain S, et al. Assistive technologies after stroke: self-management or fending for yourself? A focus group study. BMC Health Services Research 2013;13(1):1.

[13] Turchetti G, et al. Why effectiveness of robot-mediated neurorehabilitation does not necessarily influence its adoption. IEEE Reviews in Biomedical Engineering 2014;7:143–53.

[14] Sukal-Moulton T, et al. Clinical application of a robotic ankle training program for cerebral palsy compared to the research laboratory application: does it translate to practice? Archives of Physical Medicine and Rehabilitation 2014;95(8):1433–40.

[15] Hu X, et al. The effects of post-stroke upper-limb training with an electromyography (EMG)-driven hand robot. Journal of Electromyography and Kinesiology 2013;23(5):1065–74.

[16] Huang Y, et al. Translation of robot-assisted rehabilitation to clinical service: a comparison of the rehabilitation effectiveness of EMG-driven robot hand assisted upper limb training in practical clinical service and in clinical trial with laboratory configuration for chronic stroke. BioMedical Engineering Online 2018;17(1):91.

[17] Bohannon RW, Smith MB. Interrater reliability of a modified Ashworth scale of muscle spasticity. Physical Therapy 1987;67(2):206–7.

[18] Fugl-Meyer AR, et al. The post-stroke hemiplegic patient. 1. a method for evaluation of physical performance. Scandinavian Journal of Rehabilitation Medicine 1974;7(1):13–31.

[19] Carroll D. A quantitative test of upper extremity function. Journal of Chronic Diseases 1965;18(5): 479–91.

[20] Hamilton BB, et al. Interrater reliability of the 7-level functional independence measure (FIM). Scandinavian Journal of Rehabilitation Medicine 1994;26(3):115–9.

[21] Ke Z, et al. The effects of voluntary, involuntary, and forced exercises on brain-derived neurotrophic factor and motor function recovery: a rat brain ischemia model. PLoS One 2011;6(2):e16643.

[22] Lin Y, et al. Involuntary, forced and voluntary exercises are equally capable of inducing hippocampal plasticity and the recovery of cognitive function after stroke. Neurological Research 2015; 37(10):893–901.

[23] Schabrun S, Hillier S. Evidence for the retraining of sensation after stroke: a systematic review. Clinical Rehabilitation 2009;23(1):27–39.

[24] Laufer Y, Elboim-Gabyzon M. Does sensory transcutaneous electrical stimulation enhance motor recovery following a stroke? A systematic review. Neurorehabilitation and Neural Repair 2011;25(9): 799–809.

[25] Bremner AJ, Spence C. Chapter seven – the development of tactile perception. In: Benson JB, editor. Advances in child development and behavior. JAI; 2017. p. 227–68.

[26] Bolognini N, Russo C, Edwards DJ. The sensory side of post-stroke motor rehabilitation. Restorative Neurology and Neuroscience 2016;34(4):571–86.

[27] Patel N, Jankovic J, Hallett M. Sensory aspects of movement disorders. The Lancet Neurology 2014; 13(1):100–12.

[28] Gallien P, et al. Magnetoencephalography in stroke: a 1-year follow-up study. European Journal of Neurology 2003;10(4):373–82.

[29] Huang M, et al. MEG response to median nerve stimulation correlates with recovery of sensory and motor function after stroke. Clinical Neurophysiology 2004;115(4):820–33.

[30] Colombo R, et al. Design strategies to improve patient motivation during robot-aided rehabilitation. Journal of Neuroengineering and Rehabilitation 2007;4(1):3.

[31] Maclean N, et al. Qualitative analysis of stroke patients' motivation for rehabilitation. BMJ 2000; 321(7268):1051–4.

[32] Maclean N, Pound P. A critical review of the concept of patient motivation in the literature on physical rehabilitation. Social Science & Medicine 2000;50(4):495–506.

[33] Maclean N, et al. The concept of patient motivation. A Qualitative Analysis of Stroke Professionals' Attitudes 2002;33(2):444–8.

[34] Ong LM, et al. Doctor-patient communication: a review of the literature. Social Science & Medicine 1995;40(7):903–18.

[35] Chan DK, et al. Patient motivation and adherence to postsurgery rehabilitation exercise recommendations: the influence of physiotherapists' autonomy-supportive behaviors. Archives of Physical Medicine and Rehabilitation 2009;90(12):1977–82.

[36] Finkler K, Correa M. Factors influencing patient perceived recovery in Mexico. Social Science & Medicine 1996;42(2):199–207.

[37] Susanto EA, et al. Efficacy of robot-assisted fingers training in chronic stroke survivors: a pilot randomized-controlled trial. Journal of Neuroengineering and Rehabilitation 2015;12(1):42.

[38] Qian Q, et al. Early stroke rehabilitation of the upper limb assisted with an electromyography-driven neuromuscular electrical stimulation-robotic arm. Frontiers in Neurology 2017;8:447.

[39] Hu X-L, et al. Wrist rehabilitation assisted by an electromyography-driven neuromuscular electrical stimulation robot after stroke. Neurorehabilitation and Neural Repair 2015;29(8):767–76.

[40] Beebe JA, Lang CE. Absence of a proximal to distal gradient of motor deficits in the upper extremity early after stroke. Clinical Neurophysiology 2008;119(9):2074–85.

[41] Oujamaa L, et al. Rehabilitation of arm function after stroke. Literature review. Annals of Physical and Rehabilitation Medicine 2009;52(3):269–93.

PART IV

# Commercialization

15

# Commercialization of rehabilitation robotics in Hong Kong

Michael TSUI. Kam-fai

*CHIEF EXECUTIVE OFFICER, DELTASON MEDICAL LTD., MA LIU SHUI, HONG KONG*

**Chapter outline**



Worldwide, when we think of technology, we will think of Silicon Valley.

Almost all notable high-technology companies started at the Silicon Valley. Hong Kong is one of the Alpha cities. Its status in the innovation and technology industry should be as prestigious as its financial industry. Hand of Hope is a "successful project" through the collaboration of government, industry, academia, and the research sector.

**Intelligent Biomechatronics in Neurorehabilitation. https://doi.org/10.1016/B978-0-12-814942-3.00015-5**

To many stroke patients, Hand of Hope is an innovative robotic hand therapy device for rehabilitation. As its name suggests, it gives hope for patients to regain their hand mobility and function. To us, Hand of Hope means more—it is a milestone, it is the expression of effective altruism, to ease the suffering of stroke patients, by applying the knowledge as an occupational therapist and robotic engineer. I am very glad to have had the opportunity to participate in its commercialization at each step.

I started to practice as an occupational therapist in 1992, but then I joined commerce and founded several companies in Hong Kong, Macau, and China. While I was operating my company as a rehabilitation equipment importer and supplier, I also further studied an MBA and robotic engineering.

During my 15 years of experience as a rehabilitation equipment supplier, I noticed some changes that had occurred:

1. The technology gap between the Pan-China (Hong Kong, China, and Taiwan) and Western countries was reducing.
2. Western countries rely heavily on the manufacturing capability of China and Taiwan for high-technology products.

With all these apsects to my background, I have a strong belief that Hong Kong could be the technology hub of the next era. I was recognized as an entrepreneur who has implemented a successful commercialization process. I am very humbled by these recognitions and I am grateful for the experiences I have had. A large part of these "successes" was coincidences of the correct time, place, and persons. Most of this chapter is sharing of my first-hand experience. It may sound straightforward but it is a summary of many small failures and a subsequent tremendous amount of going around requesting helpful consultations from knowledgeable people. This is a manifestation of the collaborative success of government, industry, academia, and the research sector.

## Correct time (government contribution)

The Hand of Hope project started at the time when Hong Kong Government decided to provide resources to support innovative and technology projects; the government then built the world-class infrastructures like the Hong Kong Science and Technology Park and Cyber port.

In The 2013 Budget Report, the Hong Kong Government proposed to allocate a subvention of up to $12 million each to six universities in Hong Kong for a period of 3 years from 2013/14 onwards to support the work of their technology transfer offices. Our project was also specifically mentioned, we were so lucky to catch this wave (http://www.budget.gov.hk/2013/eng/budget09.html).

We have relied heavily on the funding support from the Hong Kong Science Park and Innovation Technology Commission several times. This was very important, especially for the first 4 years.

## Correct place (government contribution)

We have been registered as incubatee of the Hong Kong Science Park since 2011. The incubation program not only allowed us to use the office space in Science Park free, but also provided other financial subsidies. The Science Park staff are always very helpful. We have won a world-class award from the event Geneva International Invention Exhibition which is their initiation to organized companies to join as a pavilion. Science Park has also arranged for companies to be exposed to potential investors, it was an all round and substantial help besides the financial subsidies.

## Correct person 1 (government, academia, and research contribution)

Thanks are due to Professor Raymond Tong who approached me when he had the bright idea of making a rehabilitation robot hand. This is where the whole story started. Mr. Andrew Young, a director from Science Park, with passion and enthusiasm, insisted on taking me to Geneva to join the invention competition. I had countless help from staff from Science Park, including Jessica, Julia, Reese, Paul, HL Yiu, Carrie Ling, Simon, Oscar, Ka-wai, Anthony ... etc. who were all very kind and always ready to help. I did not need to go anywhere else when I had troubles in R&D, commercialization, and operation, as they always offered knowledge and networking to answer my queries. All I needed was to ring them for help.

## Correct person 2 (industrial contribution)

Thanks are also due to Vincent Medical Group—their organic synergy provided financial, prototyping, and industrialization support, which was a main part of the success.

## Correct person 3 (industrial contribution)

With all the above, and the minimal me and my team from Rehab-Robotics Company Limited, we formed the final minor piece of the puzzle.

## Importance of a market-oriented approach

The objective of commercialization is to lead the exploitation of scientific/R&D results, integrate R&D works with production, and disseminate the products or services. It is essential because otherwise good work with strong science base could be buried in laboratories. Many funding emphases now do not only focus on the early R&D process but also extend the support to promoting the dissemination and subsequent commercialization process of R&D projects. Other supports are encouraged to include commercialization-related activities in recent projects, and such activities could be supported under the same conditions as R&D activities.

"Successful commercialization" cannot be understood as a well defined concept. "Commercialization" is understood as converting research results into a product/services available in the market. In our case it is a direct relation between research and market success through the collaboration of government, industry, academia, and the research sector. The concept of a "market-oriented" approach has steered the process since the very beginning. It provides a better view of the market impact of R&D projects.

Thirty years ago, I graduated from Hong Kong Polytechnic University as an occupational therapist. I failed to become a good therapist, the deep empathy I had for my patients was extremely intense and distorted, and I was not mature enough to correctly deal with the clinical sessions. I finally quit my job as a full-time therapist but practiced part time until I found the new direction I wanted to take.

I then started my nonclinical career as a rehabilitation equipment importer. I sourced rehabilitation equipment from mainly the United States and Europe, countries generally believed to be more advanced than Hong Kong in clinical and technology areas. I then finished another master's degree in business administration and in robotics engineering. After being in the rehabilitation equipment importation business for more than 15 years, I learnt how to evaluate good rehabilitation equipment, I understood the market need, and I looked at the trends in rehabilitation technology. As an occupational therapist, I had a wish list of clinical needs for the future.

In 2007, Prof. Raymond Tong, a good friend from my alma mater rung me up and asked if I was interested in supporting him in a project for a rehabilitation product. I was very surprised as he was echoing my own thoughts on the industry. I said yes and very soon we decided to develop something that was needed but not yet available from the market—an assistive robot hand for helping stroke patients, later we named it the "Hand of Hope". I want to highlight here, the "market-oriented" approach that was considered since the very start of this project, well before any R&D work had been started. The commercial objective and considerations of profitability were needed to be assessed together with the technical feasibility. This was the only way to increase the chances of commercial success.

## Transfer of technologies/knowledge

Transfer of technologies/knowledge refers to the process of passing a research outcome produced in an R&D project to industry. Technologies/knowledge contained in the intellectual properties could be transferred through total acquisition or through a technology transfer usage license. Terms could be varied, but the whole idea is to allow industry to access the science/R&D outcome and further the "commercial transformation," until transformation of the research outcome into a product ready for full market commercialization.

After technology has been transferred to a company, the original R&D team may move on to other projects or leave. In either case, there are risks to losing R&D continuity. Some terms should be considered for inclusion in intellectual property (IP) acquisition/licensing.

**i)** Original R&D team moving on—in the case that the original R&D team moves on to a new project, it is very common that the team will start the new project based on the scientific outcome of the original one. The further research may render the current model undergoing commercialization obsolete. Terms should include the right to have the first claim to acquire or license the outcome from the original research.

**ii)** Original R&D team is dismissed—in the case that the original R&D team is dismissed the company should consider hiring the key researchers on a discretionary basis. The terms should include the commitment to active and regular follow-up of the project by the key researchers on a personal basis. The idea behind this is to make sure the original team bears continued and regular responsibility for a certain period even after being dismissed. This would guarantee R&D continuity.

In addition to the above, company should also consider investing resources into related additional R&D activities of up- or downstream technologies.

In our case, we signed the "Hand of Hope" technology transfer license with Hong Kong Polytechnic University, and this allowed us to use the technology for 15 years. That meant we could do further commercial/technology transformations; we then applied for further patents based on the original patent. As I was operating my equipment import business, I retained the license, new patents, and related activities in a separate start-up company called "Rehab-Robotics Company Limited."

The Hand of Hope project combined a typical technology push and market pull. The demand from the market was clear, and the technology was advanced enough to fulfill the demand. That created a natural new market for the commercialization process.

## Key factors for successful commercialization

### Balance of market demand and technologies

There is a complicated balance between technology availability and market demands. There will always be a gap between the two. Often there are demands for which we do not have appropriate technologies to fulfill, and in some case, we have technologies for which we cannot find suitable applications. Nowadays, technology is changing very rapidly. At the same time, society is also changing very rapidly, to enable new market demands to appear. We need to continuously identify the potential for new market demands and match them with available technologies. Alternatively, new technologies can sometimes drive new market demand. Civil-level commercialization relies heavily on the provision of profit within a reasonable amount of time; if the resources (time and money) needed to fill the "technology–demand" gap are too large, failure of commercialization will be the result.

## Identification of application sectors

Therefore, commercialization is the process of finding the correct application sector for innovations. The process is very dynamic, and when technology is more advanced and complex, the technology push can be greater, the trend today is for the major part of market demand to be created by technologies, so you may say that it is mainly technology-driven. However, in the sector of rehabilitation applications or generally for medical applications, technology will never be the main driving factor. As long as the human structure is remaining similar, the needs (restoration goal) would remain similar. Of course, there are exceptional cases, devices sometimes enhance human bodily function (e.g., the uses of sports protheses, where blade runners can perform sometimes better than regular sprinters), but this is not typically the case.

When people first saw the Hand of Hope, much of the audience envisaged the possibility of a robotic device to enhance human function (like the "Ironman" in the movie). We had been advised by capital investors to adjust our development for augmented human function or entertainment purposes. Identification of the application sector was therefore very important; the same technology could apply to several sectors and commercialization included a decision of how and where to apply the device. Each application sector has its own uniqueness, medical applications need regulatory registration and have their own set of ordinances, and for toy applications there is another set of regulatory requirements. The team you would need to form for commercialization would be totally different, and resources would be drained for different reasons and a favorable outcome may not be achieved if the application sectors are confused.

The duration of commercialization depends heavily on the application sector. A normal application with high technical complexity may reach the market within a couple of years, whereas the application of technologies in the rehabilitation/medical sector can take more than 10 years due to the regulatory environment.

Application complexity, if involving multidisciplines, would also further extend the commercialization duration (e.g., a radio diagnostic device would be operated by technicians, but the records would be reviewed by radiologists and surgeons, installation and other safety issues would be reviewed by the biomedical engineering department of the hospital, etc.).

## Find a key opinion leader (KOL)

It is imperative to identify a leader with strong character demonstrating vision, ambition, and willingness to take the risk in your application sector. It is not difficult to identify the enthusiastic and skillful user in your application sector in a region. The problem is how to approach them. Even with a strong belief in your technology, you need to approach them humbly and be prepared to invest your time in them. Scientists and entrepreneurs, after working in a single area for years, will also naturally become experts in this particular technology and application area.

Strongly opinioned leaders always want to share their opinions. You need to stay very open when they offer you advice and be prepared to change your design. Don't be afraid to give up the good to go for the great. Make the key users your partner and recognize their contribution whenever you have the opportunity. You can share the glory afterward.

This is particularly important in medical applications, clinical setup is normally not affordable for start-up company, which is why it is important to target some reference clinics/hospitals for cooperation. The outcomes could also be used as clinical studies, which is needed in many countries for regulatory purposes.

One of the reasons Hand of Hope gained success was because we identified opinion leaders in the countries we took our system to. We also generated clinical studies in the United States, China, and Turkey.

## Financial support

It is important to gain financial support for your commercialization project. You must understand the appetite of investors and the operation of the capital market. You also need to be aware of government and public funding which tends not to require equity.

Your project might start with a great idea and have a solid science foundation before the technology is licensed by a company. However, when the project is passed to the company from the laboratory, it is not mature enough to launch to the market, and we must estimate the duration and total expenses incurred before we can expect revenue to be generated from the market. With only a laboratory prototype, it would not be appealing for fund raising.

There are different types of capital investors. For the first few years, the funding size you could attract is always smaller, say less than 1 million USD, and you are advised to withhold most of your shares, even you are approached by an angel investor.

For our case, from 2011 to 2015, 1 million USD was the amount of money we needed, and instead of disposing of shares to an angel investor, we relied on investments from cofounders' savings and FFF (family, friends, and fools) that would not incur equity disposal. We also successfully applied for government matching funding which supported half of our payroll.

Valley of death (normally first 3–5 years) is the terms used to describe the above stage, as an average of 90% of start-ups do not pass this stage because this is a stage that generates only negative cashflow and the business model has not yet been established (https://en.wikipedia.org/wiki/Venture_capital#/media/File:Startup_Financing_Cycle.png).

During this "valley of death" stage, there is a lot of necessary preparation (Fig. 15.1). We understood that once we passed through the valley of death, we needed to approach

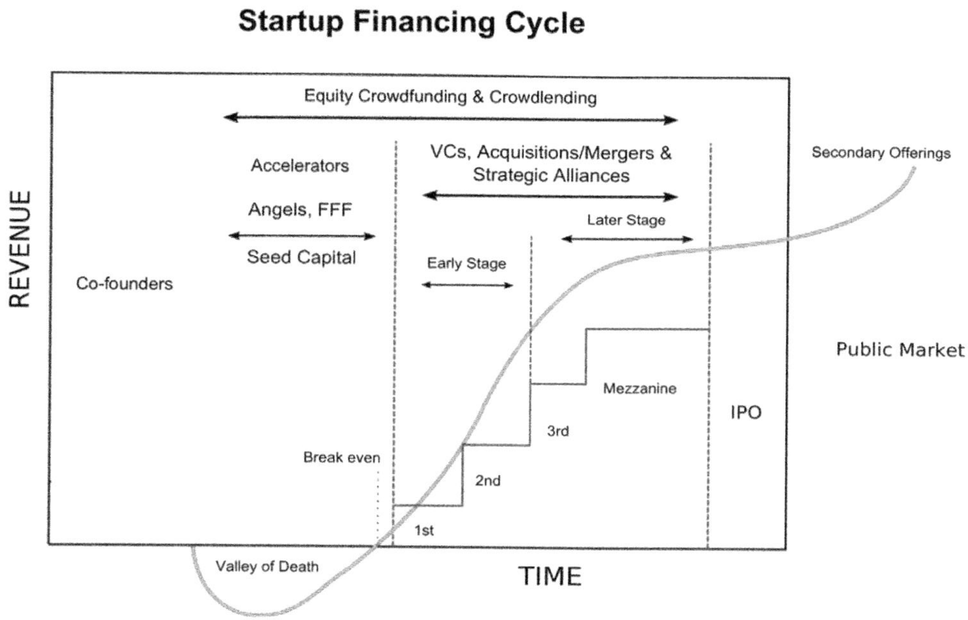


**FIGURE 15.1** Valley of death.

financial investors. Therefore the following needs to be done to make your project more appealing.

**i)** Gather pilot user references

We have already mentioned KOL identification and acquisition, if you want to recruit a KOL, you should go for a visionary, ambitious clinical leader. However, if you are passing through your valley of death, you cannot afford to be picky in obtaining pilot user references. You might want some quick advice before you formally launch the product to the market. Pilot reference users could impress your investors, especially active and paid users.

**ii)** Get some patents/trademarks

As mentioned above, there may be a need for some design changes to the product before you prepare to launch to the market. (For example, the market would require a more durable structure that can withstand heavy duty usage.) Don't forget to capitalize on your work by obtaining patents or trademark. Every logo design, outlook, mechanical or process change, can be used to gain intellectual properties. These would be needed for investors. Compare the different designs of motor layout we filed for our patent in Fig. 15.2.

**iii)** Join award winning competitions

- Award winning competitions give you the opportunity to practice your presentation skills for your project and your reaction to criticism. This helps you to

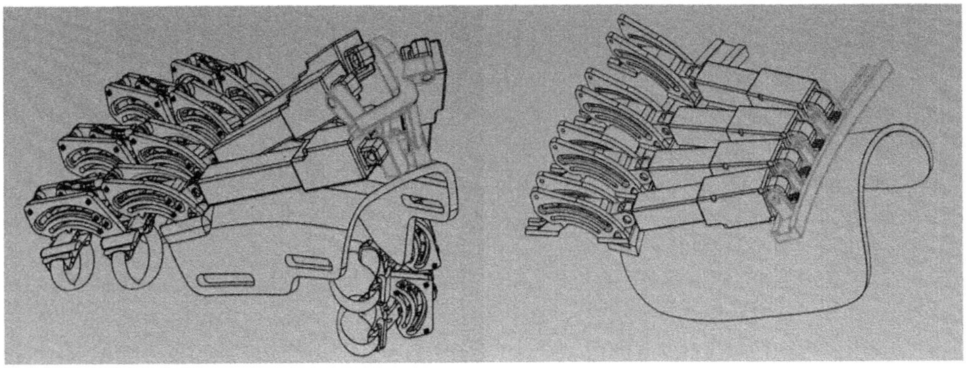

Structure of motor Bracket from laboratory, which is fragile and easy to break

Newly patented structure of modified motor bracket, more suitable for heavy duty usage.

**FIGURE 15.2** Comparison of original patent (left) and newly filed patent (right).

prepare to pitch to investors. Also, any awards you collect are a value justification for your project.

**iv)** Prepare your business plan

Before you get your business launched, spend time before diving into the project, that would be something that you probably already had in mind, but you may need to express your thoughts in writing, and you also need to gather data to support your hypothesis. For instance, information about the market size, business model, customer information, etc. You may also want to interview as many target customers as possible and put together a report to verify your assumptions. The information you gather from your interview may change your thoughts in the meantime, so be prepared to adjust the content so that your business plan reflects the reality. It's an invaluable exercise. You could very soon feel that the planning, brainstorming, and research will give you a highly defined business direction. At the same time, it will also have an impact on your product design.

**v)** Prepare financial documents

Based on your business plan, and estimated operational expenses to support the plan, prepare your revenue forecasts and cash flow.

These documents could be rather complex if you are not a financial or accounting professional. It could range from a single slide to highly complicated Excel documents that involve macros and formulas changing outcomes based on key assumptions and scenarios. Assumptions are important, so you need to work together with financial personnel who have experience and are willing to criticize. You need to cover several aspects of your financial picture: revenue projection, operational expenses, and cash flow. Typically, a 5–10-year duration would need to be covered, depending on it covering at least 3 years after the breakeven point. If you were asked for other documents, such as the balance sheet or pro-forma income statement, you could provide them after the pitch meeting.

Rehab-Robotics Company Limited spent almost 1 million USD during the valley of death period (2011–15) before almost reaching their breakeven point. During the fourth year, we had around 15 paid pilot users and some of them were key opinion leaders in their areas. We were working very closely with these initial first batch users. The favor we offered in return for their cooperation was lifelong free upgrades to any new models. They also started work from their clinics which we documented for future reference. We filed two new utility patents, one design patent and a couple of trade marks under the name "Hand of Hope" in different countries' markets. We finished the European CE registration for a medical product (Hand of Hope is classified as a class II medical device in Europe), so we were free to sell the product in Europe. We went to the Geneva International Invention Exhibition and won the Grand Prix Award, which was the biggest award, and this was the first time a Hong Kong company had won it. Subsequently we also won other local awards from the Federation of Hong Kong Industries, South China Morning Post, Hong Kong Science Park, etc. We were therefore encouraged to become quite active in entering competitions. Finally, we gathered around 15 awards when in 2015 we were ready for our investment round. The most difficult part of this was the preparation of a business plan and financial documentation. As we started with a target amount of recruiting 2.5 million USD, we needed to justify our business plan with a figure of this amount in exchange for share equity. By the end of 2015 we had exchanged 53% of our shares with a Hong Kong listed company Vincent Medical (stock code: 1612).

## Company structure and management complexity

A start-up company needs to remain small, medical equipment commercialization is highly specialized, and the process is extremely complicated. All activities are run continuously and parallel, every team member's activity constitutes a chain of subsequent steps. It implies also feedback loops incorporating feedback from internal members and external users. Activities progress and feedbacks are recorded in relevant documents (the company needs to be accredited by ISO-13485 certified bodies, and a written track of activities and its changes are required). As many of above activities take place in parallel with other key steps of the cycle (e.g., research, prototyping, industrialization, etc.), frequent updating of documents and version control are needed. A digital document control system could avoid "overwhelming numbers of papers" during a quality audit.

Fig. 15.3 shows the structure of the Rehab Robotics company and separation of duties.

### R&D team

R&D teams need to pay attention to the emergence of related new technologies and adopt to these products. The team should aim for continuous enhancement of function and performances, for instance, processing power, materials, safety, etc.

Progress of commercialization

| | Idea generation | Validation | Regulatory | Clinical Application | | Production | |
|---|---|---|---|---|---|---|---|
| R&D | Technologies paper studies | Specification Verification | | Functional Validation | | Improvement | |
| Clinical | Clinical Paper studies | Clinical Validation | | Clinical Support | | Clinical Trial | |
| Marketing | User interview Summary | Trade show | | Steering Clinical Trial | | Extend users | |
| Prototyping | | Basic | Testing | Full function | Bug Fixing | | |
| Regulatory | File IP | Regulatory Documents | | | | | |
| Industrialization | Small number Prototype support | | | | | Trial production | Mass production |

**FIGURE 15.3** Structure of Rehab Robotics company and separation of duties.

The team may need to redesign the system to improve usability and user experiences, e.g., ease of service, durability, control of potential hazards, regulatory requirement, packing and transportation, definition of serviceable parts.

## Clinical team

Technology does not fit naturally into the clinical rehabilitation regime, there are lots of design criteria that need to be considered.

There are always gaps between artificial mechanical hardware and human anatomy, e.g., an electric motor can turn 360 degrees and produce a strong torque. When it is used to assist rehabilitation physical training, the clinical team should define the movement freedom and torque of the mechanical structure so that it does not harm the joints and tissues.

Safety considerations—Patients in a rehabilitation program commonly move slower, their muscles are weaker, and their coordination ability is yet to be improved. The risks for accidents/incidents while using the rehabilitation product should be considered thoroughly. The clinical team needs to understand the symptoms of the target patient group and lay down safety requirements. Regarding designs for safety, rehabilitation products are much more stringent than other applications for civil use.

Treatment goal—The clinical team should define the treatment goal for a variety of patients (protocol of rehabilitation training). Design for technologies application must be in line with the treatment goals. The effectiveness of treatment needs to be measured, recorded, and monitored. The clinical team should review and design the report format in line with the presentation for clinical exchange. As patients recover, the technology should be flexible and amenable to adaptation to new treatment goals.

Clinical trial design—Many countries' regulatory bodies require a recognizable clinical trial report as a mandatory document for registration. It is useful for them to validate the product in the clinical environment to confirm safety and treatment effectiveness.

The design of the clinical trial protocol is essential. The clinical team needs to define the inclusion/exclusion criteria, safety precautions, treatment design for the control group to ensure the study's ethicality, decide the sample size and study duration, etc.

## Marketing team

The marketing persons in the early commercialization stage need to be very skillful. They must not only do their jobs as typical marketing persons from a mature corporate point of view, but they also need to be intensively involved in the collection of design change ideas and business intelligence. They will feedback regularly the design change ideas to the R&D team, and use business intelligence collected to formulate a business model suitable for the novel product and market. They should actively arrange meetings for R&D, the clinical team, and even the prototype team to meet with end users.

Product trials and sales: when the marketing team take out the prototype to collect users' comments. A clear naming and framing of the product should be decided. For instance, we decided to name our product "Hand of Hope" before we took our product out to the market for trial sales, because we noticed there had already been some non-EMG-driven robot hands available in the market. We wanted to characterize our product by emphasizing our differences. The name is easy to understand, and we differentiated our product in terms of our application. We stated a very clear position on the market by specifying that it is an EMG-driven robot hand for stroke rehabilitation and we gained a lot of recognition at an early stage. Patients could expect recovery through the training, and "hope" is one of the direct immediate customer benefits.

The marketing team should collect business intelligence so as to understand competition from all over the world. International competition does not only affect prices but is a constant source of potentially competitive (or even the same) technologies. In the case when infringement of intellectual properties was found, the marketing team would need to report immediately to the regulatory team for immediate action. A correct price strategy is vital for commercialization. Flexibility of price for different (geographical and economic) markets was understandable but the market today is very transparent. Users have instant exchanges in social media and professional forums. If prices are set too low it would downgrade an innovation, and yet prices would need to possess competitiveness, so that the dissemination of the innovative system would be efficient. Marketing team should disseminate the product actively through conferences, trade fairs, workshops, and publications (Fig. 15.4). We won the Grand Prix Award from the 40th International Exhibition of Inventions of Geneva (Fig. 15.5), meeting different people in trade shows (Fig. 15.6; the Hand of Hope in its current form).

A market-oriented approach is emphasized throughout the whole process, and the marketing team should constantly examine whether the assumed demand is there and

**FIGURE 15.4** Rehab Robotics Company won the Grand Prix Award from the 40th International Exhibition of Inventions of Geneva.

**FIGURE 15.5** Meeting different people in trade shows.

**FIGURE 15.6** Hand of Hopes in its current from.

strong enough. The overall economic climate and several other reasons could create a mismatch between return and investment cycles. The marketing team needs to constantly monitor these changes, even after company has successfully launched their product into the market. The information from them would greatly affect the production processes and planning. This information feedback for production chains is essential. Production costs may be reduced when production is scaled up, and in turn would feedback to and affect price strategies. The process is rather dynamic.

At the later mass production stage, the marketing team may need to restructure to incorporate routine and typical marketing persons, to ensure effective dissemination.

Key opinion leader always develop from clinical trial studies. When an important user is identified, the marketing team should line up resources, feedback and discuss with the clinical team the possibility to extend clinical studies with new important users.

## Prototyping team

Prototyping is the first step of product verification. The prototyping team needs to create a prototype according to the designs passed from the R&D team. By creating a prototype, it is possible to see the product in a different way, the team could find discover that has not been previously noticed. The prototyping team could then help the R&D team to evaluate the design defects effectively. "Prototyping/evaluation" is a continuous cycling process loop that does not stop until the displayed defects have been brought to focus and improved, therefore, the prototyping team needs to work closely with the R&D team. For different reasons as discussed below, prototypes would be produced.

**i)** Basic prototyping is produced for the "prototyping/evaluation" cycle, version control of each prototype and the corresponding evaluation report (verification and validation report) should be well documented.

**ii)** During the "prototyping/evaluation" cycle, several testing prototypes are needed. Testing prototypes are produced in a small batch scale. The related testing plans are created for different reasons. The testing plan could aim at performance, functionality, and specification evaluation. The results of these evaluations are consolidated into a report and feedback to R&D team. Tests can be carried out in house or outsourced. Some test reports are required by regulatory teams that need to be issued by an accredited laboratory to fulfill the registration requirements of some countries, testing samples are then send out to external accredited laboratories.

**iii)** Before the product goes for trial production or mass production, a final prototype is made. The product at this stage should have all design defects removed, it should also fulfill all performance, functionality, and specification requirements. The final prototype could also be used as a "golden sample" for production quality assurance.

**iv)** After full market launch, R&D may need to change the design for different reasons (e.g., a bug was found, or the marketing team collected comments from users that indicated a required version change). The prototyping team would then need to produce a prototype with this bug repaired for a new round of evaluation.

The prototyping team would advise the change of design for manufacturing and carefully consider the product manufacturability. The aim of this process is to eliminate potential manufacturing issues as early and promptly as possible. The prototyping team should bring in multidisciplinary teams every time a change is suggested (e.g., the R&D team would be looking at material types, tolerancing, and part requirements; the Industrialization team would consider tooling and setups; the clinical team would be interested in elements that could affect the final product and its functions; and the regulatory team would be concerned with the classification and registration, etc.) The fact that this collaborative approach needs to be preserved and communication is emphasized throughout the entire process.

The prototyping team should also consider ways to simplify the design so that the number of parts required can be reduced. Having fewer parts makes it easier to assemble and decreases the likelihood of a defective part holding up the production process.

After wrapping up the change, the prototyping team should help to clarify production costs, together with any residual concerns for manufacturing. The production cost estimation would also be fed back to the marketing team. The prototype team should put together costs related to fabrication and assembly, along with costs of purchase, stock, and servicing. They try to keep the cost of actual production to a minimum. They will also put in precautions that to ensure effectiveness and efficiency of the production process. For example, some special parts have an extra-long lead time, parts that need special materials for fabrication, parts that require complex tooling.

Functional and specification changes during the prototyping stage should be summarized. The R&D and clinical teams make sure any changes to design do not compromise the original functions and features.

## Regulatory team

Our regulatory team wraps up innovative ideas and checks if it is possible to file an IP (patent and trade mark). They would make a preliminary patent search and estimate the cost for pursuing a patent or trade mark. They put together a draft highlighting the novelty of ideas and prepare the meeting with the patent lawyer.

Some larger companies may have a separate legal team for IP issues. For Rehab-Robotics, as a start-up, we combined the IP and regulatory teams.

The higher the value of the project, the more potential competitive players. Other competitive players may not simply infringe the patent, they may acquire/combine with other enabling technologies/patentees, so the issue could be quite complicated. For example, we discovered a Chinese manufacturer making a similar robot hand in 2015. This company also filed for a patent in China. The claims inside their patent was not the same as our, their patent contained other patents from another industries. We finally invalidated part of their patent that copied directly from us but left part of their patent still valid.

For the creation of IP, there is no one best way how to handle it, it depends on whether an IP already exists or needs to be created. Several questions would need to be asked. For example, is it a unique technology? Is it easy to be substituted by similar or other technologies? Do the size and financial capacity of the company justify the extensive filing of a patent? In our case, we obtained the right to use the initial IP from Hong Kong Polytechnic University, and we filed patents ourselves based on our own development. Some companies choose not to file any patents because they don't want to disclose the technology work and make it relatively easier for other companies to come up with substitutes or find ways around the patent.

Product registration for different countries requires similar documents, but they are not the same. The regulatory team also coordinates with R&D to prepare the testing plan and receive the testing prototype from the prototyping team. Common tests for rehabilitation robotic devices include electrical safety and biocompatibility, a testing report would be a very commonly required document. Other documents required included the operation manual, product specification, and risk analysis report. This team would also collect the clinical trial report from the clinical team if available.

## Industrialization team

The economic structure of Hong Kong is putting less emphasis on industry. The high land cost makes it unfavorable for industrial development, and the industrial sector in Hong Kong has declined in the last 20 years.

Luckily, we have China at our back. China's industrial output is now only second to the United States. The fast growth in manufacturing capability in China benefits Hong Kong's innovation.

Despite the Chinese factory operations often not being as advanced as other international competitors, Chinese manufacturers are responsive and dependable.

To develop a rehabilitation robot in Hong Kong, we depended on factories in China from the very beginning with the prototyping stage. We appointed Chinese fabricators to help us to make our prototype. China, however, does not put too much emphasis on protecting intellectual properties, which is a concern. The best way, of course, is to set up a factory inside China and put everything under your management. However, this is not an option for a start-up. We started by splitting our work into segments for different factories. By breaking the manufacturing process into several discrete operations and manufacturing in different factories, it would prevent any single group of employees in one factory from seeing the entire process. That was a boring part of our early operation, but at least it relieved us of worries of leaking our intellectual properties in China.

The industrialization team, as mentioned above, needs to source more than one manufacturer in China. The selection of external manufacturing companies needs to be done very carefully, as we were splitting the manufacturing into different segments, each piece of work finally coming back to us for assembly. The external manufacturer would be required to provide detailed feedback in the case that they faced any design problems.

The team needed to hold regular meetings with different manufacturing vendors and collect comments from factories for feeding back to the prototyping and R&D teams. Internally we still needed to assemble the parts and so we still needed simple tooling and equipment; in order to save some cost, we purchased used equipment from internet auctions.

At a later stage, we recruited Vincent Medical group (stock code: 1612) as a synergic shareholder, and we became the subsidiary company of the list company group. As Vincent Medical is providing manufacturing capability for the "Hand of Hope" project, we are consolidating all external manufacturing to them. It is easier and safer to leave manufacturing to our shareholder. The industrialization team was responsible for manufacturing transfer from the very beginning. As the factory still stayed inside China, industrialization still needed to hold regular meetings to review different issues and requirements from manufacturing and feedback to the Hong Kong R&D and prototype teams. Of course, this was easier because they were dealing with only one manufacturer.

## Managing start-up and project commercialization

Managing a company commercializing rehabilitation product is difficult, it is a combination of all the above activities. Without proper management processes, it is not possible for commercialization to be efficient. Hand of Hope is commercialized as a class II rehabilitation product, and the company needs to be audited against ISO13485 annually. Entrepreneurs, managers, and engineers need to cooperate with a common understanding of our goals and quality processes. Management should mediate the understanding of both the market and the technical problems. The organization should typically practice with an integrated team approach. Besides actively establishing an atmosphere of openness of ideas from different perspectives, management should also be responsible for maintaining relations with all market players: key customers, press, important hardware and software partners, distributors, dealers, patient groups, universities, and certified bodies (for ISO 13485 standards). Management should also build a relationship with KOL and support the clinical trial project from them.

In the early stages, there were only three of us in the company, and we stayed in the office for more than 12 hours each day. Those were the days when everything came to us quickly, and we needed to react rapidly. We exchanged ideas in an instance. If a wrong decision was made, we just rolled it back. So although we changed our minds very often, we didn't need any plan for our decision-making, as our decision-making would not impact too many people.

When we later became part of a listed company group, we needed to submit our plans, and waited for approval from the next meeting, before the final decision was made. I learnt that when we worked with more people, and part of our staff stayed in China in the corporate factory, decision-making needs become more serious because changing our mind about something may cost the company a lot of money. It may impact many employees and involve retooling, lots of processes, and systems. When we decided to scale-up , we needed to spend more time trouble-shooting.

Therefore, I experienced a decision-making model of shifting toward doing things right more often than doing things fast. Planning has become a part of my life.

## Finale

After working in the company for almost 11 year as CEO, taking the company from a start-up to a company valued at around 50 million HKD, I am reflecting on all the decisions I have ever made. There are many right and wrong decisions, probably balanced fairly equally, but I have enjoy every part of the journey. The company evolved from an early-stage start-up to a subsidiary of a listed company. It has been an unforgettable experience.

We started with one product, Hand of Hope, and now the company is also making other products for rehabilitation. The Hand of Hope product is now in its fourth version, with many improved features. The product has been used in many countries, including Japan, the United States, Italy, Germany, Spain, Norway, Turkey, Ukraine, Russia, Saudi Arabia, Macau, Taiwan, Singapore, Malaysia, China, Israel, Hong Kong, etc. The product has been registered as a medical product in the above countries.

In 2015, Rehab-Robotics was acquired by Vincent Medical Group and became a subsidiary of the listed company group. I continued to serve the company as CEO until the middle of 2018. The company keeps improving its performamce and the Hand of Hope product is getting stable sales from different countries.

# Index

'*Note:* Page numbers followed by "f" indicate figures and "t" indicates tables.'

**D**



**E**

CPI Antony Rowe
Eastbourne, UK
November 20, 2019